Cenozoic History of the
Southern Rocky Mountains

The Geological Society of America, Inc.
Memoir 144

Cenozoic History of the Southern Rocky Mountains

Papers deriving from a symposium
presented at the
Rocky Mountain Section meeting
of The Geological Society of America
Boulder, Colorado, 1973

Edited by

BRUCE F. CURTIS

1975

Published by
THE GEOLOGICAL SOCIETY OF AMERICA, INC.
3300 Penrose Place
Boulder, Colorado 80301

The Memoir series was originally made possible
through the bequest of
Richard Alexander Fullerton Penrose, Jr.

Printed in the United States of America

Contents

Preface

Much new knowledge about the eventful Cenozoic Era in the Southern Rocky Mountains region has arisen from intensive geologic studies made during the past decade. Geologic relations long obscure and puzzling have become greatly clarified by radiometric age data, by careful petrologic analyses of sedimentary and volcanic rocks, by extension of paleontological studies, and by detailed mapping.

Ogden Tweto, a leader of these new inquiries (and of earlier ones), recognized a few years back that some parts of this emerging Cenozoic story of the Southern Rocky Mountains had been scattered in publications of varying accessibility and that many other parts were quite fresh and unpublished. He saw that the new information had accumulated in ample quantity and had developed a form coherent enough to be of great importance to geologists with interests in such diverse matters as the economic mineral deposits, regional tectonics, volcanology, stratigraphy, sedimentation, paleontology, and geomorphology of this mountain area. Plainly the time was appropriate for putting together a comprehensive account of the regional Cenozoic history, and fortunately Tweto was willing to organize this account as a symposium for the 1973 Rocky Mountain Section meetings of the Geological Society of America. The papers in this volume, contributed by many of those responsible for the new state of knowledge, derive from that very successful symposium, but are in a somewhat more detailed form. Some of the information is new since the papers were presented.

Taken together, these articles demonstrate that Cenozoic orogeny in the Southern Rocky Mountains was by no means composed of just a few scattered goings-on associated with decaying Laramide deformation, as many geologists have tended to picture it. Instead, the end of the Laramide orogenic phase about mid-Eocene time can be discerned rather clearly, along with a following quiescent interval that lasted for somewhere around 5 m.y. Following the quiescent interval, the rock records show renewed tectonism, but of a different style which involved extensive volcanism and faulting of basin-range type that created some displacements of at least several kilometers. The Laramide and the later Cenozoic tectonic phases probably were related through deep internal earth processes, but the evidence here shows that the later Cenozoic activity was, by itself, of impressive stature and quite deserving of separate recognition.

Basic to much of the history traced in these papers are the numerous absolute age determinations presented. They accurately time many events, and therefore an attempt has been made in the text to correlate these dates with both general and provincial geologic time scales. Some time scale discrepancies remain among the papers because of current uncertainties about Tertiary epochal boundaries. The most conspicuous case in point is that of the Pliocene-Miocene boundary. Currently, interpretations of its age range from 5 m.y. to 12 m.y. One man's

Pliocene thus may be another's Miocene, but as the authors have indicated their particular interpretations of the time scale, it is rather easy for a reader to resolve such apparent conflicts.

The papers are arranged in approximate order of the Cenozoic episodes. Tweto first presents an inclusive summary of the many complex Laramide events that ended by late Eocene time and gave way to quiescent conditions when a widespread erosion surface of subdued relief developed. Epis and Chapin discuss the well-documented 10,000 km² central area of this surface and the evidence that it probably extended over a much broader region. Steven examines the chiefly andesitic widespread volcanism that occurred approximately in Oligocene time and deposited a cover upon substantial portions of the erosion surface. Clark presents an overview of the Oligocene fluvial sedimentary deposits on the Great Plains, especially those lying northeast of the Front Range. He points out sediment sources not only in the volcanic rocks but also in some still-exposed highland masses. A second episode of volcanism that followed in Neogene time produced extensive basalt flows while block faulting occurred close by. Lipman and Mehnert examine the record of this activity in the vicinity of the Rio Grande rift depression and propose an interpretation of its pattern in terms of large-scale tectonics and magma generation. Larson, Ozima, and Bradley consider the Neogene volcanic and volcaniclastic rocks in northwestern Colorado, and relate them to the tectonic record, changing climates, and an evolving drainage pattern. Izett discusses many clear examples of Neogene faulting and folding in northern Colorado, which are of impressive magnitude and which strongly affected development of drainage patterns. He notes that even more intense deformation occurred farther south. Taylor sets forth the evidence for some of this strong late Tertiary tectonism, discerning maximum activity in early Miocene and latest Miocene or early Pliocene time. Scott next presents the erosional history of the eastern range in the Southern Rocky Mountains. He develops evidence that the extensive erosion surfaces observed at various elevations in the mountains are fragments of the late Eocene surface, overlain in places by channel deposits of Oligocene and later ages, and much broken by Neogene faulting. Finally, Blackstone summarizes the Cenozoic history of the region surrounding the northern ends of the Southern Rocky Mountain ranges in southern Wyoming. Past studies and the information from his own extensive detailed mapping are interpreted as reflecting a sequence of Cenozoic events remarkably similar to that discerned farther south. It should occasion no surprise that some differing explanations, especially of high-level conglomerates and erosion surfaces, appear among papers dealing with various areas. This attests to the complexity of the geologic problems and to the opportunities for further study—which are many.

This newly defined history certainly constitutes a deeply interesting story, and it also has important implications both concerning local mineral deposits and geologic features and concerning large-scale tectonic and depositional patterns. Exploring these matters in coming years should afford scientific pleasure to many geologists who use this volume.

BRUCE F. CURTIS

Department of Geological Sciences
University of Colorado
Boulder, Colorado 80302

Geological Society of America
Memoir 144
© 1975

Laramide (Late Cretaceous–Early Tertiary) Orogeny in the Southern Rocky Mountains

OGDEN TWETO

U.S. Geological Survey
Federal Center
Denver, Colorado 80225

ABSTRACT

At the beginning of Laramide orogeny, a blanket of undisturbed Cretaceous and minor older Mesozoic sedimentary rocks 1,500 to 3,000 m thick covered the Southern Rocky Mountain province, and the last of a series of Cretaceous seas was starting to withdraw northeastward across the region. Beneath the blanket was an older and inhomogeneous terrane that in some places consisted of the eroded stumps of late Paleozoic mountain ranges made up of Precambrian rocks, and in other places of piles of sedimentary rocks thousands of meters thick on the sites of late Paleozoic basins. At the onset of Laramide orogeny in Late Cretaceous time, most of the buried mountain ranges were re-elevated, and adjoining Laramide basins, in part inherited from the late Paleozoic basins, began to subside and receive orogenic sediments. In addition, two anticlines of mountain-range proportions—the Sawatch and Uinta—rose from the sites of late Paleozoic basins.

Orogeny began in the southwest part of the province before marine deposition ended in the northeast. The late Paleozoic San Luis highland of southwestern Colorado and northern New Mexico was re-elevated in late Campanian or middle Montana Cretaceous time, as indicated by orogenic sediments in the San Juan basin to the south and the Raton basin to the east. The Sawatch anticline, which diverges northward from the rejuvenated San Luis highland, rose at about the same time, as indicated by the age (72 to 70 m.y.) of fault-controlled porphyries on the flanks. The Front, Park-Sierra Madre, and Medicine Bow Ranges, on the site of the late Paleozoic Front Range highland, rose after the marine Fox Hills Sandstone was deposited over their sites 67.5 m.y. ago. Uplift and erosion were rapid. By 66 to 65 m.y., and before the close of Cretaceous time, 3,000 m of sedimentary rocks had been eroded from at least parts of these ranges, and streams were carrying detritus from Precambrian rocks to bordering basins.

Once started, uplift of mountain units continued through Paleocene and into

1

Eocene time, as indicated by nearly continuous Upper Cretaceous to Eocene sedimentary sequences in the interiors of bordering basins. Uplifts grew laterally as they rose vertically. Consequently, the major uplifts of today are areally larger than those that supplied the first orogenic sediments, and their border structures are younger than those sediments. The crystalline rock body of the interior of the Laramide San Luis highland supplied sediments to adjoining basins in Cretaceous time, but the sedimentary rock flank of the uplift at the site of the Sangre de Cristo Range was not uplifted and deformed until Paleocene and early Eocene time. Similarly, an interior part of the Front Range supplied Precambrian rock detritus to bordering basins very late in Cretaceous time, but the flank structures of the range, which involve the early orogenic sediments, developed in Paleocene and Eocene time. The Laramie Range prong of the northern Front Range probably did not begin to rise until Paleocene time. The White River Plateau, the last major Laramide uplift unit to appear in the province, began to rise in early Eocene time, but it did not attain its present outline and flank structure until late Eocene time.

Laramide volcanism and intrusion were almost entirely confined to a broad northeast-trending belt that cuts diagonally across major tectonic units of the province. The Colorado mineral belt constitutes an inner zone of this igneous belt. Andesitic volcanism occurred at several localities within the igneous belt at an early stage in Larmide orogeny but after the first uplift of mountain units. Intrusion of granodioritic porphyries in stocks and smaller bodies began at about the same time as volcanism but continued longer, at least through Paleocene time. Though magmatism may have been the cause of some structural features within the igneous belt, the belt itself has no evident Laramide structural control and seems to be independent of the tectonic elements it crosses. The only unifying structural feature within the belt is a system of discontinuous and overlapping Precambrian shear zones. This system probably furthered the rise of magma bodies into the upper crust from batholiths at depth, but it had no role in the generation of those batholiths. Concurrent magmatism and tectonism must have shared the same cause, but they proceeded independently in the upper crust.

INTRODUCTION

The Laramide system of mountain ranges extends along the summit of the Rocky Mountains far northward in British America, and southward into Mexico . . . In the United States it occupies the summit region of the mountains, between the line of the Wasatch Archean and the Front Range . . . The rocks involved were those of all Paleozoic and Mesozoic time, Cambrian beds making the bottom, and the Laramie, or the uppermost formation of the Cretaceous, the top.

In these words, Dana (1895, p. 359) introduced the term "Laramide," which subsequently has come into wide and varying usage. The name was clearly derived from "Laramie formation," a term that, in Dana's time, had been applied widely as a synonym of "Lignitic group" to coal-bearing strata above fossiliferous marine Cretaceous rocks through much of the Rocky Mountain West. Two factors unknown to Dana have served to make a rather vague definition more ambiguous: (1) the term "Laramie formation" as originally applied in many areas was later found to include strata ranging in age from pre-Laramie Cretaceous to Eocene, and (2) various units of the Rocky Mountains were found to differ appreciably in date of origin. In 1910, the U.S. Geological Survey restricted the application of the term "Laramie Formation" to the Denver basin, and in 1939 the restricted unit was designated as entirely Cretaceous in age. Meanwhile, the descendant term "Laramide" has flourished, but owing to the misconceptions inherent in the original definition, it has come to have many connotations. Hence, the term is only a convenient wastebasket except when the usage is defined. In this paper, which refers only to the Southern Rocky Mountain portion of the Rocky Mountain system, in Colorado and adjoining parts of New Mexico and Wyoming, the term "Laramide" is applied to orogenic events that occurred between late Campanian Cretaceous and late Eocene time. Other parts of the Rocky Mountains had different time frames of orogeny, and even within the restricted area considered here, the movements of major tectonic elements started and ended at different times within the limits just named.

Though generally recognized as Laramide in origin, the Southern Rocky Mountains are not solely a Laramide creation. Several of the main ranges are wholly or in part elements of late Paleozoic uplifts that were rejuvenated in the Laramide orogeny, and several of the bordering sedimentary and structural basins had late Paleozoic expressions. Many of the major faults of the mountain units are Precambrian faults that were rejuvenated in Laramide and later time. Once formed, Laramide uplifts were greatly modified by later igneous activity and rifting or block-faulting. These later processes produced many of the features of the modern mountains.

The purpose of the papers in this volume is to summarize the evidence—much of it new in the last decade—for the post-Laramide development of the Southern Rocky Mountains. This paper on the Laramide will set the stage. The papers will establish, we believe, that the Laramide orogenic episode, characterized by large-scale warping, deep erosion of uplifts, and deposition of orogenic sediments in basins, died out in the latter part of the Eocene. In Oligocene time, magmatism superseded tectonism and was accompanied by sedimentation high onto the flanks of the eroded Laramide uplifts. In Miocene time, tectonism recurred on a large scale but in quite different form than in the Laramide period; this was a time of rifting and block-faulting, of an abrupt change in the character of magmatism, and of sedimentation in newly created and partially rejuvenated basins.

This paper is based on hundreds of sources of information. References are given where specific credit is appropriate or as examples, but no attempt is made to cite all sources. I can only acknowledge that I stand on the backs of all the geologists who have worked in the region in a little over 100 years. Summaries of the structural development of large parts of the region were made by Hills (1891) and by Burbank and Lovering (1933), and for the Front Range by Lovering (1929), Lovering and Goddard (1938), and Warner (1956).

GENERAL FEATURES

The Southern Rocky Mountains rise abruptly above the Great Plains along a front that extends northward from north-central New Mexico through Colorado and into southeastern Wyoming (Fig.1). On the west, the mountains border the Colorado Plateau along a sinuous front that in places has little or no topographic relief. Sedimentary and structural basins complementary to the mountain ranges are mainly in the Plains and Plateau provinces, though some of the smaller basins are intermontane. The larger basins are outlined by contours on the basement-rock surface in Figure 1. Surface elevations in the Plains area near the mountain front are 1,400 to 1,800 m. In the Plateau area in western Colorado, surface elevations are generally 1,800 to 2,400 m, deep valleys excepted. Large parts of the mountain areas are above 3,000 m in elevation, and many peaks and crest lines are in the range of 3,600 to 4,200 m. Most of the altitude and relief expressed in these figures resulted from post-Laramide uplift and differential erosion in late Tertiary time. As will be shown, the mountain province was at or very near sea level at the beginning of Laramide orogeny, and it was much closer to sea level than to present altitude after the orogeny had drawn to a close in the latter part of Eocene time.

The relation of the Southern Rocky Mountains to other mountain provinces and the evolution of the western Cordillera has been summarized by King (1959, p. 89–131) and by Gilluly (1963).

PALEOTECTONIC SETTING

The Southern Rocky Mountain province is on a part of the continental interior platform that has had positive tendencies since Precambrian time. Consequently, the record of the Paleozoic through Mississippian time is largely one of unconformity.

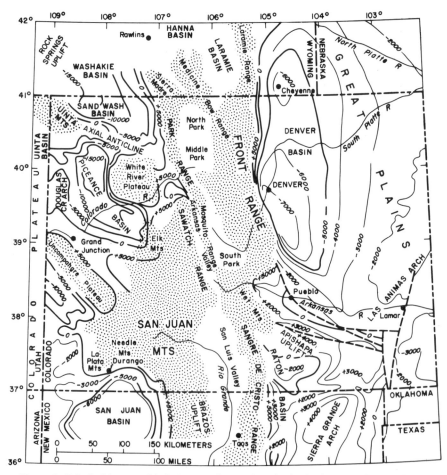

Figure 1. Principal topographic features and tectonic units in the Southern Rocky Mountain province. Major mountain units shaded. Selected contours (in feet) on Precambrian basement surface shown in the larger basins (after MacLachlan and Kleinkopf, 1969). Subsurface faults south and east of Pueblo shown by heavy dashed lines.

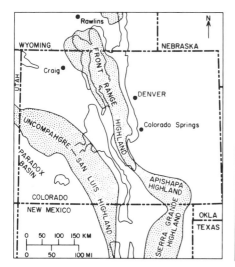

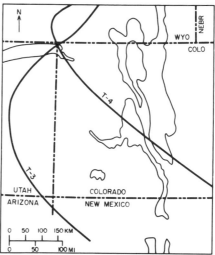

Figure 2. Late Paleozoic uplifts (shaded) in Southern Rocky Mountain province (modified from Mallory, 1960, 1967). Outlines of main bodies of Precambrian rocks as now exposed shown for reference (see Figs. 1 and 4).

Figure 3. General positions of strandlines at maximum stages of last two Late Cretaceous marine transgressions (strandlines after Weimer, 1960, Figs. 9 and 10; and McGookey, 1972, Figs. 44 and 45). Outlines of main bodies of Precambrian rocks as now exposed are shown for reference (see Figs. 1 and 4). T-3, early Campanian; T-4, late Campanian-early Maestrichtian. Final northeastward regression beyond map area, early middle Maestrichtian.

The rock sequence from Cambrian through Mississippian is only about 150 m thick in the central part of the region and 300 to 350 m in the western part. In Pennsylvanian time, major uplifts—generally known as the Ancestral Rockies—and complementary basins formed in the region (Fig. 2). At that time, the thin cover of pre-Pennsylvanian rocks was quickly eroded from the uplifts, leaving Precambrian rocks exposed much more widely than they are today. Coarse arkosic sediments derived from the Precambrian rocks accumulated during Pennsylvanian and Permian time to thicknesses of 3,000 to 4,500 m in the elongate sedimentary basin or trough between the Front Range-Apishapa and Uncompahgre-San Luis highlands and in the basin southwest of the Uncompahgre-San Luis highland. Accumulations were smaller to the east of the uplifts, but reached 1,500 m in the area between Denver and Colorado Springs.

The late Paleozoic uplifts were gradually reduced by erosion and buried by sediments during Late Permian and early Mesozoic time. By the beginning of Cretaceous sedimentation, a nearly planar surface existed that was underlain partly by huge bodies of Precambrian rocks beneath a thin cover of sedimentary rocks, and partly by basins filled with thousands of meters of sedimentary rocks.

Cretaceous sedimentation produced a cover of intertonguing marine and continental rocks 1,500 to 3,000 m thick over the region. These rocks were derived from sources far to the west, where orogeny occurred earlier than in the Southern Rocky Mountain province. Several cycles of marine transgression and regression are recorded in the Cretaceous rocks (Reeside, 1957; Weimer, 1960, 1970). Over much of the province, strandlines trended northwest; marine waters transgressed southwestward and regressed northeastward (Fig. 3). This pattern probably reflects the developmental pattern of Laramide orogeny, which within the region considered here began at about the time the last marine cycle changed from transgression to regression. As discussed below, Laramide uplifts began to rise in the southwestern part of the region before marine deposition had ended in the northeastern part, and retreat of the last Cretaceous sea was concurrent with the general northeastward advance of the Laramide orogenic front in this region.

MOUNTAIN UNITS

Existing mountain units and structural-sedimentary basins of the Southern Rocky Mountains (Fig. 1) are of various origins and characters. Distribution of rocks in them is shown in Figure 4.

The Front Range, Medicine Bow, Park–Sierra Madre, and Wet Mountains are early Laramide uplifts on the site of the late Paleozoic Front Range highland (Fig. 2). Therefore, these mountains had a comparatively thin sedimentary cover that was removed during initial stages of uplift, and they now consist largely of Precambrian rocks (Fig. 4). The mountains are flanked by upturned sedimentary rocks and, in many places, by border faults (Fig. 5). Some of the border faults

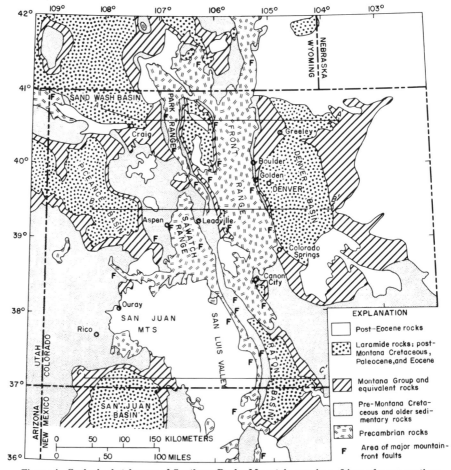

Figure 4. Geologic sketch map of Southern Rocky Mountain province. Lines of cross sections (Fig. 5) are indicated. See Figure 7 for Laramide igneous rocks.

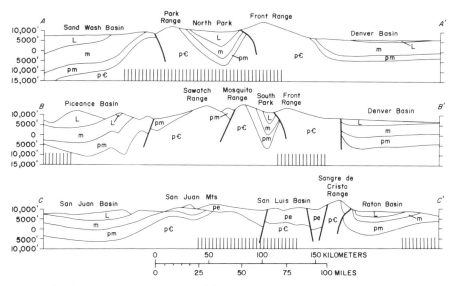

Figure 5. Diagrammatic cross sections of Southern Rocky Mountain province. Locations of sections indicated in Figure 4. pe, post-Eocene rocks; L, Laramide rocks (Eocene, Paleocene, and post-Montana Cretaceous); m, rocks of Montana age; pm, pre-Montana Mesozoic and Paleozoic sedimentary rocks; pЄ, Precambrian rocks. Belts of short vertical lines extend across areas that were parts of late Paleozoic uplifts.

are reverse faults of Laramide origin, and some are steep normal faults of later origin.

The Sawatch Range consists largely of Precambrian rocks in the core of a huge anticline of early Laramide origin that rose axially in a trough of late Paleozoic sedimentation. The Mosquito Range is a part of the east flank of the Sawatch anticline, separated from the crestal area by the deep Arkansas Valley graben, a northern element of the Neogene basin-and-range San Luis–Rio Grande rift system. The Elk Mountains are a subsidiary faulted anticline or structural bench on the west flank of the Sawatch anticline; they consist of folded and thrust-faulted Paleozoic and Mesozoic sedimentary rocks extensively intruded by Oligocene igneous rocks.

The Sangre de Cristo Range is the faulted east flank of an early Laramide uplift at the site of the late Paleozoic San Luis highland (Fig. 2). Much of this uplift was obliterated by subsidence of the San Luis Valley segment of the Rio Grande rift system in late Tertiary time. The Sangre de Cristo Range consists of Precambrian rocks along its faulted western side and elsewhere chiefly of deformed upper Paleozoic rocks.

The San Juan Mountains are an Oligocene volcanic edifice resting in part on the truncated Laramide-Paleozoic uplift just mentioned and in part on the Needle Mountains uplift to the south. The Needle Mountains are the exposed part of a Laramide domal uplift that projects south or southwest from the rejuvenated San Luis–Uncompahgre uplift into the adjoining sedimentary basin. The Brazos uplift in northern New Mexico and the Uncompahgre Plateau also are elements of the late Paleozoic San Luis–Uncompahgre uplift (Fig. 2) that were re-elevated in Laramide time. The Uncompahgre Plateau was elevated still further in the late Tertiary. The La Plata Mountains are a Laramide intrusive center in upper Paleozoic rocks.

The White River Plateau is a flat-topped domal uplift of late Laramide age;

it consists mainly of Paleozoic rocks, but Precambrian rocks of the core are exposed in deep canyons, and a part of the dome is covered by upper Tertiary basalts. Flanking Mesozoic and lower Cenozoic rocks are sharply upturned in the Grand Hogback along the western and southwestern flanks.

The Uinta Mountain range, only the eastern tip of which is included in Figure 1, is—like the Sawatch Range—a huge anticlinal uplift; major Laramide faults border the anticline on the north and in places on the south; at its eastern end, the anticline is split by an axial graben of post-Laramide age. The Laramide Douglas Creek arch, which has little or no topographic expression, projects southward from near the eastern end of the Uinta anticline.

The Apishapa uplift and the Sierra Grande and Las Animas arches east of the mountain front (Fig. 1) are subsurface features that have little or no topographic expression. They are late Paleozoic—or older—highs that were moderately reactivated anticlinally in Laramide and perhaps later time. They are truncated by high-plains surfaces and by the Ogallala Formation of late Miocene and Pliocene(?) age.

BASIN UNITS

In addition to the mountains, Laramide orogeny produced several major sedimentary and structural basins (Fig. 1). The sedimentary rocks of these basins (Fig. 6) provide a more detailed record of the orogeny than do the mountains. Main features of the basins are outlined here, and the stratigraphic units that record the Laramide orogeny are discussed in the following section. Like the mountains, several of the basins inherited their structure in part from late Paleozoic features. Collectively, the basins are also parts of the very large Late Cretaceous basin of sedimentation—the Rocky Mountain geosyncline—which was deepest at the northern edge of the province and in the adjoining Middle Rocky Mountain province (Reeside, 1944; Gilluly, 1963, p. 146).

The San Juan structural and sedimentary basin occupies an area of about 20,000 km^2 in the Colorado Plateau province of northwestern New Mexico and the adjoining part of Colorado. The basin has steep sides around the northern half of its periphery and gentle sides around the southern half; the basement-rock floor slopes gently northward to a low at about 2,100 m below sea level near the Colorado-New Mexico boundary (Peterson and others, 1965). The basin contains about 1,500 m of Laramide orogenic sediments (Fig. 6) underlain by Mesozoic and Paleozoic rocks. The deep part of the basin coincides with a late Paleozoic sedimentary trough along the southwest side of the ancient Uncompahgre highland. This trough was an arm of the larger and deeper Paradox sedimentary basin to the northwest (Fig. 2). Thus, the basin is in part a heritage from the late Paleozoic.

The Raton structural and sedimentary basin of southern Colorado and northern New Mexico (Fig. 1) is in large part a rejuvenated late Paleozoic basin. The Paleozoic basin lay between the Apishapa-Sierra Grande uplifts on the east and the San Luis highland (beneath the present San Luis Valley) on the west; the depositional axis was along the eastern flank of the present Sangre de Cristo Range (Fig. 1). The Laramide basin, including a northwest arm known as Huerfano Park, has the Sangre de Cristo Range as a western boundary and the same eastern and northern boundaries as the Paleozoic basin had. Orogenic sediments of Late Cretaceous to middle Eocene age in the basin total as much as 3,600 m in thickness (Johnson and Wood, 1956), but owing to nonuniform deposition and many local unconformities, much less than this is present at any one place. These sediments constitute the most complete record of Laramide orogeny anywhere along the east front of the Southern Rockies.

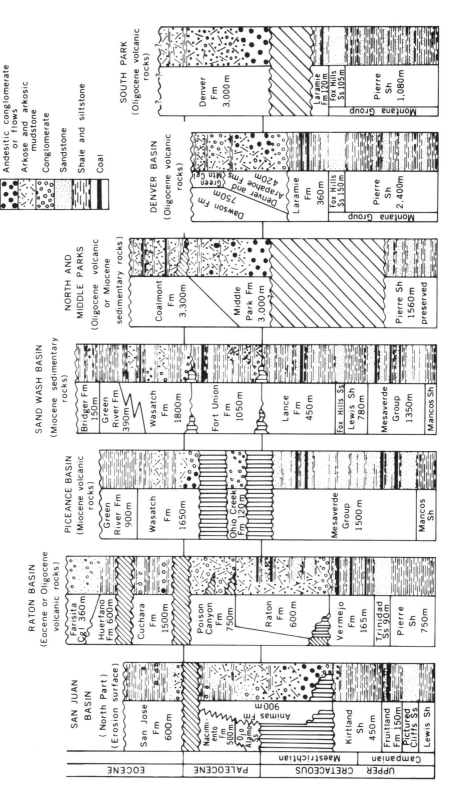

Figure 6. Stratigraphic columns of the Laramide interval in the principal basins of the Southern Rocky Mountain province. Lithologic columns are generalized for entire basins and emphasize features bearing on Laramide orogeny. Thicknesses indicated are approximate maximums. Horizontal correlation is approximate. Lewis Shale and Mesaverde Group of Sand Wash basin are not correlative with the type Lewis Shale and underlying type Mesaverde Group of the San Juan basin (see Gill and others, 1970, p. 2-5, for summary).

The Piceance basin (Fig. 1) occupies a large area in the Colorado Plateau portion of northwestern Colorado. The southwest flank of the basin is the flank of the rejuvenated late Paleozoic Uncompahgre uplift, and the northeast flank, which lies in the late Paleozoic sedimentary basin, is the Grand Hogback, a steep Laramide monocline. The Piceance basin reaches a basement-rock depth of 5,500 m below sea level, or about 7,900 m below the land surface on Eocene rocks (MacLachlan and Kleinkopf, 1969). In its northern part, the basin contains 2,400 to 2,700 m of lower Eocene and Paleocene rocks (Donnell, 1961). These are underlain by more than 3,000 m of Upper Cretaceous rocks and, at greater depth, by a southwestward-thinning wedge of upper Paleozoic rocks.

The Sand Wash basin of extreme northwestern Colorado is essentially a southern arm of the larger Washakie basin of southern Wyoming (Fig. 1). It is separated from the Piceance basin to the south by a curving anticlinal feature between the White River and Uinta uplifts. The basement-rock surface beneath the Sand Wash basin is as deep as 5,200 m (17,000 ft) below sea level, and in the Washakie basin it reaches 6,400 m (Love, 1960, p. 205). These basins contain at least 3,000 m of Paleocene and Eocene sedimentary rocks that shingle against the adjoining Park–Sierra Madre and Uinta uplifts in zones of unconformity and overlap, and an equal or greater thickness of Upper Cretaceous rocks (Fig. 5, section A–A′).

North and Middle Parks (Fig. 1) constitute one Laramide structural and sedimentary basin, and South Park constitutes another. Both basins are in a synclinal belt between the regional anticlines of the Front Range on the east and the Park and Sawatch Ranges on the west. Both basins contain a few thousand meters of conglomerates and arkoses of Laramide age, and both extend in depth to basement lows at about 2,500 m below sea level (MacLachlan and Kleinkopf, 1969) from surface elevations of 2,500 to 3,000 m. The North Park–Middle Park basin is on the site of the late Paleozoic Front Range highland (Fig. 2) and thus had little or no pre-Laramide expression. South Park basin straddles the boundary between the late Paleozoic Front Range highland and the sedimentary trough to the west of the highland, as shown by the eastward wedge-out of thousands of meters of Paleozoic rocks within the area of the park. Laramide sedimentary and volcanic rocks are unconformable on older formations in the basin, but a nearly complete Cretaceous stratigraphic section is preserved locally along the east side.

As a structural feature, the Denver basin occupies an enormous area—about 130,000 km[2]—in the Great Plains province of northeastern Colorado and adjoining parts of Wyoming and Nebraska. As a Laramide sedimentary basin, it was much smaller. Laramide orogenic sediments (Denver and Dawson Formations) now occupy an area of about 10,500 km[2] in the west-central part of the basin, and it is questionable whether they ever extended over an appreciably larger area, except possibly as a very thin blanket. However, thin quasi-orogenic sediments (Laramie Formation) may have covered the entire area. The basin is asymmetric, with a broad, gentle eastern flank and an abrupt western flank at the mountain front. Though it reaches a depth of 2,300 m below sea level at the basement-rock surface (Fig. 1), it is shallow in respect to its great width. The north part of the basin (Fig. 5, section A–A′) is filled principally with rocks of the Upper Cretaceous Montana Group (Pierre Shale and Fox Hills Sandstone).[1] The south part (Fig. 5, section B–B′)

---

[1]The term "Montana Group" is used here in the way it has been applied in this region for several decades, though Gill and Cobban (1973) now propose to restrict the term to central and eastern Montana. Further, these authors would include the Hell Creek Formation— an approximate equivalent of the Lance and Laramie Formations (Fig. 6)—in the Montana Group as thus geographically restricted.

contains, in addition, moderately thick Laramide and upper Paleozoic sedimentary sequences. The late Paleozoic depositional axis corresponds closely with that of the Denver basin (Martin, 1965, Fig. 8), indicating that this part of the basin was in part inherited from that time.

Other basins in the region need be noted only briefly. The Laramie basin, at the northern edge of the province, is described by Blackstone in this volume. Other basins in southern Wyoming have been described by Love (1960) and Bradley (1964). The deep, narrow basin on the southwest side of the Uncompahgre Plateau (Fig. 1) is in the late Paleozoic Paradox basin and is characterized by salt anticline tectonics (Cater, 1970); rocks of Laramide age are not preserved in it. The basin of San Luis Valley is a product of late Tertiary faulting and is deeply filled with upper Tertiary sediments and volcanic rocks. It is significant to Laramide history only in that major Laramide and late Paleozoic uplifts have disappeared beneath it. It is discussed in several of the papers in this volume.

SEDIMENTARY RECORD

A major part of the record of the Laramide orogeny—particularly of the early phases—is in the sedimentary rocks and associated unconformities in the basins bordering the mountains. In all the basins, Upper Cretaceous rocks of Montana age, derived from distant western sources, are succeeded upward by orogenic sediments derived from local sources. The marine rocks constitute the highest stratigraphic datum common to all of the basins, though the age or stratigraphic level of this datum varies systematically from basin to basin in accord with the pattern of marine transgression and regression discussed previously. The last of the Cretaceous marine transgressions affected only the northeastern half of the region, whereas the next preceding transgression covered almost the entire region (Fig. 3). In the extreme southwest, in the San Juan basin, the youngest marine unit is the regressive Pictured Cliffs Sandstone (Fig. 6) of late Campanian age (Gill and Cobban, 1966). Sandstone of this same regressive stage in the Raton basin is in the somewhat younger Trinidad Sandstone, and in the Piceance basin, probably in marine units in the lower part of the Mesaverde Group. In the northern and northeastern parts of the region, in the Denver, South Park, and Sand Wash basins and the basins of southern Wyoming, the youngest marine unit is the lower Maestrichtian Fox Hills Sandstone. This sandstone is increasingly younger in age eastward across southern Wyoming and northern Colorado (Gill and others, 1970, p. 43).

In all the basins except North Park–Middle Park, the youngest marine sandstones grade upward into, and intertongue with, coal-bearing brackish- and fresh-water sandstones and shales. These rocks constitute the Laramie Formation of the Denver basin and South Park, the Lance Formation of the Sand Wash basin and basins in southern Wyoming, the Vermejo Formation of the Raton basin, the Fruitland Formation and Kirtland Shale of the San Juan basin, and the upper part of the undivided Mesaverde Group in the Piceance basin. The generally fine-grained rocks of these units were derived in part from the same sources as the underlying rocks and in part from the reworking of Cretaceous rocks on rising Laramide uplifts.

The fine-grained sediments of the Laramie Formation and its analogs in the various basins are overlain conformably or with local angular unconformity by orogenic sediments that mark the onset of major Laramide uplift. These younger sediments are in general conglomeratic and arkosic, and in several of the basins they contain abundant andesitic materials and local andesitic flows (Fig. 6). The timing of this episode of uplift and sedimentation is discussed in a following section, but in general, and in accord with the pattern of underlying rock units, the basal orogenic units are oldest in the southwest and youngest in the northeast.

Many authors have interpreted the presence of feldspars, micas, and rock grains in certain Cretaceous sandstones as evidence that the sands were derived in part from local uplifts. However, sandstones containing several percent of these materials occur throughout the region and throughout the Upper Cretaceous rock sequences, as do clean quartz sands. Sabins (1964) has noted that the composition of the Cretaceous sandstones was affected markedly by environment of deposition: Marine sands are cleaner than the continental, and fine-grained sands are cleaner than the coarse grained; fine-grained marine sands consist almost entirely of quartz, whereas intertonguing, continental sands of the same derivation contain a large nonquartz component. In general, no compositional pattern relating any part of the intertonguing marine and continental Cretaceous sequence to local sources of sediment is evident in the region. In contrast, rocks above this intertonguing sequence display abundant evidence of local derivation. These rocks constitute the Laramide orogenic sequence.

Though the volume of Laramide sediments in the various basins is large, it represents only a fraction of the material eroded from the uplifts in Laramide time. In some basins, the orogenic sediments consist almost entirely of materials derived from Precambrian crystalline rocks or uppermost Cretaceous volcanic rocks, and in others, they contain a large fraction of these materials. The erosional products of the fine-grained Cretaceous rocks are missing or are present in amounts far disproportionate to the enormous volumes of these rocks that once covered the areas of uplift. Though the Cretaceous rocks should have been in general the first to be removed from the uplifts, it is unlikely that erosion proceeded layer by layer. Presumably, once uplift and erosion were under way, valleys or canyons were cut to the Precambrian basement long before any given highland was wholly stripped of its sedimentary cover. Thus, Cretaceous and other sedimentary rocks probably contributed detritus to the drainage systems concurrently with the basement rocks. But the basins were not perfect traps; they caught a larger proportion of the coarse materials from resistant rocks than of the fine materials from the sedimentary rocks. The remainder of fine-grained materials was swept beyond the basins and out of the system.

Similar losses of sediment are recorded in other Rocky Mountain basins. For example, Keefer (1970, p. 33) calculated that the volume of sediment deposited in the Wind River basin of central Wyoming was only about half of the volume eroded from the bordering uplifts.

San Juan Basin

Sediments of Laramide age are in general coarsest along the northern and eastern sides of the San Juan basin, and their compositions show that they were derived from highlands in those two directions. A northern part of these highlands may have been a source of sediment as early as Pictured Cliffs time (Dane, 1946; Baltz, 1967, p. 19), though this is disputed by Fassett and Hinds (1971, p. 37). Local unconformities between the Pictured Cliffs and the Fruitland Formation (Fassett and Hinds, 1971, p. 12) and within the Kirtland Shale (Baltz, 1967, p. 29) in the southeastern part of the basin record the beginning of uplift east of that area. The earliest positive indications of widespread uplift and accompanying volcanism in bordering highlands are in the upper part of the Kirtland Shale (Fig. 6). In places on the northern and eastern sides of the basin, sandstones near the top of the Kirtland contain siliceous pebbles and andesitic debris, have a high feldspar content, and display gradients in grain size that indicate a contribution of materials from a source in a northerly to easterly direction (Dane, 1946; Zapp, 1949; Baltz, 1967; O'Sullivan and others, 1972). The Kirtland is poorly fossiliferous but is generally classed as late Montana in age (Reeside, 1924, p. 24) or very late Campanian (Gill and Cobban, 1966, Pl. 4). Fossil pollen in a locality on the north side of the San Juan Mountains indicates that the Campanian-Maestrichtian boundary lies within the Kirtland Shale (Dickinson and others, 1968, p. 140). However, on the south side of the San Juan basin, 240 km south of this locality, pollen indicates that the uppermost beds there assigned to the Kirtland are Paleocene in age (Fassett and Hinds, 1971, p. 22–23).

The coarse-grained and andesitic Animas Formation (Fig. 6) records major uplift and volcanism in the area north and northeast of the San Juan basin. Through most of the northern part of the basin, the Animas bevels across very slightly tilted Kirtland and Fruitland strata (Fassett and Hinds, 1971, p. 38), but in the northwestern part its contact with the Kirtland is gradational (Hayes and Zapp, 1955). The McDermott Member at the base of the Animas (formerly McDermott Formation) consists largely of andesitic conglomerate but has a basal bed that contains only siliceous clasts (Zapp, 1949). Clastic rocks of the main body of the Animas are mixtures of andesitic debris and materials derived from sedimentary and Precambrian granitic rocks. The McDermott Member is of Montana age (Reeside, 1924, p. 28), and the remainder of the Animas is of post-Montana Cretaceous and Paleocene age.

The Animas Formation grades southward into finer-grained rocks assigned to the Paleocene Ojo Alamo Sandstone and overlying Nacimiento Formation. The Ojo Alamo is separated from underlying rocks by an unconformity that represents a greater hiatus than that at the base of the Animas (Fig. 6). Sandstones of the Ojo Alamo are arkosic and include pebbly beds and conglomerates characterized by clasts derived from sedimentary rocks. These materials evidently were derived in part from eastern and northeastern sources (Baltz, 1967, p. 33) and in part from western sources (Fassett and Hinds, 1971, p. 38). The Nacimiento Formation is largely shale and siltstone, but in the northern part of the basin it contains much sandstone and some conglomerate. Abundant feldspar from granitic rocks and scattered volcanic pebbles in these rocks indicate a source in highlands to the north and east, just as for the Animas Formation. The predominant shale and siltstone in the southern part of the basin were derived from Cretaceous rocks in uplifts to the south and west (Baltz, 1967, p. 39).

The lower Eocene San Jose Formation lies disconformably on the Nacimiento in much of the San Juan basin and in angular unconformity along the eastern and western sides. Unconformities also occur within the San Jose (Baltz, 1967, p. 54). The San Jose consists of intertonguing sandstone, conglomerate, and shale units that contain abundant feldspar derived from granitic rocks. Some of the conglomerates consist principally of quartzite derived from the Precambrian terrane of the Brazos uplift of northern New Mexico (Fig. 1). This and other highland areas on the site of the ancient San Luis highland (Fig. 2) were the source of most of the coarse materials in the San Jose, and Cretaceous and older sedimentary rocks father south were a source of shales (Baltz, 1967).

In summary, the Laramide sedimentary sequence of the San Juan basin (Fig. 6) records uplift and erosion of a highland to the north and east through the period extending from late Campanian Cretaceous into early Eocene time. The San Juan basin subsided concurrently with sedimentation, and its border structures developed gradually through the same period; subsidence ceased after deposition of the San Juan Formation.

Raton Basin

Downwarping of the Raton basin evidently began in Montana time during deposition of the Trinidad Sandstone and the conformably overlying Vermejo Formation (Fig. 6). Isopach maps of the Trinidad and Vermejo, based on subsurface data (Johnson and Wood, 1956, Figs. 4 and 5), indicate that these units are more than twice as thick in the axial part of the basin as on the flanks. Matuszczak (1969, Fig. 1) has presented an alternative map, based on projected sedimentational trends, that does not show this basin control of Trinidad thickness. Compositions of the Trinidad and Vermejo are similar to those of the Pictured Cliffs and Fruitland-Kirtland in the San Juan basin. These compositions would permit, but not require, a component of locally derived materials—such as reworked older Cretaceous sedimentary rocks—in the Trinidad and Vermejo.

The Raton Formation of Late Cretaceous and Paleocene age marks a sudden influx of locally derived sediments into the Raton basin. The Raton lies with marked to slight erosional unconformity on the Vermejo around the borders of the basin (Lee, 1917; Pillmore, 1969) and is in angular unconformity with older formations along the southwestern flank of the basin. Though in general fine grained, the Raton is arkosic, and a widespread conglomerate at its base is characterized by materials derived from many types of sedimentary and Precambrian rocks. The Raton has a complex relation to the overlying Paleocene Poison Canyon Formation (Johnson and Wood, 1956, p. 715). The Poison Canyon consists mainly of granitic conglomerate and coarse arkose, and through much of the basin, the Raton grades upward into these materials. Toward the southwest, however, the Raton grades laterally into the coarser facies of the Poison Canyon, and near the mountain front about 40 km south of the Colorado-New Mexico boundary, only the coarser facies (Poison Canyon) is present. There, this unit lies with angular unconformity on strata as old as Pierre Shale. In the northern part of the Raton basin, the Raton Formation is overlain gradationally by a lower unit of the Poison Canyon, but a coarsely conglomeratic upper unit of the Poison Canyon extends in angular unconformity across eroded edges of the Raton and underlying formations down to the Pierre Shale (Johnson and Wood, 1956, p. 715).

The composition, facies changes, and unconformable relations of the Raton and Poison Canyon indicate sources of detritus in highlands southwest and northwest of the Raton basin. The southwestern source was the same rejuvenated San Luis highland that was the source of sediments in the San Juan basin, discussed above. The present Sangre de Cristo Range could not have been a source because its eastern slope and crest are still mantled with upper Paleozoic rocks, and the generally fine-grained formations once present above the Paleozoic rocks could not have been a source of coarse arkosic detritus. On the western flank of the Sangre de Cristo Range, the Paleozoic rocks themselves lap against the Precambrian rocks of the Paleozoic San Luis highland. The northwestern source of materials in the Raton and Poison Canyon evidently was in the Wet Mountains, a rejuvenated element of the eastern or Front Range late Paleozoic highland rather than of the western or San Luis highland (Fig. 2). The source could not have been in the present Sangre de Cristo Range not only for the reasons just given, but also because boulders in the Poison Canyon are coarser than those in the Paleozoic conglomerates present at this general latitude. In a small outlier preserved near Canon City, on the east flank of the Wet Mountains and north of the Raton basin, the Vermejo and a thin Raton Formation are overlain with marked erosional unconformity by conglomerate and arkosic sandstone of the Poison Canyon. The conglomerate contains abundant Precambrian debris, part of which is in coarse angular fragments, as well as volcanic rocks and materials from various sedimentary formations, including coal and soft sandstone from the Vermejo (Washburne, 1910). Clearly, the Wet Mountain front existed at essentially its present location during Poison Canyon deposition.

The Poison Canyon and older rocks of the Raton basin are overlain with regionally angular unconformity by as much as 1,500 m of sandstone and claystone of the lower Eocene Cuchara Formation. The Cuchara overlaps various formations from Poison Canyon down to Pierre Shale, and it was evidently derived in part from these rocks as well as from various older rocks (Johnson and Wood, 1956, p. 717). Volcanic materials also are present in the Cuchara (Burbank and Goddard, 1937, p. 958). The Cuchara is unconformably overlain by the lower and middle Eocene Huerfano Formation, the lower part of which consists of red beds almost identical to the finer grained red beds in the upper Paleozoic strata of the Sangre de Cristo Range and evidently derived from that range. The upper part of the Huerfano (older usage) was denoted the Farisita Conglomerate by Johnson and Wood (1956, p. 718). This coarse conglomerate consists largely of Precambrian debris, some of it in subangular fragments and some of it in 2-m boulders. The unit is confined to the northern end of the Huerfano Park arm of the Raton basin, between the Sangre de Cristo Range and the Wet Mountains, and it must have been derived from the Precambrian terrane of the Wet Mountains. The Farisita is overlain unconformably by volcanic rocks of very late Eocene or early Oligocene age.

In summary, the Raton basin contains a sequence of orogenic sediments and associated unconformities that span the time from some part of the Montana Cretaceous at least to the middle Eocene. Sediments in the lower part of this sequence were derived from granitic highlands to the southwest and northwest that were elevated in Late Cretaceous time at the sites of late Paleozoic uplifts. The upper part of the sequence was derived principally from sedimentary rocks in the nearby Sangre de Cristo Range. This range began to develop structurally in Paleocene time, as indicated by the unconformity at the base of the Poison Canyon along the flank of the range, but it did not rise enough at this time to prevent delivery of the granitic materials in the Poison Canyon from a source area west of the present range, and it evidently did not become a major source of sediment in the Raton basin until Eocene time.

Piceance Basin

All Cretaceous rocks above the Mancos Shale in the Piceance basin are assigned to the Mesaverde Formation or Group. This unit, which consists of as much as 1,500 m of sandstone and shale, is predominantly of continental origin but contains marine tongues in its lower part. The lower part is also coal bearing. Though the Mesaverde is normally classed as Montana in age, as recognized in this basin it probably includes a post-Montana Lance equivalent (Fig. 6). Most of the Mesaverde rocks were derived from the same distant western sources as other Cretaceous rocks of the province, but the upper part probably contains a fraction reworked from older Cretaceous rocks on local uplifts.

Conglomeratic sandstone of the Paleocene Ohio Creek Formation constitutes the oldest coarse orogenic deposit in the basin. The Ohio Creek is regionally disconformable on the Mesaverde; J. R. Donnell (1973, oral commun.) states that it truncates 300 m of Mesaverde westward toward the Uncompahgre and Douglas Creek uplifts, indicating prior rise of these structures. However, as defined at the southern end of the basin by Gaskill and Godwin (1963), the Ohio Creek includes strata formerly assigned to the Mesaverde and is in gradational contact with the restricted Mesaverde. It seems likely that one of the erosional unconformities noted by Gaskill and Godwin (1963) within their Ohio Creek marks a major hiatus. Pebbles in the Ohio Creek are principally chert in most parts of the Piceance basin, but in the southern part they are mainly quartzite and quartz. Near the Colorado River, granite pebbles are also present. Many of the chert pebbles are fossiliferous, and in general they closely resemble the cherts of various Paleozoic formations of central Colorado. The Ohio Creek was evidently derived from the Sawatch Range east and southeast of the Piceance basin. Areas to the south and west of the basin are on the site of the late Paleozoic Uncompahgre highland and therefore contained no Paleozoic rocks. These areas could have supplied small chert pebbles reworked from Mesozoic formations such as the Dakota and Morrison, but not the pebbles several inches in diameter found in the Ohio Creek. The Ohio Creek is a thin unit—in most places, less than 30 m thick—and it evidently constitutes a lag concentrate of the coarse materials in sediments that moved across the site of the Piceance basin to depositional basins farther north.

The main orogenic unit in the Piceance basin is the Wasatch Formation, which has a maximum thickness of about 1,650 m. As classified in this area, the Wasatch is early Eocene and Paleocene in age; a basal unit about 150 m thick probably is a Fort Union equivalent (Donnell, 1961, 1969). The Wasatch lies disconformably on the Ohio Creek Formation through most of the area but laps onto the Mesaverde on the Douglas Creek arch, where the Ohio Creek is absent. In the southern part of the basin, the Wasatch consists largely of arkosic sandstones and conglomerates characterized by clasts of andesitic and dacitic volcanic rocks and Precambrian granitic rocks (Gaskill and Godwin, 1966). Elsewhere in the basin, lower and upper members of the Wasatch consist largely of claystones, and a middle member is arkosic sandstone which is locally conglomeratic (Donnell, 1969). The amounts of conglomerate in the Wasatch and of volcanic materials in the conglomerates decrease northward in the basin. The amount of sandstone decreases westward in the central and northern part of the basin (McDonald, 1972, p. 254). Sources of the Wasatch sediments were to the east and south, probably mainly in the Sawatch Range and the part of the Laramide Uncompahgre uplift now buried beneath younger volcanic rocks. The White River uplift apparently began to rise during Wasatch time, as discussed later; this uplift probably was also a source of sediment. Though the Wasatch thins markedly on the flanks of the Douglas Creek arch, there is no evidence that either this arch or that part of the Uncompahgre uplift to the south of it were significant sources of Wasatch sediments.

The oil-shale-bearing lower and middle Eocene Green River Formation, which has a maximum preserved thickness of about 900 m, lies conformably upon, and intertongues with, the Wasatch Formation. The Green River consists of lacustrine marlstones and shales and peripheral fluvial sandstones deposited in and around a lake that occupied the sites of both the Piceance basin and the Uinta basin to the west in Utah. As shown by the distribution of sandstones and by grain-size gradients (Cashion, 1967; Bradley, 1931; Donnell, 1961), the Green River sediments were supplied from many different directions, and they also include much air-fall volcanic ash. The generally fine grain of the Green River sediments, together with the great extent and long duration of the lake in which the sediments were deposited, indicates a period of relative tectonic stability in late early to early middle Eocene time. However, stability was not absolute. The thickening of the Green River toward the basin center indicates that the Piceance basin continued to subside during this time, and unconformities within the Green River on the east side of the basin indicate continuing local deformation. After disappearance of the Green River lake, westward-flowing streams supplied granitic and minor volcanic debris from the mountains of north-central Colorado to the upper Eocene Uinta Formation of the Uinta basin (Stagner, 1941). Similar material probably was deposited in the Piceance basin but has since been eroded.

Sand Wash Basin

The sedimentary record of the Sand Wash basin is similar to that of the Piceance basin, though there are differences in some rock units and nomenclature. The Sand Wash basin, unlike the Piceance, was in the area of the youngest Cretaceous marine transgression (Fig. 3). The continental Mesaverde Group is overlain by the marine Lewis Shale, and this by the marine Fox Hills Sandstone. The Fox Hills grades upward into the brackish- and fresh-water Lance Formation of latest Cretaceous age. These four units combined are probably equivalent to the Mesaverde of the Piceance basin (Fig. 6). The Lance Formation is similar lithologically to the Mesaverde; it probably was derived partly from the same western sources as the Mesaverde, and partly from reworking of Cretaceous rocks in rising uplifts, particularly the Park Range.

The Fort Union Formation, of Paleocene age, is separated from the Lance by an erosional unconformity in the eastern and western parts of the basin, but in the central part, the two units seem to be in conformable or gradational contact. On the north flank of the Uinta Mountains, the Fort Union lies with angular unconformity upon units of the Mesaverde Group (Hansen, 1965, p. 109; Colson, 1969, p. 124). As described by Colson (1969), the Fort Union of the Sand Wash basin consists of arkose and conglomerate derived from the Park–Sierra Madre Range to the east, reworked sedimentary rocks derived from the Uinta uplift to the west, and feldspar- and chert-rich sandstone derived from a southern source. Porphyritic and fossiliferous chert pebbles in a basal conglomerate in the eastern part of the basin (Ritzma, 1955, p. 36) probably also were from the southern source, though they are mixed with granitic detritus from the Park Range. The southern source was almost certainly the same as that of the Ohio Creek Formation in the Piceance basin, namely the Sawatch Range.

The Wasatch Formation records further uplift of mountain units as well as continued sinking of the basin. This formation, as thick as 1,800 m, is in conformable or gradational contact with the Fort Union in the western part of the basin and in unconformable contact in the eastern part. The Wasatch consists of fluvial sandstones and claystones in a lower main body and in an overlying series of tongues that alternate with tongues of the lacustrine Green River Formation. In the eastern part of the basin, sandstones in both the Wasatch and the Green River are arkosic; some are coarse arkoses consisting almost entirely of granitic materials, and some of these are conglomeratic (Colson, 1969; Theobald, 1970). These materials were evidently derived from the Precambrian terrane of the Park–Sierra Madre Range. On the western side of the basin, the Wasatch contains coarse conglomerates characterized by clasts of quartzites of the Precambrian Uinta Mountain Group and of various Paleozoic formations (Ritzma, 1955; Hansen, 1965). Farther west along the north side of the Uinta Mountains, the entire Wasatch–Green River–Bridger sequence passes into conglomerate, which clearly was derived from the Uinta uplift (Bradley, 1964, p. 54).

The middle and upper(?) Eocene Bridger Formation is preserved in the Sand Wash basin only as a remnant less than 150 m thick lying conformably on the Green River Formation; it is unconformably overlain by the Miocene Browns Park Formation. It consists of lacustrine and fluvial mudstones and sandstones that contain abundant altered volcanic ash. The source of the ash probably lay to the north (Bradley, 1964, p. 54), rather than in the Southern Rocky Mountains.

North Park–Middle Park Basin

In the North Park–Middle Park basin, orogenic sediments of latest Cretaceous(?) to Eocene age lie with marked unconformity on older rocks ranging from the Pierre Shale down to the Precambrian. The upper part of the Pierre Shale, the Fox Hills Sandstone, and the Laramie Formation or equivalent are absent in the basin; the Pierre and Fox Hills were almost certainly present before early Laramide uplift and erosion, but the Laramie may not have been deposited there.

Much of the 1,560 m of Pierre preserved in the basin is a sandstone-bearing zone that correlates closely with the sandy zone in the middle Pierre of the Denver basin (Izett and others, 1971). This sandy zone in the Pierre on the two sides of the Front Range has been interpreted to indicate uplift of the Front Range in middle Pierre time and the derivation of sediments in adjoining basins from that source thereafter (Lovering, 1929; Griffitts, 1949). Lovering and Goddard (1950, p. 58) postulated that by Fox Hills time the Precambrian rocks in the Front Range had become a source of sediments.[2] However, the sandstones of the Pierre and the Fox Hills are elements of Cretaceous sedimentational patterns that extend far beyond the environs of the Front Range and beyond the Southern Rocky Mountain region (Reeside, 1957; Weimer, 1960, 1970). Moreover, these sandstones show no compositional or thickness patterns that relate them to the Front Range. As shown by Izett and others (1971, Fig. 2), the sandy zone of the Pierre and the numerous ammonite zones within it project directly from the Denver basin through the section preserved in North Park–Middle Park to equivalent positions in the Mesaverde Group in the Sand Wash basin. It is unlikely that the abbreviated section of the Pierre in North and Middle Park reflects either nondeposition or a Front Range source.

On the east side of Middle Park, andesitic breccias and flows, conglomerates, and arkoses of the Middle Park Formation lie with angular unconformity on Pierre Shale and older sedimentary rocks and lap over Precambrian rocks (Tweto, 1957; Izett, 1968). In this formation, which may be as much as 3,000 m thick (Kinney, 1970), andesitic materials predominate in the lower part, and a fraction derived from Precambrian rocks increases upward. In the western part of the park, the very andesitic lower part of the formation is absent, and arkoses higher in the formation lie on an irregular surface cut over Precambrian and various Mesozoic rocks. The basal andesite breccia and conglomerate member of the east side of the park is questionably of very Late Cretaceous age (Izett and others, 1962; Izett, 1968); the remainder of the formation is Paleocene and possibly early Eocene in age (Kinney, 1970).

In North Park, strata largely equivalent to the Middle Park Formation are called the Coalmont Formation. The Coalmont consists principally of conglomerates, arkosic sandstones, and mudstones or claystones derived from Precambrian rocks; carbonaceous shales and some coal occur in the upper part (Hail, 1965). The formation

---

[2]This inference was evidently based on the assignment by Lovering and Goddard (1950, p. 40) of a conglomerate at Colorado Springs to the Fox Hills. No other mention of conglomerate in the Fox Hills has been made in a voluminous literature on the Colorado Springs area and on the Cretaceous rocks. Lovering and Goddard possibly misidentified as Fox Hills the Dawson strata that unconformably overlap the Fox Hills northwest of Colorado Springs (discussed under "Orogenic Sequence and Structural Development").

does not contain the abundant andesite debris that characterizes the Middle Park Formation, but andesitic pebbles are present in a lower member in the southwestern part of the park (Hail, 1968) and in about the lower one-third of the formation in the southeastern part (Kinney, 1970). The Coalmont lies on a very uneven surface cut on Mesozoic and Precambrian rocks and contains a widespread unconformity between its middle and upper members. Its thickness is therefore difficult to establish but may approach 3,300 m. The strata beneath the unconformity within the Coalmont are Paleocene, and the strata above are Eocene—probably early Eocene (Hail, 1968, p. 46).

The Middle Park and Coalmont Formations are overlain unconformably by Oligocene igneous or Miocene sedimentary rocks.

Denver Basin

In the northern part of the Colorado portion of the Denver basin, the Pierre Shale thins eastward from a thickness of 2,500 m or more near the mountain front to 600 to 900 m near the eastern border of Colorado (Mather and others, 1928, p. 109-115; Reeside, 1944). Though the thinning is distributed through all parts of the formation, it is most evident in lower units (Blair, 1951; Gill and others, 1972, Fig. 23). The gradient in thickness of the Pierre is almost identical with the opposing slope gradient of the basement-rock surface on the eastern flank of the basin as determined from numerous oil test holes that reached Precambrian rocks (MacLachlan and Kleinkopf, 1969). Thus, the gentle westward slope of the basement surface and pre-Pierre rock units on the broad east flank of the basin is almost exactly compensated by westward increase in thickness of the Pierre. The thickness gradient of the Pierre or equivalent rocks continues westward beyond the Front and Park Ranges to a depositional center in the Sand Wash basin and basins to the north, as diagrammed by Haun and Weimer (1960, Fig. 5). These relations indicate that the eastern flank of the northern part of the Denver basin is of Pierre age and is a segment of a regional basin of Upper Cretaceous (Montana) sedimentation. The steep western flank is a younger tectonic feature, created by uplift of the Front Range as a welt on the flank of the Cretaceous basin.

In the central and southern parts of the Denver basin, the Pierre is thinner than in the northern part, and eastward thinning is less pronounced. In this area, the floor of the Late Cretaceous sedimentational basin was modified by later rise of the Las Animas arch, over which the Upper Cretaceous formations are gently warped. This part of the basin, then, is bounded on both east and west by post-Montana uplifts. This part was also a basin of latest Cretaceous and Paleocene orogenic sedimentation, as discussed below.

The marine Fox Hills Sandstone, the upper unit of the Montana Group, is overlain gradationally in the Denver basin by brackish-water and continental sandstones, shales, and coals of the Upper Cretaceous Laramie Formation. The contact near the mountain front west of Denver has been described as an erosional unconformity by Moody (1947), but many other workers in the area do not even agree on where the contact lies in a sequence of lenticular sandstones. The lower part of the Laramie probably was derived in large part from the same sources as the underlying marine rocks, but much of the formation contains an admixture of materials that probably came from the rising mountains in Colorado. From about the latitude of Denver southward, sandstones in the upper part of the Laramie contain pebbles of quartz, chert, and various sedimentary rocks in scattered lenses (Scott, 1972; 1963). Some of the sandstones in the upper part of the Laramie near Boulder are reported to be arkosic (Spencer, 1961).[3] Heavy-mineral suites in the Laramie differ from place to place; north of Boulder the suites in the Laramie and Fox Hills are almost identical, whereas near Golden they differ markedly (Goldstein, 1950). Clay-mineral assemblages differ from zone to zone within the Laramie in the Golden area (Gude, 1950, Fig. 1). Shale and clay that compose a large part of the Laramie Formation on the plains east of Denver (Dane and Pierce, 1936) and in the northern part of the Denver basin (Mather and others, 1928) probably were derived from Cretaceous shales in the Front Range. The coarser and arkosic materials in the upper Laramie farther south near the mountain front may have come from more deeply eroded uplifts farther southwest, such as the Sawatch Range.

In the west-central part of the Denver basin, coarse orogenic sediments of the Upper Cretaceous and Paleocene Arapahoe, Denver, and Dawson Formations overlie the Laramie Formation. At the mountain front near Golden, the Arapahoe Formation, characterized by sandstones and conglomerates derived from sedimentary and Precambrian rocks, is overlain gradationally by the Denver Formation, characterized by abundant andesitic debris and intercalated lava flows (LeRoy, 1946; Van Horn, 1957; Scott, 1972). The Arapahoe grades laterally into andesitic facies and is distinguished from the Denver Formation only in a small area. Both formations intertongue southward with the Dawson Formation, which is characterized by arkose derived from the Precambrian Pikes Peak Granite in the Front Range to the west. Conglomerates in the lower part of the Dawson contain sedimentary-rock pebbles as well as granitic pebbles. Andesitic conglomerates are present locally (Goldman, 1910, p. 320; Varnes and Scott, 1967, p. 14).

An erosional unconformity at the base of the Arapahoe or Denver in the north and the Dawson in the south is evident along the mountain front, but farther east, where the Denver- or Dawson-Laramie contact is problematic (Dane and Pierce, 1936, p. 1320; Soister, 1965), it may not exist. A lengthy literature, summarized by Brown (1943), has been devoted to this unconformity as a hiatus, as an indicator of tectonic events, and, especially, as a candidate for the conspicuous dividing plane (once sought so obsessively) between the Cretaceous and Tertiary. The unconformity was classed as only a minor break in sedimentation by Lovering (1929) and Brown (1943), and as discussed below, the time constraints on early Laramide events indicated by radiometric dates do not permit a major hiatus. The Cretaceous-Tertiary boundary is now closely established on the basis of vertebrate fossils at an inconspicuous horizon in the Denver Formation about 240 m above the Laramie, and in the Dawson, 150 to 180 m above the base (Gazin, 1941; Brown, 1943).

In a small area immediately east of the mountain front south of Golden, coarse sandstones of the Denver Formation that consist almost entirely of andesitic and Precambrian detritus derived from the Front Range are overlain disconformably by the Paleocene Green Mountain Conglomerate. This coarse conglomerate contains conspicuous clasts of quartzite, sandstone, and chert as well as of Precambrian rocks and andesite, and it reflects an influx of detritus from a source that may have been west of the Front Range. The Green Mountain is of Paleocene age and is possibly equivalent to strata near the top of the Dawson Formation farther south (Scott, 1972). In a small area southeast of Denver, the Dawson is disconform-

---

[3]Lovering (1929, p. 90) reported that some Laramie sands contain "as much as 85 percent of fresh feldspar." That statement is in error; the samples studied by an assistant were from the Denver Formation rather than the Laramie (T. S. Lovering, 1938 and later occasions, oral commun.).

ably overlain by ash-flow tuff or by Castle Rock Conglomerate, both of early Oligocene age (Izett and others, 1969).

The Arapahoe and Denver Formations and Green Mountain Conglomerate have a total thickness of about 600 m, and the Dawson reaches about 750 m. Both the Denver and Dawson Formations become markedly finer grained eastward, and they probably thinned in this direction as well as to the north and south of present exposures. A map showing altitude and configuration of the Laramie-Fox Hills aquifer (Romero and Hampton, 1972) gives a measure of the thickness distribution of the Denver and Dawson, as the thickness of the Laramie above the aquifer probably is fairly constant. The map indicates that the Denver and Dawson fill a spoon-shaped depression which corresponds approximately in outline with the closed −6,000-ft contour on the basement surface shown in Figure 1. As judged from surface exposures, strata within this depression are nearly horizontal, which implies that all but the uppermost parts of the Denver and Dawson Formations must taper to a vanishing edge within the "spoon" and in the area of present exposures. Thin upper units could have extended much farther.

South Park Basin

The Laramide sedimentary record in South Park (Fig. 6) generally parallels that of the Denver basin, but a few important differences exist. The Laramie Formation, which is preserved only in a small area in the northeastern part of the park and is very poorly exposed, has been reported to contain volcanic tuff and agglomerate as well as beds of highly arkosic sandstone (Stark and others, 1949, p. 56, 167). These materials are characteristic of the overlying Denver Formation, and they might be parts of that unit, since the stratigraphic section in which these materials are reported (Stark and others, 1949, p. 167) was measured in an area that is cut by many small faults and is very poorly exposed (D. G. Wyant and R. J. Weimer, 1972, oral communs.).

The Denver Formation lies with angular unconformity on tilted and locally sharply folded Laramie and older sedimentary rocks and laps onto Precambrian rocks, much as the Middle Park Formation in Middle Park. The Denver may be as much as 3,000 m thick (D. G. Wyant, 1972, oral commun.). The lower part consists largely of andesitic conglomerates, breccias, tuffs, and lava flows, and the upper part consists mainly of Precambrian granitic detritus. The formation also contains detritus from sedimentary rocks and pebbles of various intrusive porphyries (Stark and others, 1949, p. 57, 137; Lovering and Goddard, 1938, p. 44). Intrusive plugs and large dikes near the north border of South Park were evidently the source of the volcanic flow rocks. Sources of nonandesitic clastic materials in the lower part of the formation were to the west, in the Mosquito or Sawatch Ranges (Stark and others, 1949, p. 135). Many of the porphyry pebbles can be identified with intrusive porphyries in the Leadville area of the Mosquito Range and with intrusive bodies near the Continental Divide north of the park. Increasing amounts of granitic detritus of Front Range provenance upward and eastward in the Denver (Sawatzky, 1964, p. 136) indicate that the Front Range became a major source of sediment in later Denver time.

The Denver is overlain unconformably by Oligocene volcanic rocks.

IGNEOUS RECORD

Volcanism and intrusion accompanied Laramide orogeny in parts of the Southern Rocky Mountain region but were absent or trivial elsewhere. The record of volcanism lies chiefly in the rather imposing piles of andesitic sediments present in some of the basins just discussed; the volume of actual volcanic rocks accompanying or contained in these sediments is small in comparison. The volcanic sediments and associated flow rocks lie in a broad northeast-trending belt defined by the Animas Formation of the northern San Juan basin, the Cimarron Ridge Formation (Dickinson and others, 1968) on the north side of the San Juan Mountains, the Wasatch Formation in the southern part of the Piceance basin, the Denver Formation in South Park, the Middle Park Formation in Middle Park, and the Denver Formation in the west-central part of the Denver basin (Fig. 7). Intrusive stocks and associated sills and dikes are concentrated in a somewhat narrower belt within the broad volcanic belt, though a few lie outside. The narrow belt is the well-known Colorado mineral belt (Burbank and Lovering, 1933; Lovering and Goddard, 1938, 1950; Tweto and Sims, 1963; Tweto, 1968). Intrusive bodies and associated mineral deposits in this belt were long thought to be Laramide in age, though a few exceptions were recognized. With the advent of radiometric dating, supported by many detailed geologic studies, it became clear that the mineral belt contains at least three interspersed populations of intrusive bodies and ore deposits. The Laramide population, mainly in the 70- to 50-m.y.-age range, is largely concentrated in a fairly sharply defined and narrow inner zone in two segments, one extending from the east front of the Front Range to the west side of the Sawatch Range, and the other extending along the west side of the San Juan Mountains. A middle Tertiary population (~40 to 25 m.y.) is more widespread, has the effect of widening the mineral belt, and is represented in several areas outside the mineral belt as generally defined. A third and late Tertiary population (~15 to 10 m.y.) is smaller than the other two and is widely diffused through the mountain region.

Volcanic rocks and sediments in all areas of occurrence are at or very near the bottoms of the orogenic sedimentary sequences. They reflect onset of volcanism—in various places and at somewhat different times—in the general environs of the Colorado mineral belt shortly after first uplift of the mountains. All the volcanic sediments contain admixed detritus from Precambrian rocks, even at their bases, indicating uplift and erosion prior to the volcanic episode. Paleozoic conglomerates could not have been the source of the granitic detritus, for, owing to late Paleozoic uplift, they did not exist in most of the source areas of Laramide

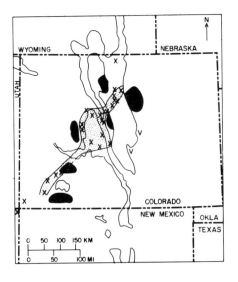

Figure 7. Laramide igneous and mineral belts in Colorado. X, intrusive bodies of known or inferred Laramide age; solid black areas, andesitic sedimentary rocks and flows; V, minor occurrences of andesitic detritus; shaded strip, mineral belt as defined by Laramide intrusive bodies and productive mining districts. Outlines of main bodies of Precambrian rocks as now exposed are shown for reference (see Figs. 1 and 4).

sediments. Only one major episode of volcanism is recorded in the sediments of each of the affected basins. In the Denver and San Juan basins and on the north side of the San Juan Mountains, and possibly in Middle Park, this episode is dated paleontologically as beginning in very Late Cretaceous time.

Radiometric dates in two areas serve to define the time of volcanism more closely. At Table Mountain at Golden, west of Denver, dacitic pumice about 120 m above the base of the Denver Formation and 10 m above the Cretaceous-Tertiary boundary as defined by vertebrate fossils (Gazin, 1941) was dated by Evernden and others (1964, p. 187) at 64.8 m.y. (K-Ar). Pumice from a bed 10 m higher has recently been dated by Obradovich (in Scott, 1972) at 64.3 m.y., thus closely corroborating the date of Evernden and co-workers. As the coarse clastic sediments of the Denver Formation are inferred to have accumulated rapidly, the Cretaceous-Tertiary boundary in this locality probably is little older than the dated rocks, perhaps ≤65 m.y. This age is in general accord with that of 64 ± 1 m.y. assigned to the boundary by Lambert (1971, p. 33), largely on the basis of dates from localities in Montana and Alberta. A basalt flow within the Denver Formation 72 m above the pumice horizon dated by Evernden and co-workers was dated by these same workers at 58.7 m.y. This age probably is too young—a dike that has been generally recognized as a probable feeder for the lava flows has yielded a K-Ar age of 63 ± 2.5 m.y. (E. E. Larson in Scott, 1972).

On the northwestern side of the San Juan Mountains, rhyodacitic flows, breccias, and tuffs of the Cimarron Ridge Formation lie unconformably on Kirtland Shale and older units and are overlain unconformably by Telluride Conglomerate of Wasatch age (Dickinson and others, 1968). A large dike, one of several evident sources of the volcanic rocks, has a K-Ar age of 66.9 ± 4.0 m.y. Age determinations on various mineral separates from the volcanic rocks are in the general range of 61 to 71 m.y. and average 66 m.y. Tuffaceous shales near the base of the Cimmaron Ridge contain Maestrichtian pollen (Dickinson and others, 1968). Igneous activity of the type recorded in this locality is inferred to have occurred widely along or near the western side of the present San Juan Mountains, giving rise to the andesitic sediments of the Animas Formation. Some of the intrusive rocks in the La Plata Mountains (Fig. 1) and the Rico area to the north have yielded K-Ar ages of 63 to 65 m.y. (Young, 1972; Armstrong, 1969), but others have yielded spuriously greater ages that are not consistent with the geologic record. As judged by the occurrence of volcanic pebbles in the upper part of the Kirtland Shale (O'Sullivan and others, 1972, p. 54), igneous activity probably began at some still-unidentified locality in the western San Juan region as early as 69 to 70 m.y. ago.

In the Sawatch, Mosquito, and Front Ranges, intrusive rocks of Laramide age occur in thousands of dikes and sills and in numerous stocks, most of which are small—less than 5 km maximum dimension. These igneous bodies are heavily concentrated in the mineral belt (Fig. 7). Most of them are in Precambrian or Paleozoic rocks, and they cannot be closely dated geologically. Available radiometric dates (Young, 1972) indicate that intrusion was mainly in the period 70 to 60 m.y. ago but continued on diminishing scale to about 50 m.y. The early intrusions are especially pertinent to the correlation of Laramide events.

A porphyry sill on the west side of the Sawatch Range, in the Aspen area, has been dated at 72.2 ± 2.2 m.y., and two determinations on aplitic rock in a small pluton nearby gave ages of 70.0 ± 2.3 and 67.4 ± 2.2 m.y. (Obradovich and others, 1969). Granodiorite in a small stock in Precambrian rocks in the northern part of the Sawatch Range has been dated in two determinations at 69.5 and 68.9 m.y. as recalculated by Obradovich in terms of currently accepted constants from Pearson and others (1962). The Pando Porphyry, the earliest and most abundant of the porphyries in the Leadville area on the west flank of the Mosquito Range, is dated at 69.6 m.y. as similarly recalculated, and the Lincoln Porphyry, one of the younger porphyries, is dated at 64.2 m.y. Several other porphyries are geologically bracketed between the Pando and Lincoln Porphyries (Tweto, 1960). A stock on the east side of the Mosquito Range east of Leadville is dated by two determinations at 64 m.y. (C. E. Hedge in Young, 1972). A large stock in the southern Mosquito Range is dated at 70.4 ± 2.1 (biotite) or 69.4 ± 2.1 (hornblende) (McDowell, 1966). The Mosquito Range is structurally a part of the eastern flank of the huge anticline of the Sawatch Range, separated from the topographic Sawatch Range by a deep graben of late Tertiary age. In both the Aspen area on the west flank of the Sawatch anticline and the Leadville area on the east flank, the earliest porphyries were intruded after rise of the anticline and faulting of its flanks (Tweto, 1960; Obradovich and others, 1969, p. 1751). The large 70-m.y.-old stock in the southern Mosquito Range, just mentioned, cuts the Weston fault, a major fault that extends much of the length of the range.

Early stocks in the Front Range are near or east of the crest of the range in the general latitude of Boulder (Lovering and Goddard, 1950, Pl. 2). The Audubon monzonite stock (Wahlstrom, 1940) is dated at about 68 m.y. (Hart, 1961, p. 193). The Caribou mafic monzonite stock, a few kilometers to the south (Smith, 1938), is closely related compositionally to the more mafic facies of the Audubon stock and is inferred to be of the same general age. Lovering and Goddard (1938) suggested that the Audubon stock had a volcanic superstructure which was a source of pebbles in the Denver Formation. Assuming volcanic superstructures, both the Audubon and the Caribou stocks qualify in age, composition, and location as sources of volcanic materials in the Middle Park Formation as well as the Denver Formation. Lavas and volcanic breccias in the lower parts of both formations came from local dikes and plugs on the two sides of the range, but the great volumes of andesitic detritus overlying and distant from these local centers must have come in large part from other sources, such as volcanic fields higher on the Front Range. It is likely that there were many source volcanic centers, of various sizes and compositions, in the Front Range. Cross (in Emmons and others, 1896, p. 201, 315) emphasized the great variety among the generally andesitic pebbles in the Denver Formation and their entirely extrusive character. In the upper part of the Middle Park Formation, though, the andesitic rocks are accompanied by abundant pebbles of light-colored prophyries of intrusive origin and also by some of olivine basalt (Tweto, 1957, p. 23).

Several younger Laramide stocks are also present on the eastern flank of the Front Range. The Eldora stock, dated at about 54 m.y. (Hart, 1964), is an example. These stocks are too young to have contributed significantly to the Denver or Middle Park Formations. The Montezuma stock, the main porphyritic body on the western flank of the range, is of middle Tertiary age.

The stocks and numerous other hypabyssal porphyry bodies of the mineral belt have long been interpreted on geologic grounds as expressions of an underlying batholith or string of batholiths (Crawford, 1924; Lovering and Goddard, 1938; Tweto and Sims, 1963; Tweto, 1968). Gravity data support this interpretation. In the Front, Mosquito, and Sawatch Ranges, the mineral belt coincides with a gravity valley pocked by deep gravity lows (Behrendt, 1968; Behrendt and Bajwa, 1972; Tweto and Case, 1972). The western San Juan segment of the mineral belt does not have such expression, probably because of the dominant gravity effects of the overlapping mid-Tertiary caldera complexes and underlying intrusive bodies. The almost complete restriction of Upper Cretaceous volcanic rocks and sediments to a broad but definite belt centered over the intrusive belt suggests that the inferred batholiths of the mineral belt were also the source of the early volcanic rocks. The entire belt of volcanic and intrusive rocks cuts obliquely across a succession of major mountain ranges. Though seemingly the northeast-trending line of batholithic intrusion and the north-northwest-trending mountain ranges were independent features, it is hard to believe that their essentially simultaneous appearance in early Laramide time was pure coincidence. The Laramide intrusive belt widens markedly in the area of the Sawatch anticline, as indicated by dated stocks near the north and south ends of the anticline; possibly, intrusion at depth was a contributing cause of the rise of the anticline, though other factors must have controlled orientation of the anticline. Similarly, intrusion may have caused early Laramide uplift in a part of the ancient Uncompahgre–San Luis highland now buried beneath volcanic rocks of the San Juan Mountains. However, this would leave unexplained the rise of other parts of the old highland southward into New Mexico, where there is no evidence of contemporary magmatism. In the Front Range, no case whatever can be made for shallow intrusion as a cause for the range. Nevertheless, intrusion probably did cause the part of the range in the volcanic-intrusive belt to rise higher, and perhaps earlier, than the rest of the range. This part of the range was a principal source of early Laramide orogenic sediments, and today is the highest part.

As indicated, magmatism reached a climax in the early Laramide and dwindled thereafter. From a longer view, the dwindling reflected only a period of dormancy of the source-magma bodies at depth. In Oligocene time, these and other bodies rose higher into the crust, causing magmatism on a vastly greater scale than that of the Late Cretaceous and Paleocene (Steven, 1975).

EARLY LARAMIDE TIMING CONSTRAINTS

The marine Cretaceous rocks place a close—though not absolute—limit on the start of Laramide orogeny in any given area. Gill and Cobban (1966, p. 34-37) related ammonite zones in these richly fossiliferous rocks to a numerical time scale devised from potassium-argon ages of bentonite beds. The upper part of this scale as currently revised by J. D. Obradovich and W. A. Cobban (1973, oral commun.) is incorporated in Figure 8 along with ages of early igneous rocks. Radiometric ages of both the igneous rocks and the ammonite zones are subject to many uncertainties and caveats, as generally recognized, and also to continuing revision. The dates currently available are used here at face value as a basis for discussion.

As indicated in Figure 8, the early part of the igneous sequence overlaps the late part of the ammonite sequence in the region as a whole. The time interval between marine conditions expressed by ammonites and early phases of orogeny measured by intrusion was short. To the extent that the radiometric dates for ammonite zones and igneous rocks are both correct, they indicate a very rapid progression of events in the early stages of orogeny. Cretaceous rocks preserved on the west side of the Sawatch Range near Aspen extend with certainty as high as the zone of *Exiteloceras jenneyi* (72 m.y., in Fig. 8) and possibly to the zone of *Baculites compressus* (71.5 m.y.) (Freeman, 1972). These fossils are near the top of the Mancos Shale, which here extends higher stratigraphically than in the Piceance basin to the west (Fig. 6). The Mancos is overlain by remnants of Mesaverde nonmarine sandstones. Several igneous bodies emplaced after rise and faulting of the Sawatch anticline are in the 70- to 60-m.y.-age range (Fig. 8). The two sets of age data indicate, therefore, that only 2 to 3 m.y. were available for deposition of the Mesaverde and the subsequent—or partly concurrent(?)—rise of the Sawatch anticline to a structural height great enough to have caused large displacements on antithetic flank faults.

The Front Range and adjoining basins display an even tighter time schedule. As discussed earlier, the marine Fox Hills Sandstone, dated at about 67.5 m.y. (Fig. 8), is inferred to have been deposited over the site of the entire range. Early stocks in the Front Range are dated at about 68 ± 2 m.y., suggesting—but not proving—appreciable uplift by that time. The evidence of the Denver and Arapahoe Formations, however, requires Precambrian rocks to have been exposed

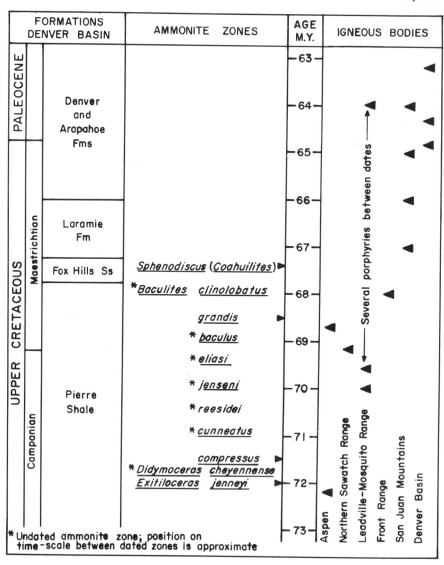

Figure 8. Relation of early Laramide igneous rock dates to time scale based on Western Interior ammonite sequence and to formations in the Denver basin. Ammonite scale from Gill and Cobban (1966) as revised by Obradovich and Cobban (1973, oral commun.). Igneous rock dates from sources noted in text. Analytical uncertainties of dates on ammonite zones, 1 to 2 m.y., and of dates on igneous rocks, 2 to 3.5 m.y.

to erosion in the Front Range prior to 65 m.y.—perhaps at 66 m.y. Thus, in ~1.5 m.y., the Front Range rose structurally an amount at least equal to the thickness of the covering sedimentary rocks—about 3,000 m—plus an unknown amount required for the range to have supplied abundant detritus from Precambrian rocks and to have supported streams of high gradient to transport this coarse detritus. These figures indicate a minimum rate of 2 mm/yr of uplift and erosion. In the same period, major volcanism occurred on the range, and large quantities of volcanic rocks were eroded from it. On the west side of the range, moreover, sedimentary rocks were deformed and eroded locally to the Precambrian basement before deposition of the Coalmont, Middle Park, and Denver Formations began in North, Middle, and South Parks.

OROGENIC SEQUENCE AND STRUCTURAL DEVELOPMENT

Data presented in preceding sections indicate that orogeny began at somewhat different times in different parts of the Southern Rocky Mountain province, starting with the southwest part. This is interpreted to reflect the advance of an irregular orogenic front rather than progress of orogeny by fits and starts. Once uplift of a mountain unit or downwarping of a basin started, the process continued. No evidence exists of distinct breaks in the orogenic processes, though there were times when, or places where, they proceeded more rapidly than in others. Except near the edges of the sedimentary basins, the Laramide orogenic formations are conformable with each other, and with few exceptions (Fig. 6), they record almost continuous sedimentation through the Laramide time span from Late Cretaceous into Eocene time—at least 25 m.y. Through much if not all of this time, the mountain units grew laterally as they rose, and they were eroded nearly or quite as rapidly as they rose. Consequently, the mountains of today are areally larger than the uplifts that supplied the first orogenic sediments, and their border structures are younger than those sediments. The once-popular idea that Laramide orogeny consisted of two distinct phases, one of uplift in Late Cretaceous time and one of deformation in late Paleocene-early Eocene time, is not supported by the evidence in this part of the Rocky Mountains.

Because of the long-continued growth and erosion of the mountain ranges, structural features of the early, smaller uplifts are obscure. The early uplifts are presumed to have been generally anticlinal in geometry, whatever the mechanics of generation may have been. In my opinion, the crystalline cores of these uplifts rose by movements on many faults in the Precambrian rocks, and by slight adjustments on multitudes of fractures that had existed since Precambrian time in these rocks, rather than by flexure.

Mountain units are classified according to age of first major uplift in Figure 9. The uplifts are discussed below in order of inferred development first within the generally older southern part of the region and then within the somewhat younger northern part.

San Juan-San Luis Uplift

The upper part of the Kirtland Shale and the overlying Animas Formation in the San Juan basin and the Raton and Poison Canyon Formations in the Raton basin record the presence of a highland in the area of the present San Juan Mountains, San Luis Valley, and Brazos uplift in Montana time. In general, this highland coincided in location with the southeastern or San Luis part of the late Paleozoic Uncompahgre-San Luis highland; it is here referred to as the Laramide or rejuvenated San Luis highland. A northern part of the rejuvenated highland may have existed

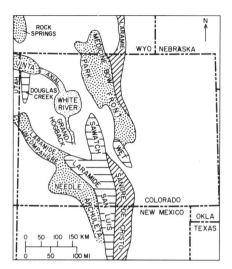

Figure 9. Major uplifts in Southern Rocky Mountain province classified according to time of first rise. Horizontal rules: Cretaceous, >70 m.y.; coarse dots: Cretaceous, 70 to 65 m.y.; diagonal rules: Paleocene; fine dots: Eocene.

as early as Pictured Cliffs time, as suggested by Dane (1946) and Baltz (1967). The base of the Pictured Cliffs is in the zone of *Didymoceras cheyennense* (Gill and Cobban, 1966, Pl. 4), and from this it may be inferred that the highland began to rise in late Campanian time, about 72 m.y. ago (Fig. 8). Local unconformities between the Pictured Cliffs and Fruitland and within the Kirtland along the east side of the San Juan basin, noted earlier, and the more widespread unconformity at the base of the Animas Formation indicate tectonic movements of the southern part of the highland at an early date. This is indicated also by the angular unconformity at the base of the Cretaceous portion of the Poison Canyon Formation in the southwestern part of the Raton basin.

By the time of andesitic volcanism, the area of uplift had expanded into the sedimentary terrane south and west of the crystalline rock body of the San Luis highland, in the area of the western San Juan Mountains. This uplifted area, the San Juan dome of Burbank (1930), is now largely covered by Oligocene volcanic rocks or obliterated by caldera development, but parts of it are exposed as a western monoclinal front (Burbank, 1940; Dickinson and others, 1968) and as the Needle Mountains. An early version of the modern Uncompahgre Plateau—which lies along the southwestern side of the late Paleozoic Uncompahgre highland—also developed at this time (Burbank, 1940; Dickinson and others, 1968). The modern plateau, however, is largely a product of later movements (Cater, 1970, p. 67).

Details of later history of the main or northern part of the San Luis highland are obscure because this part is now either covered with volcanic rocks or depressed beneath the San Luis Valley. It is clear, however, that the highland persisted and supplied sediments to the San Juan and Raton basins into Eocene time. Near the Colorado-New Mexico border, a sharply folded belt—the Archuleta anticlinorium—lies between the Laramide San Luis highland and the San Juan basin and merges northwestward with the Needle Mountains uplift. Farther south, the old highland and the San Juan basin have a common boundary, marked by steep monoclinal folds and faults. All these structures developed through Paleocene and early Eocene time, and in part they postdate the lower Eocene San Jose Formation (Dane, 1948; Wood and others, 1948; Baltz, 1967).

Throughout its length, the eastern flank of the Laramide San Luis highland is preserved as the younger Sangre de Cristo Range, discussed separately below.

Sawatch and Mosquito Ranges and Elk Mountains

As already indicated, the anticline of the Sawatch and Mosquito Ranges existed by about 70 m.y., and perhaps by 72 m.y., ago. Thus, the anticline is about the same age as the rejuvenated San Luis highland, and on its southwest flank, the anticline merged with this highland (Fig. 9). At the time the anticline began to rise, a sedimentary sequence 4,500 to 5,400 m thick covered the area. The igneous rocks by which the anticline is dated are exposed mainly in the lower 600 m of this sequence and in the Precambrian core of the anticline. By the time the intrusive rocks were emplaced, erosion of the crestal part of the anticline must have been under way, presumably in the fine-grained Cretaceous rocks. The products of this early erosion are not recognized in the nearby sedimentary basins but are probably incorporated in units such as the upper Mesaverde of the Piceance basin, the Vermejo Formation of the Raton basin, and the uppermost Pierre of South Park and the Denver basin. The oldest detrital materials that seem definitely linked to a Sawatch source are in the Ohio Creek Formation in the Piceance basin and in the lower part of the Denver Formation of South Park. The materials in these formations indicate that by Paleocene time, most of the original sedimentary cover had been eroded from the higher parts of the Sawatch anticline, and that lower Paleozoic, Precambrian, and Laramide intrusive rocks were exposed. The abundant volcanic debris in the lower part of the Wasatch Formation in the southern Piceance basin indicates extensive volcanism in the range prior to late Paleocene time. Sites of this volcanism are not identified but could have been at any of several localities through the length of the range.

The Sawatch anticline grew laterally as well as vertically throughout these erosional stages. Though rumpled by subsidiary folds and displaced by major faults, the east flank of the anticline extends 55 to 65 km across northern South Park to the foot of the Front Range. The Fox Hills and Laramie Formations, and to a degree even the Denver Formation, are involved in the eastward tilt of this flank, Yet, if radiometric dates are correct, the axial portion of the anticline must have existed in the area of the present Sawatch Range before the Fox Hills, Laramie, and Denver were deposited in South Park. Growth of the anticline, then, extended through a period from pre-Fox Hills to post-Denver time. In the latter part of Denver time, however, the Front Range rose high enough at this latitude to become a major source of sediment for the Denver Formation of South Park. In post-Denver time, Precambrian rocks of the Front Range were thrust westward over the Denver and older formations in South Park along the Elkhorn thrust fault (Stark and others, 1949; Sawatzky, 1964).

On the western flank of the Sawatch anticline, the structural bench or sharply asymmetric anticline of the Elk Mountains probably developed in primitive form at an early stage in the rise of the main anticline. A major fault in these mountains, the Elk Mountain thrust, is attributed by Bryant (1966) to gravity sliding off the flank of the rising Sawatch Range, probably in Paleocene time. A major border fault of the Sawatch Range, the Castle Creek fault, displaces the thrust. The steep western front of the Elk Mountains, facing the Piceance basin, is a southern extension of the Grand Hogback, an Eocene monoclinal fold that is discussed below in connection with the White River uplift. In the Elk Mountains, this fold was further modified by emplacement of Oligocene plutons.

Wet Mountains Uplift

The Wet Mountains probably rose as an anticlinal uplift at or nearly the same time as the Sawatch anticline to the northwest. The two north-northwest-trending anticlines, in echelon arrangement, are now separated by the younger Sangre de Cristo Range, constructed in part out of the intervening syncline. Within the small outlier of uppermost Cretaceous and Paleocene sedimentary rocks preserved on the east side of the Wet Mountains near Canon City, the Vermejo Formation changes rapidly eastward from predominantly sandstone to predominantly shale (Washburne, 1910, p. 344), suggesting that the Vermejo sediments of this area may have been derived from older Cretaceous rocks in the adjoining Wet Mountains. The change described by Washburne is most evident in the lower two-thirds of the Vermejo, beneath a marine sandstone that, according to Lee (1917, p. 168), contains a Fox Hills fauna. Abundant course detritus from Precambrian rocks in the Poison Canyon Formation on both sides of the Wet Mountains indicates that extensive parts of the uplift had been stripped of sedimentary cover by Paleocene time. Similar detritus in younger formations, together with unconformities (Fig. 6), indicates continued uplift of the Wet Mountains until at least middle Eocene time.

A major high-angle reverse fault—the Wet Mountain fault—borders the Wet Mountains on the northeast side, and a major parallel fault—the Ilse—cuts longitudinally through the Precambrian core of the range. The Ilse is a Precambrian fault that underwent movement at many later times, including Laramide and mid-Tertiary; the Wet Mountain fault probably had a parallel history (G. R. Scott and R. B. Taylor, 1973, oral commun.). The date of first Laramide movement on the Wet Mountain fault is not firmly established, but coarse, angular clasts in the Poison Canyon just across the fault from Precambrian rocks suggest that the fault existed in Paleocene time.

Sangre de Cristo Range

The Sangre de Cristo Range is bounded on the west through its full 400-km length by middle to late Tertiary faults that separate it from the deeply subsided San Luis and Rio Grande Valleys. The range consists of two main structural elements: (1) a subordinate northern segment that is structurally continuous with the Mosquito Range and, like that range, is a part of the east flank of the Sawatch anticline, and (2) the main body of the range, which is constructed from the eastern flank of the rejuvenated or Laramide San Luis highland, discussed above. Though they have no clear-cut boundary, the two parts differ somewhat and are discussed separately, the main body first.

The relation of the main body of the Sangre de Cristo Range to the late Paleozoic and Laramide San Luis highland is displayed directly only near the southern end of the range in New Mexico. There, for 80 km, Precambrian rocks of the western part of the range are separated from Paleozoic rocks of the eastern part by the Picuris-Pecos fault (Miller and others, 1963). Compositional and depositional features of a thick sequence of Pennsylvanian rocks east of the fault indicate that the fault is essentially the border of the late Paleozoic San Luis highland. As interpreted by Miller and co-workers, this vertical, north-trending fault originated as a Precambrian strike-slip fault with many kilometers of right-lateral displacement. Large dip-slip movements, with upthrow to the west, occurred in both Pennsylvanian and Late Cretaceous times; the fault became the eastern border structure of a San Luis highland on each occasion. Additional movements probably occurred on parts of the fault during development of the Rio Grande rift or graben in Tertiary time. The fault disappears northward beneath Tertiary cover in the Rio Grande Valley, but it probably continues northward along the border of the buried San Luis highland into the San Luis Valley in Colorado.

The main body of the Sangre de Cristo Range is characterized structurally by a system of Laramide thrust faults and folds in a thick sequence of predominantly Paleozoic sedimentary rocks and by a younger block-fault system that has elevated Precambrian rocks to great topographic and structural heights in some areas. Laramide thrusting, folding of thrusts, and overturning were toward the east and were largely confined to the sedimentary rocks (Burbank and Goddard, 1937; Briggs and Goddard, 1956; Johnson, 1959). This deformation clearly was related to the rise of the San Luis highland to the west and the sinking of the Raton basin to the east. The central part of the San Luis highland began to rise in Cretaceous time, as noted previously, but deformation did not extend to the flank of the present Raton basin until Paleocene time. The angular unconformity at the base of the Poison Canyon Formation, the lowest of several in the Laramide orogenic sequence, records this stage. Burbank and Goddard (1937) documented later stages of deformation extending into Huerfano (middle Eocene) time. The deformational climax is correlated by them with the unconformity between the Poison Canyon and the lower Eocene Cuchara Formation. Increasing amounts of detritus from sedimentary rocks upward in the Cuchara and Huerfano Formations indicate that by Eocene time the San Luis highland had expanded topographically into the sedimentary terrane of the Sangre de Cristo Range. Whether a Sangre de Cristo Range distinct from the San Luis highland existed at this time is uncertain. A probable Cuchara or Huerfano equivalent is preserved in places on the west side of the Sangre de Cristo Range. It consists of Precambrian materials and lies on Precambrian rocks, and it was interpreted by Upson (1941) as having an eastern source, in the crestal area of the Sangre de Cristo Range. However, the source could as well have been in the Precambrian terrane of the San Luis highland immediately to the west, at the present site of the San Luis Valley. Most likely, the main body of the Sangre de Cristo Range did not become an entity separate from the Laramide San Luis highland until mid-Tertiary time.

The northern part of the Sangre de Cristo Range consists of Precambrian rocks and—mainly on the east slope—of overlying Paleozoic rocks that dip sharply eastward (Gabelman, 1952; Litsey, 1958). The Paleozoic rocks are cast into open folds but are only mildly deformed as compared with the sedimentary rocks farther south in the range. Numerous faults are present, including some major reverse faults, but the flat and folded thrust faults found farther south in the range are absent. This part of the range is north of the latitude of the Raton basin, and its structural pattern, as distinct from that to the south, probably reflects the absence of a deep sedimentary basin along its east side. Instead, the east side is buttressed against the Precambrian rock body of the Wet Mountains, in large part inherited from the late Paleozoic Front Range highland. Northwest-trending faults between the northern Sangre de Cristo and Wet Mountains are in part reactivated late Paleozoic faults (DeVoto and Peel, 1972), and these in turn probably are reactivated Precambrian faults.

Front, Park, and Associated Ranges

Uplift of the large body of Precambrian rocks now exposed in the Front, Laramie, Medicine Bow, Park, and Sierra Madre Ranges (Fig. 4) is inferred to have begun in the area of Middle and North Parks. The timing constraints discussed earlier seem to require that uplift and deformation start in this area very shortly after deposition of the Fox Hills Sandstone, probably in latest Cretaceous time. A first requirement was uplift in the Front and Park Ranges to create a basin of Middle Park–Coalmont sedimentation. Concurrent with this uplift, the site of the basin was itself uplifted, strongly deformed by folding and faulting, and eroded in places to the Precambrian basement rocks. Products of this stage of intense erosion may lie chiefly in the deep sedimentary basins of southern Wyoming. The Medicine Bow Formation of these basins, equivalent in major part to the Laramie and Lance Formations, is several thousand feet thick and locally contains conglomerate (Knight, 1953, p. 67) that could well have been derived from the North Park area (D. L. Blackstone, 1972, oral commun.). After the early uplift and erosion in the North Park–Middle Park basin, deposition began of the lower volcanic unit of the Middle Park Formation, dated with some reservation as very Late Cretaceous.

In the South Park area, the Sawatch uplift existed as a western border of a sedimentary basin well before Denver time. Before Denver deposition began, the Front Range was necessarily uplifted enough to create a basin, though not enough at this time to become a major source of sediment. Concurrently, the site of the basin was deformed and eroded, though not so severely as in Middle and North Parks.

Though the Front Range was evidently low with reference to South Park during early Denver time, it must have been high with respect to the Denver basin at this general latitude. The Dawson Formation in this basin, of the same age as the Denver, contains much coarse arkose and many conglomerate lenses near the mountain front and grades rapidly into finer grained rocks eastward. These features indicate a persistent highland of at least moderate relief a short distance to the west through latest Cretaceous and some part of Paleocene time.

The location of the interface between highland, or sediment-source, and depositional basin in Denver-Dawson time is not known, but in most places, it was somewhere west of the present east margin of the Front Range. Along most of the segment of the mountain front where Denver or Dawson rocks are preserved, no indications of angular unconformity, wedge-out, or overstepping are seen. The line or belt along which older formations must have been truncated in order to expose Precambrian rocks, and the depositional edge of the Denver or Dawson, was therefore farther west. However, in an area a few miles northwest of Colorado Springs, angular unconformity exists at the present mountain front between a part of the Dawson Formation and older rocks. At this locality, noted by Harms (1965, p. 986), the base of the Dawson is conformable with the underlying Laramie Formation, but an upper part, above an unconformity within the Dawson (G. R. Scott, 1973, oral commun.), oversteps more steeply dipping Laramie and Fox Hills strata (Varnes and Scott, 1967, Pl. 1; Boos and Boos, 1957, Fig. 10).

From these relations, I infer that the line of Late Cretaceous uplift in the Front Range was near the present mountain front in the south and that it diverged deeper into the present range northward. At the latitude of southern North Park, only the Medicine Bow prong of the northern Front Range may have been involved in the Late Cretaceous uplift.

The Laramie Range prong probably developed as the Front Range uplift expanded toward its present front in Paleocene time. The earliest positive evidence of the existence of the Laramie Range uplift within the area considered here is arkose and fine conglomerate in the upper Paleocene Hanna Formation near the flank of the range (D. L. Blackstone, 1972, written commun.). These materials imply earlier uplift and unroofing of the Precambrian rocks. This probably occurred in Paleocene time, when most of the ranges of Wyoming were stripped of their sedimentary cover and the resultant detritus was deposited as the Fort Union Formation and equivalents (Love and others, 1963; Denson and Chisholm, 1971).

Unlike the rest of the Front Range, the northernmost part, extending into the Laramie Range, was not inherited from the late Paleozoic Front Range highland but is a Laramide addition. This part of the range also differs from the main body in trend (Fig. 1) and in character of the front, reflecting the different basement control and mechanics of origin. South of the bend in the range near Boulder, the mountain front is a steep monocline associated in many places with steep Laramide reverse faults as well as with younger faults. To the north, in contrast, frontal faults are absent, sedimentary rocks extend higher onto the range, and the fold structure is dominated by an en echelon series of northwest-trending folds and faults. The folds are asymmetric, and both the folds and faults step northeastern blocks upward. Without attempting analysis here, it is worth noting that the change in the character of the range occurs in the belt of major north-northwest-trending breccia reef faults (Lovering and Goddard, 1950), the predominant fracture structures of the north-central part of the Front Range. These faults are of Precambrian origin (Tweto and Sims, 1963) but they underwent complex movements in Laramide and other times. Where they extend into sedimentary rocks, almost all are upthrown to the northeast.

In the Denver basin, the Denver and Dawson Formations are folded along with underlying formations in the monoclinal upturn at the mountain front. This folding, therefore, and also steep reverse faulting that accompanied or followed it in many places is post-Denver-Dawson in age. Little or no evidence exists within this basin to date the mountain-front deformation more accurately, but from the evidence of other basins, the deformation is presumed to be of early to middle Eocene age. Folds and thrust faults along the western side of the range are also post-Denver in age. In North Park, where the Coalmont Formation is involved in the mountain-front folds and faults, deformation must have been younger than early Eocene.

At some time during the late Laramide evolution of the Front Range, the adjoining plains region to the east was elevated. From near the latitude of Boulder northward into central-eastern Wyoming, Paleocene rocks are absent, and Eocene rocks are absent through an even larger area that extends southward to the Raton basin. Lower Oligocene rocks are present in places through this region and are inferred to have covered most or all of it. Uplift could have begun in the area of the northern Denver basin as early as Paleocene time, preventing deposition, though conceivably Paleocene rocks were deposited and then eroded during Eocene uplift. Farther north, in central-eastern Wyoming, deposition continued through Paleocene time, and regional drainage was eastward (Love and others, 1963, p. 202). By early middle Eocene time, drainage in this area was reversed by uplift in the plains area (Love and others, 1963, Fig. 8). This uplift evidently affected much of the plains province, for except in a rather small area in the northern part of the Raton basin, Eocene rocks are absent east of the mountain front from Wyoming into New Mexico.

The history of the Park–Sierra Madre Range generally parallels that of the Front Range. Detritus from sedimentary rocks eroded from the range during first uplift probably went to the Lance Formation of the Sand Wash basin and to basins in southern Wyoming, though some may have been carried out of the region. By Paleocene time, the range was supplying Precambrian materials to the Coalmont Formation in North Park and to the Fort Union Formation in the Sand Wash basin. This supply continued through Wasatch and into Green River time. The main area of supply must have been at the latitude of North Park and northward across the Colorado-Wyoming boundary, where the Precambrian rock bodies of the Sierra Madre and Medicine Bow Ranges merge (Fig. 4). Farther south in Colorado, the Park Range contains only a narrow strip of Precambrian rocks, and in places sedimentary rocks extend to or across the crest. This part could not have been a major source of Precambrian detritus in Laramide time. Though faulted in places, the flanks of the Park Range are essentially anticlinal southward almost to the Colorado River. Farther south, a southern segment of the range, known as the Gore Range, is bounded on the west by a major fault—the Gore fault. This north-northwest-trending fault existed in Precambrian time, was active in late Paleozoic time when it marked the western boundary of the Front Range highland (Fig. 2), was active in Laramide time, and was markedly reactivated in later Tertiary time (Tweto and others, 1970, p. 10).

Uinta Mountains

Only the eastern end of the enormous anticline of the Uinta Mountains extends into the area discussed here. The anticline is asymmetric, steep on the north, and the north flank is broken by a major reverse fault, the Uinta fault. On the south side of the anticline, large but subsidiary folds parallel the main anticline; a pair of lengthy longitudinal normal faults define a prominent horst in the subsidiary fold belt. Angular unconformity between the Fort Union Formation and the underlying Mesaverde Group on the north side of the mountains indicates that uplift of the Uinta anticline began in post-Montana Cretaceous time, and structural relations indicate that the Uinta fault also dates to Cretaceous time (Hansen, 1965). As indicated by the composition and unconformable relations of the Wasatch Formation, discussed previously, pronounced uplift occurred in early Eocene time, and deformation of the Wasatch and Green River Formations indicates that uplift continued at least through early middle Eocene time. Hansen (1965) has noted that at least 2,550 m of Cretaceous strata were eroded from the anticline prior to Fort Union (Paleocene) time; by Wasatch (early Eocene) time, Precambrian rocks were exposed in the range, implying at least local erosion of an additional 1,800 to 1,900 m of strata.

The eastern part of the Uinta anticline lies athwart a line of uplift that consists of the north-trending Douglas Creek arch on the south side of the anticline and a north-northwest-trending structural uplift on the north side. East-trending subsidiary folds of the Uinta system are superposed on the north-trending Douglas Creek arch and are clearly younger (Gow, 1950; Ritzma, 1955). The arch existed at least as early as Late Cretaceous time, as units of the Mesaverde Group thin over the crest. The Lance and Fort Union formations are absent on the Douglas Creek arch, and the Wasatch Formation thins markedly over the crest, indicating that uplift continued into Wasatch time. Uplift ended prior to Green River time, however, as the Green River was deposited uninterruptedly across the arch. Thus, though an arch transverse to the Uinta axis existed before the Uinta anticline formed, the transverse arch continued to rise along with the Uinta anticline until Wasatch time.

White River Plateau

The broad flat-topped dome of the White River Plateau is largely a product of Eocene uplift and was the last of the major Laramide mountain units to form. In the Piceance basin, adjoining the White River Plateau, strata equivalent to a part of the Fort Union Formation (but assigned to the basal Wasatch) show no evidence of a sediment source in this uplift. Further, the Fort Union of the eastern part of the Sand Wash basin to the north contains materials inferred to have been derived from the Sawatch Range; as the White River uplift now lies between these localities (Fig. 1), this also strongly suggests that the uplift did not exist in Paleocene time. In early Eocene time, in contrast, the Piceance basin began to subside as the sediments of the thick Wasatch Formation accumulated there, and the White River dome began to rise concurrently. The Wasatch thins perceptibly near its eroded edge along the present flank of the dome (Donnell, 1961, Pl. 48), even though it definitely extended eastward beyond the present flank structure. The first materials eroded from the rising dome were the fine-grained Cretaceous rocks; this may account for the fact that the Wasatch in this general latitude is markedly finer in grain than in the southern part of the Piceance basin.

On its western and southwestern sides, the dome of the White River Plateau is bordered by steeply upturned strata of the Grand Hogback. This abrupt monoclinal fold continues southward along the flank of the Piceance basin to the Elk Mountains. To the northwest of the White River dome, the monocline merges into the fold system on the southwest flank of a major anticline—the Axial anticline—that extends between the White River and Uinta uplifts and separates the Piceance and Sand Wash basins (Fig. 9). Strata of the Green River Formation, including upper members of middle Eocene age, are involved in the Grand Hogback fold, though they are not so steeply tilted as underlying formations. The fold, then, is at least in part younger than early middle Eocene, and it might be almost entirely so. Donnell (1961, p. 862) has noted that sandstones in the lower and uppermost members of the Green River along the east side of the Piceance basin suggest a source of sediment not far to the east of the Grand Hogback, but that the oil-shale-bearing Parachute Creek Member evidently extended considerably beyond the site of the Grand Hogback. Thus, though a central area of the White River dome probably rose high enough to supply sediments in Wasatch time, the dome did not attain its present outline and flank structure until after Green River time. In the Elk Mountains, the Grand Hogback monocline is cut and deformed by an Oligocene pluton dated at 34.1 m.y. (Obradovich and others, 1969). Hence, the monocline is bracketed in age between middle Eocene and fairly early Oligocene time.

The Axial anticline, to the northwest of the White River Plateau, probably began to rise in Wasatch time and certainly existed by Green River time, when it separated the two major Green River lakes (Bradley, 1930). Much of its structural relief is due, however, to post–Green River uplift, as noted by Sears (1924), who correlated it with late (post-Bridger) uplift of the Uinta anticline. Presumably, this is the same stage of uplift recorded by the Grand Hogback.

CLOSE OF THE LARAMIDE OROGENY

No precise and all-inclusive date can be assigned to the end of Laramide orogeny in the Southern Rocky Mountain province. Just as orogeny started at somewhat different times in different parts of the region, so it ended at somewhat different times. In general, the ending, if it can be so designated, seems to have occurred in middle to late Eocene time, depending on locality.

In the San Juan region, Paleocene structures such as those of the Needle Mountains, the monoclinal west flank of the San Juan dome, and the Uncompahgre uplift were beveled before the close of Wasatch or San Jose time; a thin clastic sequence, equivalent to some part of the upper San Jose Formation and variously known as Telluride Conglomerate or Blanco Basin Formation, was deposited over the truncated surface. The little-deformed Telluride and Blanco Basin were in turn regionally beveled beneath a surface of low relief before lower Oligocene volcanic rocks were deposited widely over this surface (Steven, 1975).

In the area of the Sangre de Cristo Range and Raton basin, the fine-grained sediments of the Huerfano Formation suggest low relief in the Sangre de Cristo–San Luis source area in middle Eocene time. However, lower strata of the Huerfano record the dying stages of thrusting, as noted by Burbank and Goddard (1937). Presumably, this thrusting reflects gravity movement of weak sedimentary rocks into the still-subsiding Raton basin. The Farisita Conglomerate (or upper Huerfano of some usages) records uplift of the Wet Mountain block in late-middle or later Eocene time. This unit was beveled, along with the Wet Mountain crystalline block, before volcanic rocks of very late Eocene or earliest Oligocene age were deposited over the region (Taylor, 1975; Epis and Chapin, 1975).

In the Front Range region, the chief handle on the close of Laramide orogeny is the widespread surface of low relief that was in existence before early Oligocene time in the southern half of the range (Epis and Chapin, 1975; Taylor, 1975). This surface was beveled across the region after thrust faults of post-Denver age had formed along the east side of South Park, and it presumably developed through much of Eocene time (Epis and Chapin, 1968). In North Park, as noted, strata of early Eocene age in the Coalmont Formation are deformed along the flanks of the Medicine Bow and Park Ranges. Lower Oligocene strata of the White River Formation lie on a surface of moderate relief that truncates the Laramide structures and extends high onto the Medicine Bow Range (Steven, 1960), which is true also on the eastern side of the range (Knight, 1953).

Northwestern Colorado and adjoining areas remained tectonically active at least through middle Eocene time, as indicated by deformation of the Green River Formation along the Grand Hogback, Axial anticline, and on the flanks of the Uinta Mountains. Further, the Bridger Formation is moderately deformed along the south edge of the Sand Wash basin (Sears, 1924). The Uinta Mountains continued to supply sediments to adjoining basins through the remainder of post-Bridger Eocene time and possibly into the Oligocene, but deformation had apparently ended except for slight sagging on the borders of the basins.

After Laramide orogeny, the Southern Rocky Mountain region stood somewhat higher than at the beginning of orogeny—when it was at sea level—but at a much lower level than it does today. From several lines of evidence, Bradley (1930, 1931) estimated that the lakes in which the Green River Formation was deposited stood at 300 to 600 m above sea level. Altitudes in other basins probably were similar to this. Mountain areas unquestionably stood higher, probably at altitudes of 1,000 to 2,000 m. Similarly, Keefer (1970, p. 32) estimated that the Wind River basin of central Wyoming had an altitude of about 450 m at the end of Laramide orogeny and that the highest mountains reached 2,400 m.

TECTONIC AND MAGMATIC PATTERNS

The Laramide uplifts and basins of the Southern Rocky Mountain province have many features in common with major mountain and basin units to the north in Wyoming and central Montana (for example, Keefer, 1970; Love, 1970). This similarity must reflect the location of all of these tectonic units on the interior platform of the continent and some causative mechanism common to all. Much has been written on the relations of the mountains and their structures to regional stress patterns but little on the reasons for, or sources of, these stresses. Two hypotheses presented recently do get at the causes. Lipman and others (1972) presented a model relating Laramide orogeny and Cenozoic magmatism to a subduction plate suspended in the mantle at depths ranging from 400 km at the Rocky Mountain front to about 200 km farther west, at the general longitude of the Colorado Plateau. In two papers, Gilluly (1971, 1973) relates magmatism, orogeny, and epeirogeny in this interior region to inhomogeneities in the crust and mantle and to irregularities in the low-velocity zone over which the continent has drifted. The inhomogeneities are attributed by Gilluly to additions of sialic material to crust and mantle throughout the period of continental drifting, but he rejects, as a cause for orogeny and magmatism, a suspended subduction zone so far from the continental margin.

The distinguishing characteristic of the Southern Rocky Mountain province is its long history of buoyancy. Not only were most of the Laramide mountain units rejuvenated from late Paleozoic uplifts, but there were earlier Paleozoic uplifts that affected this region to a greater degree than adjoining areas. A good case can even be made for a differential buoyancy during the latter part of Precambrian time, though this is not to be undertaken here. It may be re-emphasized, however, that many of the faults that controlled Paleozoic and Laramide uplifts were also major faults in Precambrian time. The long history of buoyancy of this region, and of differential buoyancy of large tracts within the region, must reflect an inhomogeneity built into the continental plate at an early date. This inherent buoyancy was very likely nurtured at times by processes resulting from movement of the plate as visualized either by Gilluly or by Lipman and co-workers.

The evident control of many uplifts by paleotectonic and structural features dating from the Paleozoic and the Precambrian lends little support to the once-popular model of external horizontal forces as the cause of all Laramide structure in the region. In the several ranges inherited from late Paleozoic uplifts, sharp folds and thrust faults exist only in narrow belts on the flanks; sedimentary strata a few miles away are almost horizontal. These uplifts are generally described as anticlinal because strata on the flanks dip outward, but most of them were flat-topped and are better described as paired monoclines. In my opinion—not shared by all—the monoclines are drape structures on the flanks of recurrently rising crystalline rock bodies. When the impetus came—whatever the cause—for these bodies to rise, they did so primarily by movements on a network of pre-existing fractures. In places, old and new fractures were integrated into prominent faults, but it is probably inaccurate to conceive of an uplift as a fault-bounded block that rose bodily as a unit. Rather, the history of growth of mountain ranges detailed above indicates that uplift began along a central axis and expanded outward as it extended upward. Mountain-border faults represent a late stage in the process.

Unlike the Laramide tectonism, the Laramide magmatism had essentially no precursor in Phanerozoic time. The only pre-Cretaceous Phanerozoic igneous rocks known in the province are small plutons of mafic and alkalic rocks of very early Cambrian age north of the San Juan Mountains and in the Wet Mountains (Olson and Marvin, 1971; Parker and Sharp, 1970), and ultramafic diatremes of post–Middle Silurian, pre–Late Devonian age in the Front Range (Chronic and others, 1969). Volcanic materials in Pennsylvanian strata were reported by Koschmann and Wells (1946, p. 70), but having seen all of the rocks and localities concerned, I cannot agree to the interpretation that Pennsylvanian volcanism is established. Bentonites in the Cretaceous shales and volcanic detritus in older Mesozoic formations obviously were introduced from outside the province.

Thus, after being essentially absent since Precambrian time, magmatism occurred suddenly near the end of Cretaceous time. Perhaps this was related to migration of a subduction zone into the region (Lipman and others, 1972), or to mantle-crustal processes independent of subduction (Gilluly, 1971, 1973), but in either case, the magmatism must have shared the same root cause as the orogeny. The problem, from the purview of this paper, is not so much why or how magmas were generated, but why magmatic activity took the pattern that it did—that is, of a rather sharply defined belt diagonal to all major tectonic elements in an extensive region that elsewhere is nearly devoid of contemporary igneous rocks.

Neither the igneous belt nor the more restricted mineral belt within it contains any lengthy longitudinal Laramide structural features, such as faults, to give structural continuity. The only internal structure that is quasi-continuous through the length of the belt is a system of echelon Precambrian shear zones. These shear zones evidently influenced localization of the igneous belt (Tweto and Sims, 1963; Tweto and Case, 1972), but it is doubtful that they caused magma to generate on the batholithic scale indicated by the gravity data. If a subduction zone migrated into the region (Lipman and others, 1972) and magmas rising from it had reached crustal heights by Late Cretaceous time, the shear zones might have provided access to higher levels and account for the northeast-trending belt. But why, then, didn't other Precambrian shear zones of this same trend provide similar access? Alternatively, if mantle processes caused the eventual appearance of granodioritic batholiths high in the crust, why did these elongate batholiths string out in a northeast line at the very time that major mountain ranges were rising athwart this line on north-northwest trends?

Whatever the answers, it seems evident that although magmatism and tectonism were related in basic cause, the orientations of magmatic and tectonic features were determined by separate factors, quite possibly at different depth levels in the lithosphere.

ACKNOWLEDGMENTS

For beneficial reviews of earlier drafts of this paper, I thank James Gilluly, W. R. Keefer, J. C. Reed, Jr., and T. A. Steven. I also thank W. A. Cobban for guidance on Cretaceous correlations and J. D. Obradovich for use of unpublished data on the ages of ammonite zones. Finally, many other colleagues, several of whom are credited in the text, generously supplied information and advice.

REFERENCES CITED

Armstrong, R. L., 1969, K-Ar dating of laccolithic centers of the Colorado Plateau and vicinity: Geol. Soc. America Bull., v. 80, no. 10, p. 2081-2086.

Baltz, E. H., 1967, Stratigraphy and regional tectonic implications of part of Upper Cretaceous and Tertiary rocks, east-central San Juan basin, New Mexico: U.S. Geol. Survey Prof. Paper 552, 101 p.

Behrendt, J. C., 1968, The most negative Bouguer anomaly in the conterminous United States [abs.]: EOS (Am. Geophys. Union Trans.), v. 49, no. 1, p. 329.

Behrendt, J. C., and Bajwa, L. Y., 1972, Bouguer gravity map of Colorado: U.S. Geol. Survey open-file map.

Blair, R. W., 1951, Subsurface geologic cross sections of Mesozoic rocks in northeastern Colorado: U.S. Geol. Survey Oil and Gas Inv. Chart OC-42.

Boos, C. M., and Boos, M. F., 1957, Tectonics of eastern flank and foothills of Front Range, Colorado: Am. Assoc. Petroleum Geologists Bull., v. 41, no. 12, p. 2603-2676.

Bradley, W. H., 1930, The varves and climate of the Green River epoch: U.S. Geol. Survey Prof. Paper 158-E, p. 87-110.

——1931, Origin and microfossils of the oil shale of the Green River Formation of Colorado and Utah: U.S. Geol. Survey Prof. Paper 168, 58 p.

——1964, Geology of Green River Formation and associated Eocene rocks in southwestern Wyoming and adjacent parts of Colorado and Utah: U.S. Geol. Survey Prof. Paper 496-A, 86 p.

Briggs, L. I., and Goddard, E. N., 1956, Geology of Huerfano Park, Colorado, in Guidebook to the geology of the Raton basin, Colorado: Denver, Rocky Mtn. Assoc. Geologists, p. 40-45.

Brown, R. W., 1943, Cretaceous-Tertiary boundary in the Denver basin. Colorado: Geol. Soc. America Bull., v. 54, no. 1, p. 65-86.

Bryant, Bruce, 1966, Possible window in the Elk Range thrust sheet near Aspen, Colorado, *in* Geological Survey Research 1966: U.S. Geol. Survey Prof. Paper 550-D, p. D1-D7.

Burbank, W. S., 1930, Revision of geologic structure and stratigraphy in the Ouray district of Colorado, and its bearing on ore deposition: Colorado Sci. Soc. Proc., v. 12, no. 6, p. 151-282.

——1940, Structural control of ore deposition in the Uncompahgre district, Ouray County, Colorado: U.S. Geol. Survey Bull. 906-E, p. 189-265.

Burbank, W. S., and Goddard, E. N., 1937, Thrusting in Huerfano Park, Colorado, and related problems of orogeny in the Sangre de Cristo Mountains: Geol. Soc. America Bull., v. 48, no. 7, p. 931-976.

Burbank, W. S., and Lovering, T. S., 1933, Relation of stratigraphy, structure, and igneous activity to ore deposition of Colorado and southern Wyoming, *in* Ore deposits of the Western States (Lindgren volume): New York, Am. Inst. Mining and Metall. Engineers, p. 272-316.

Cashion, W. B., 1967, Geology and fuel resources of the Green River Formation, southeastern Uinta basin, Utah and Colorado: U.S. Geol. Survey Prof. Paper 548, 48 p.

Cater, F. W., 1970, Geology of the salt anticline region in southwestern Colorado: U.S. Geol. Survey Prof. Paper 637, 80 p.

Chronic, John, McCallum, M. E., Ferris, C. S., and Eggler, D. H., 1969, Lower Paleozoic rocks in diatremes, southern Wyoming and northern Colorado: Geol. Soc. America Bull., v. 80, no. 1, p. 149-156.

Colson, C. T., 1969, Stratigraphy and production of the Tertiary formations in the Sand Wash and Washakie basins, *in* Wyoming Geol. Assoc. Guidebook 21st Ann. Field Conf., 1969, p. 121-128.

Crawford, R. D., 1924, A contribution to the igneous geology of central Colorado: Am. Jour. Sci., 5th ser., v. 7, p. 365-388.

Dana, J. D., 1895, Manual of geology [4th ed.]: New York, American Book Co., 1087 p.

Dane, C. H., 1946, Stratigraphic relations of Eocene, Paleocene, and latest Cretaceous formations of eastern side of San Juan basin, New Mexico: U.S. Geol. Survey Oil and Gas Inv. Chart 24.

——1948, Geologic map of part of eastern San Juan basin, Rio Arriba County, New Mexico: U.S. Geol. Survey Oil and Gas Inv. Prelim. Map 78.

Dane, C. H., and Pierce, W. G., 1936, Dawson and Laramie Formations in southeastern part of Denver basin, Colorado: Am. Assoc. Petroleum Geologists Bull., v. 20, no. 10, p. 1308-1326.

Denson, N. M., and Chisholm, W. A., 1971, Summary of mineralogic and lithologic characteristics of Tertiary sedimentary rocks in the Middle Rocky Mountains and the northern Great Plains, *in* Geological Survey Research 1971: U.S. Geol. Survey Prof. Paper 750-C, p. C117-C126.

DeVoto, R. H., and Peel, F. A., 1972, Pennsylvanian and Permian stratigraphy and structural history, northern Sangre de Cristo Range, Colorado, *in* DeVoto, R. H., ed., Paleozoic stratigraphy and structural evolution of Colorado: Colorado School Mines Quart., v. 67, no. 4, p. 283-320.

Dickinson, R. G., Leopold, E. B., and Marvin, R. F., 1968, Late Cretaceous uplift and volcanism on the north flank of the San Juan Mountains, Colorado, *in* Epis, R. C., ed., Cenozoic volcanism in the Southern Rocky Mountains: Colorado School Mines Quart., v. 63, no. 3, p. 125-132.

Donnell, J. R., 1961, Tertiary geology and oil-shale resources of the Piceance Creek basin between the Colorado and White Rivers, northwestern Colorado: U.S. Geol. Survey Bull. 1082-L, p. 835-891.

——1969, Paleocene and lower Eocene units in the southern part of the Piceance Creek basin, Colorado: U.S. Geol. Survey Bull. 1274-M, 18 p.

Emmons, S. F., Cross, Whitman, and Eldridge, G. H., 1896, Geology of the Denver Basin in Colorado: U.S. Geol. Survey Mon. 27, 556 p.

Epis, R. C., and Chapin, C. E., 1968, Geologic history of the Thirtynine Mile volcanic field, central Colorado, *in* Epis, R. C., ed., Cenozoic volcanism in the Southern Rocky Mountains: Colorado School Mines Quart., v. 63, no. 3, p. 51-85.

Epis, R. C., 1975, Geomorphic and tectonic implications of the post-Laramide, late Eocene erosion surface in the Southern Rocky Mountains, *in* Cenozoic history of the Southern Rocky Mountains: Geol. Soc. America Mem. 144, p. 45-74.

Evernden, J. F., Savage, D. E., Curtis, G. H., and James, G. T., 1964, Potassium-argon dates and the Cenozoic mammalian chronology of North America: Am. Jour. Sci., v. 262, p. 145-198.

Fassett, J. E., and Hinds, J. S., 1971, Geology and fuel resources of the Fruitland Formation and Kirtland Shale of the San Juan basin, New Mexico and Colorado: U.S. Geol. Survey Prof. Paper 676, 76 p.

Freeman, V. L., 1972, Geologic map of the Woody Creek quadrangle, Pitkin and Eagle Counties, Colorado: U.S. Geol. Survey Geol. Quad. Map GQ-967.

Gabelman, J. W., 1952, Structure and origin of the northern Sangre de Cristo Range, Colorado: Am. Assoc. Petroleum Geologists Bull., v. 36, no. 8, p. 1574-1612.

Gaskill, D. L., and Godwin, L. H., 1963, Redefinition and correlation of the Ohio Creek Formation (Paleocene) in west-central Colorado, *in* Short papers in geology and hydrology: U.S. Geol. Survey Prof. Paper 475-C, p. C35-C38.

——1966, Geologic map of the Marcellina Mountain quadrangle, Gunnison County, Colorado: U.S. Geol. Survey Geol. Quad. Map GQ-511.

Gazin, C. L., 1941, Paleocene mammals from the Denver basin, Colorado: Washington Acad. Sci. Jour., v. 31, no. 7, p. 289-295.

Gill, J. R., and Cobban, W. A., 1966, The Red Bird section of the Upper Cretaceous Pierre Shale in Wyoming: U.S. Geol. Survey Prof. Paper 393-A, 69 p.

——1973, Stratigraphy and geologic history of the Montana Group and equivalent rocks, Montana, Wyoming, and North and South Dakota: U.S. Geol. Survey Prof. Paper 776, 37 p.

Gill, J. R., Merewether, E. A., and Cobban, W. A., 1970, Stratigraphy and nomenclature of some Upper Cretaceous and lower Tertiary rocks in south-central Wyoming: U.S. Geol. Survey Prof. Paper 667, 53 p.

Gill, J. R., Cobban, W. A., and Schultz, L. G., 1972, Stratigraphy and composition of the Sharon Springs Member of the Pierre Shale in western Kansas: U.S. Geol. Survey Prof. Paper 728, 50 p.

Gilluly, James, 1963, The tectonic evolution of the western United States: Geol. Soc. London Quart. Jour., v. 119, p. 133-174.

——1971, Plate tectonics and magmatic evolution: Geol. Soc. America Bull., v. 82, no. 9, p. 2383-2396.

——1973, Steady plate motion and episodic orogeny and magmatism: Geol. Soc. America Bull., v. 84, no. 2, p. 499-514.

Goldman, M. I., 1910, The Colorado Springs coal field, Colorado: U.S. Geol. Survey Bull. 381-C, p. 317-340.

Goldstein, August, Jr., 1950, Mineralogy of some Cretaceous sandstones from the Colorado Front Range: Jour. Sed. Petrology, v. 20, no. 2, p. 85-97.

Gow, Kenneth, 1950, Douglas Creek gas field, *in* Utah Geol. Soc. Guidebook to the Geology of Utah, no. 5, p. 139-146.

Griffitts, M. O., 1949, Zones of Pierre Formation, Colorado: Am. Assoc. Petroleum Geologists Bull., v. 33, no. 12, p. 2011-2028.

Gude, A. J., 3d, 1950, Clay minerals of Laramie Formation, Golden, Colorado, identified by X-ray diffraction: Am. Assoc. Petroleum Geologists Bull., v. 34, no. 8, p. 1699-1717.

Hail, W. J., Jr., 1965, Geology of northwestern North Park, Colorado: U.S. Geol. Survey Bull. 1188, 133 p.

——1968, Geology of southwestern North Park and vicinity, Colorado: U.S. Geol. Survey Bull. 1257, 119 p.

Hansen, W. R., 1965, Geology of the Flaming Gorge area, Utah-Colorado-Wyoming: U.S. Geol. Survey Prof. Paper 490, 196 p.

Harms, J. C., 1965, Sandstone dikes in relation to Laramide faults and stress distribution in the southern Front Range, Colorado: Geol. Soc. America Bull., v. 76, no. 9, p. 981-1002.

Hart, S. R., 1961, Mineral ages and metamorphism, *in* Kulp, J. L., ed., Geochronology of rock systems: New York Acad. Sci. Ann., v. 91, p. 192-197.

——1964, The petrology and isotopic-mineral age relations of a contact zone in the Front Range, Colorado: Jour. Geology, v. 72, no. 5, p. 493-525.

Haun, J. D., and Weimer, R. J., 1960, Cretaceous stratigraphy of Colorado, *in* Weimer, R. J., and Haun, J. D., eds., Guide to the geology of Colorado: Geol. Soc. America, Rocky Mtn. Assoc. Geologists, and Colorado Sci. Soc., p. 58-65.

Hayes, P. T., and Zapp, A. D., 1955, Geology and fuel resources of the Upper Cretaceous rocks of the Barker Dome-Fruitland area, San Juan County, New Mexico: U.S. Geol. Survey Oil and Gas Inv. Map OM-144.

Hills, R. C., 1891, Orographic and structural features of Rocky Mountain geology: Colorado Sci. Soc., Proc., v. 3, p. 363-458.

Izett, G. A., 1968, Geology of the Hot Sulphur Springs quadrangle, Grand County, Colorado: U.S. Geol. Survey Prof. Paper 586, 79 p.

Izett, G. A., Taylor, R. B., and Hoover, D. L., 1962, Windy Gap Member of the Middle Park Formation, Middle Park, Colorado, *in* Short papers in geology, hydrology, and topography: U.S. Geol. Survey Prof. Paper 450-E, p. E36-E39.

Izett, G. A., Scott, G. R., and Obradovich, J. D., 1969, Oligocene rhyolite in the Denver basin, Colorado, *in* Geological Survey Research 1969: U.S. Geol. Survey Prof. Paper 650-B, p. B12-B14.

Izett, G. A., Cobban, W. A., and Gill, J. R., 1971, The Pierre Shale near Kremmling, Colorado, and its correlation to the east and the west: U.S. Geol. Survey Prof. Paper 684-A, 19 p.

Johnson, R. B., 1959, Geology of the Huerfano Park area, Huerfano and Custer Counties, Colorado: U.S. Geol. Survey Bull. 1071-D, p. 87-119.

Johnson, R. B., and Wood, G. H., Jr., 1956, Stratigraphy of Upper Cretaceous and Tertiary rocks of Raton basin, Colorado and New Mexico: Am. Assoc. Petroleum Geologists Bull., v. 40, no. 4, p. 707-721.

Keefer, W. R., 1970, Structural geology of the Wind River basin, Wyoming: U.S. Geol. Survey Prof. Paper 495-D, 35 p.

King, P. B., 1959, The evolution of North America: Princeton, N.J., Princeton Univ. Press., 189 p.

Kinney, D. M., 1970, Preliminary geologic map of the Rand quadrangle, North and Middle Parks, Jackson and Grand Counties, Colorado: U.S. Geol. Survey open-file map.

Knight, S. H., 1953, Summary of the Cenozoic history of the Medicine Bow Mountains, Wyoming, *in* Wyoming Geol. Assoc. Guidebook, 8th Ann. Field Conf., 1953, p. 65-76.

Koschmann, A. H., and Wells, F. G., 1946, Preliminary report on the Kokomo mining district, Colorado: Colorado Sci. Soc. Proc., v. 15, no. 2, p. 49-112.

Lambert, R. St. J., 1971, The pre-Pleistocene Phanerozoic time-scale—A review, *in* Part I of the Phanerozoic time-scale—A supplement: Geol. Soc. London Spec. Pub. no. 5, p. 9-34.

Lee, W. T., 1917, Geology of the Raton Mesa and other regions in Colorado and New Mexico: U.S. Geol. Survey Prof. Paper 101, p. 9-221 [1918].

LeRoy, L. W., 1946, Stratigraphy of the Golden-Morrison area, Jefferson County, Colorado: Colorado School Mines Quart., v. 41, no. 2, 115 p.

Lipman, P. W., Prostka, H. J., and Christiansen, R. L., 1972, Cenozoic volcanism and plate-tectonic evolution of the Western United States. I—Early and middle Cenozoic: Royal Soc. London Philos. Trans., v. 271, p. 217-248.

Litsey, L. R., 1958, Stratigraphy and structure of the northern Sangre de Cristo Mountains, Colorado: Geol. Soc. America Bull., v. 69, no. 9, p. 1143-1178.

Love, J. D., 1960, Cenozoic sedimentation and crustal movement in Wyoming (Bradley volume): Am. Jour. Sci., 258-A, p. 204-214.

——1970, Cenozoic geology of the Granite Mountains area, central Wyoming: U.S. Geol. Survey Prof. Paper 495-C, 154 p.

Love, J. D., McGrew, P. O., and Thomas, H. D., 1963, Relationship of latest Cretaceous and Tertiary deposition and deformation to oil and gas in Wyoming, *in* Backbone of the Americas: Am. Assoc. Petroleum Geologists Mem. 2, p. 196-208.

Lovering, T. S., 1929, Geologic history of the Front Range, Colorado: Colorado Sci. Soc. Proc., v. 12, no. 4, p. 59-111.

Lovering, T. S., and Goddard, E. N., 1938, Laramide igneous sequence and differentiation in the Front Range, Colorado: Geol. Soc. America Bull., v. 49, no. 1, p. 35-68.

——1950, Geology and ore deposits of the Front Range, Colorado: U.S. Geol. Survey Prof. Paper 223, 319 p.

MacLachlan, J. C., and Kleinkopf, M. D., eds., 1969, Configuration of the Precambrian surface of Colorado: Mtn. Geologist, v. 6, no. 4, p. 193-197 and insert map.

Mallory, W. W., 1960, Outline of Pennsylvanian stratigraphy of Colorado, *in* Weimer, R. J., and Haun, J. D., eds., Guide to the geology of Colorado: Geol. Soc. America, Rocky Mtn. Assoc. Geologists, Colorado Sci. Soc., p. 23-33.

——1967, Pennsylvanian and associated rocks in Wyoming: U.S. Geol. Survey Prof. Paper 554-G, 31 p.

Martin, C. A., 1965, Denver basin: Am. Assoc. Petroleum Geologists Bull., v. 49, no. 11, p. 1908-1925.

Mather, K. F., Gilluly, James, and Lusk, R. G., 1928, Geology and oil and gas prospects of northeastern Colorado: U.S. Geol. Survey Bull. 796-B, p. 65-124.

Matuszczak, R. A., 1969, Trinidad Sandstone interpreted, evaluated, in Raton basin, Colorado-New Mexico: Mtn. Geologist, v. 6, no. 3, p. 119-124.

McDonald, R. E., 1972, Eocene and Paleocene rocks of the southern and central basins, *in* Geologic atlas of the Rocky Mountain region: Denver, Rocky Mtn. Assoc. Geologists, p. 243-256.

McDowell, F. W., 1966, Potassium-argon dating of Cordilleran intrusives, *in* Gast, P. W., Investigations in isotope geochemistry: U.S. Atomic Energy Comm., NYO 1669-11, Appendix A, 79 p.

McGookey, D. P. [chmn.], 1972, Cretaceous System, *in* Geologic atlas of the Rocky Mountain region: Denver, Rocky Mtn. Assoc. Geologists, p. 190-228.

Miller, J. P., Montgomery, Arthur, and Sutherland, P. K., 1963, Geology of part of the southern Sangre de Cristo Mountains, New Mexico: New Mexico Bur. Mines and Mineral Resources Mem. 11, 106 p.

Moody, J. D., 1947, Upper Montana Group, Golden area, Jefferson County, Colorado: Am. Assoc. Petroleum Geologists Bull., v. 31, no. 8, p. 1454-1471.

Obradovich, J. D., Mutschler, F. E., and Bryant, Bruce, 1969, Potassium-argon ages bearing on the igneous and tectonic history of the Elk Mountains and vicinity, Colorado—A preliminary report: Geol. Soc. America Bull., v. 80, no. 9, p. 1749-1756.

Olson, J. C., and Marvin, R. F., 1971, Rb-Sr whole-rock age determinations of the Iron Hill and McClure Mountain carbonatite-alkalic complexes, Colorado—Discussion: Mtn. Geologist, v. 8, no. 4, p. 221.

O'Sullivan, R. B., Repenning, C. A., Beaumont, E. C., and Page, H. G., 1972, Stratigraphy of the Cretaceous rocks and the Tertiary Ojo Alamo Sandstone, Navajo and Hopi Indian Reservations, Arizona, New Mexico, and Utah: U.S. Geol. Survey Prof. Paper 521-E, 65 p.

Parker, R. L., and Sharp, W. N., 1970, Mafic-ultramafic igneous rocks and associated carbonatites of the Gem Park complex, Custer and Fremont Counties, Colorado: U.S. Geol. Survey Prof. Paper 649, 24 p.

Pearson, R. C., Tweto, Ogden, Stern, T. W., and Thomas, H. W., 1962, Age of Laramide porphyries near Leadville, Colorado, *in* Short papers in geology, hydrology, and topography: U.S. Geol. Survey Prof. Paper 450-C, p. C78-C80.

Peterson, J. A., Loleit, A. J., Spencer, C. W., and Ullrich, R. A., 1965, Sedimentary history and economic geology of San Juan basin: Am. Assoc. Petroleum Geologists Bull., v. 49, no. 11, p. 2076-2119.

Pillmore, C. L., 1969, Geology and coal deposits of the Raton coal field, Colfax County, New Mexico: Mtn. Geologist, v. 6, no. 3, p. 125-142.

Reeside, J. B., Jr., 1924, Upper Cretaceous and Tertiary formations of the western part of the San Juan basin, Colorado and New Mexico: U.S. Geol. Survey Prof. Paper 134, p. 1-70.

——1944, Maps showing thickness and general character of the Cretaceous deposits in the western interior of the United States: U.S. Geol. Survey Oil and Gas Inv. Prelim. Map 10.

——1957, Paleoecology of the Cretaceous seas of the western interior of the United States, *in* Ladd, H. S., ed., Paleoecology: Geol. Soc. America Mem. 67, v. 2, p. 505-541.

Ritzma, H. R., 1955, Early Cenozoic history of the Sand Wash basin, northwest Colorado, *in* Ritzma, H. R., and Oriel, S. S., eds., Guidebook to the geology of northwest Colorado: Salt Lake City, Utah, Intermtn. Assoc. Petroleum Geologists, and Denver, Colo., Rocky Mtn. Assoc. Geologists, p. 36-40.

Romero, J. C., and Hampton, E. R., 1972, Maps showing the approximate configuration and depth to the top of the Laramie-Fox Hills aquifer, Denver basin, Colorado: U.S. Geol. Survey Misc. Geol. Inv. Map I-791.

Sabins, F. F., Jr., 1964, Symmetry, stratigraphy, and petrography of cyclic Cretaceous deposits in San Juan Basin: Am. Assoc. Petroleum Geologists Bull., v. 48, no. 3, p. 292-316.

Sawatzky, D. L., 1964, Structural geology of southeastern South Park, Park County, Colorado: Mtn. Geologist, v. 1, no. 3, p. 133-139.

Scott, G. R., 1963, Bedrock geology of the Kassler quadrangle, Colorado: U.S. Geol. Survey Prof. Paper 421-B, p. 71-125.

——1972, Geologic map of the Morrison quadrangle, Jefferson County, Colorado: U.S. Geol. Survey Misc. Inv. Map I-790 A.

Sears, J. D., 1924, Geology and oil and gas propects of part of Moffat County, Colorado, and southern Sweetwater County, Wyoming: U.S. Geol. Survey Bull. 751-G, p. 269-319.

Smith, W. C., 1938, Geology of the Caribou stock in the Front Range, Colorado: Am. Jour. Sci., 5th ser., v. 36, no. 213, p. 161-196.

Soister, P. E., 1965, Geologic map of the Platteville quadrangle, Weld County, Colorado: U.S. Geol. Survey Geol. Quad. Map GQ-399.

Spencer, F. D., 1961, Bedrock geology of the Louisville quadrangle, Colorado: U.S. Geol. Survey Geol. Quad. Map GQ-151.

Stagner, W. L., 1941, The paleogeography of the eastern part of the Uinta basin during Uinta B (Eocene) time: Pittsburgh, Carnegie Mus. Annals, v. 28, p. 273-308.

Stark, J. T., and others, 1949, Geology and origin of South Park, Colorado: Geol. Soc. America Mem. 33, 188 p.

Steven, T. A., 1960, Geology and fluorspar deposits, Northgate district, Colorado: U.S. Geol. Survey Bull. 1082-F, p. 323-432.

——1975, Middle Tertiary volcanic field in the Southern Rocky Mountains, *in* Cenozoic history of the Southern Rocky Mountains: Geol. Soc. America Mem. 144, p. 75-94.

Taylor, R. B., 1975, Neogene tectonism in south-central Colorado, *in* Cenozoic history of the Southern Rocky Mountains: Geol. Soc. America Mem. 144, p. 211-226.

Theobald, P. K., Jr., 1970, Preliminary geologic map of the north half of the Craig quadrangle, Moffat County, Colorado: U.S. Geol. Survey open-file rept.

Tweto, Ogden, 1957, Geologic sketch of southern Middle Park, Colorado, *in* Rocky Mtn. Assoc. Geologists Guidebook, 9th Ann. Field Conf., North and Middle Park Basins, Colorado, 1957: p. 18-31.

——1960, Pre-ore age of faults at Leadville, Colorado, *in* Short papers in the geological sciences: U.S. Geol. Survey Prof. Paper 400-B, p. B10-B11.

——1968, Geologic setting and interrelationships of mineral deposits in the mountain province of Colorado and south-central Wyoming, *in* Ridge, J. D., ed., Ore deposits of the United States, 1933-1967 (Graton-Sales volume): New York, Am. Inst. Mining Metall. Petroleum Engineers, v. 1, p. 551-588.

Tweto, Ogden, and Case, J. E., 1972, Gravity and magnetic features as related to geology in the Leadville 30-minute quadrangle, Colorado: U.S. Geol. Survey Prof. Paper 726-C, 31 p. [1973].

Tweto, Ogden, and Sims, P. K., 1963, Precambrian ancestry of the Colorado mineral belt: Geol. Soc. America Bull., v. 74, no. 8, p. 991-1014.

Tweto, Ogden, Bryant, Bruce, and Williams, F. E., 1970, Mineral resources of the Gore

Range-Eagles Nest Primitive Area and vicinity, Summit and Eagle Counties, Colorado: U.S. Geol. Survey Bull. 1319-C, 127 p.

Upson, J. E., 1941, The Vallejo Formation—New early Tertiary red-beds in southern Colorado: Am. Jour. Sci., v. 239, no. 8, p. 577-589.

Van Horn, Richard, 1957, Bedrock geology of the Golden quadrangle, Colorado: U.S. Geol. Survey Geol. Quad. Map GQ-103.

Varnes, D. J., and Scott, G. R., 1967, General and engineering geology of the United States Air Force Academy site, Colorado: U.S. Geol. Survey Prof. Paper 551, 93 p.

Wahlstrom, E. E., 1940, Audubon-Albion stock, Boulder County, Colorado: Geol. Soc. America Bull., v. 51, no. 12, p. 1789-1820.

Warner, L. A., 1956, Tectonics of the Colorado Front Range, in Am. Assoc. Petroleum Geologists Rocky Mtn. Sec. Geol. Record, 1956: p. 129-144.

Washburne, C. W., 1910, The Canon City coal field, Colorado: U.S. Geol. Survey Bull. 381-C, p. 341-378.

Weimer, R. J., 1960, Upper Cretaceous stratigraphy, Rocky Mountain area: Am. Assoc. Petroleum Geologists Bull., v. 44, no. 1, p. 1-20.

——1970, Rates of deltaic sedimentation and intrabasin deformation, Upper Cretaceous of Rocky Mountain region, in Morgan, J. P., ed., Deltaic sedimentation, modern and ancient: Soc. Econ. Paleontologists and Mineralogists Spec. Paper 15, p. 270-292.

Wood, G. H., Kelley, V. C., and MacAlpin, A. J., 1948, Geology of southern part of Archuleta County, Colorado: U.S. Geol. Survey Oil and Gas Inv. Prelim. Map 81.

Young, E. J., 1972, Laramide-Tertiary intrusive rocks of Colorado: U.S. Geol. Survey open-file rept., 206 p.

Zapp, A. D., 1949, Geology and coal resources of the Durango area, La Plata and Montezuma Counties, Colorado: U.S. Geol. Survey Oil and Gas Inv. Map 109.

MANUSCRIPT RECEIVED BY THE SOCIETY MAY 8, 1974

Geological Society of America
Memoir 144
© 1975

Geomorphic and Tectonic Implications of the Post-Laramide, Late Eocene Erosion Surface in the Southern Rocky Mountains

Rudy C. Epis

Colorado School of Mines
Golden, Colorado 80401

AND

Charles E. Chapin

New Mexico Bureau of Mines
Socorro, New Mexico 87801

ABSTRACT

A late Eocene erosion surface of low relief, which extended throughout south-central Colorado, provides a post-Laramide, pre-Oligocene, regional structural datum. The age and geomorphic character of the surface are documented for an area of more than 10,400 km² in the southern Front Range, Rampart Range, South Park, Thirtynine Mile volcanic field, southern Mosquito Range, upper Arkansas River valley, southern Sawatch Range, and adjacent Great Plains to the east. The surface truncated middle Eocene and older rocks, which were deformed during the Laramide orogeny, and deeply beveled crystalline Precambrian rocks across wide areas. Size, shape, and distribution of the overlying Wall Mountain Tuff and associated gravel units show that the surface sloped gently southward and eastward and merged with the western Great Plains.

Correlation of deposits on the surface indicates that it was uplifted 1,500 to 3,000 m and disrupted by block faulting of basin-and-range style in Miocene and later time. Many erosion surfaces now at various levels in the mountainous terrain of the area are faulted segments of this late Eocene surface.

It is suggested that erosion surfaces of similar geomorphic character and age developed over a much larger region of the Southern Rocky Mountains province and of the adjoining Basin and Range province to the south and southwest.

INTRODUCTION

Earlier studies of the Thirtynine Mile volcanic field of central Colorado showed that the volcanic pile was underlain by an erosion surface of relatively low relief (Cross, 1894; Chapin and Epis, 1964; Epis and Chapin, 1968; Steven and Epis, 1968). Additional field, petrographic, and radiometric-age data have provided a better understanding of the stratigraphic sequence of units in the volcanic field and of their regional correlation in outlying areas (Epis and Chapin, 1974). It is now possible to document the age and geomorphic character of the prevolcanic surface across an area of more than 10,000 km^2 and to show that the surface was extensively uplifted and block faulted in Miocene and later time.

LOCATION

The area under consideration contains some of the major geomorphic elements of the Southern Rocky Mountains (Figs. 1 and 2). These include the southern Front Range, Rampart Range, South Park, southern Mosquito Range, upper Arkansas River valley, southern Sawatch Range, northern San Luis Valley, northern Sangre de Cristo Mountains, Wet Mountain Valley, Wet Mountains, and adjacent Great Plains to the east. Most of the documentary evidence and discussion of this paper relates to the central portion of the region where rocks of the Thirtynine Mile volcanic field occur.

AGE OF THE SURFACE

The age of the prevolcanic surface can be bracketed between the ages of the oldest deposits resting on it and the youngest deposits beneath it. Most important are the age and correlation of a widespread ash-flow tuff at the base of the Thirtynine Mile volcanic pile (Figs. 3 and 9). This tuff, previously referred to as ash flow 1 (Epis and Chapin, 1968) or the Agate Creek Tuff (DeVoto, 1964, 1971) and now termed the Wall Mountain Tuff (Epis and Chapin, 1974), occurs mainly as discontinuous patches along paleovalleys in the prevolcanic surface. Six K-Ar age

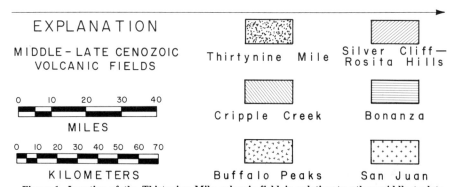

Figure 1. Location of the Thirtynine Mile volcanic field in relation to other middle to late Cenozoic volcanic rocks. Important localities mentioned in the text are included. Modified from Burbank and others (1935), Chapin and Epis (1964), and Epis and Chapin (1968).

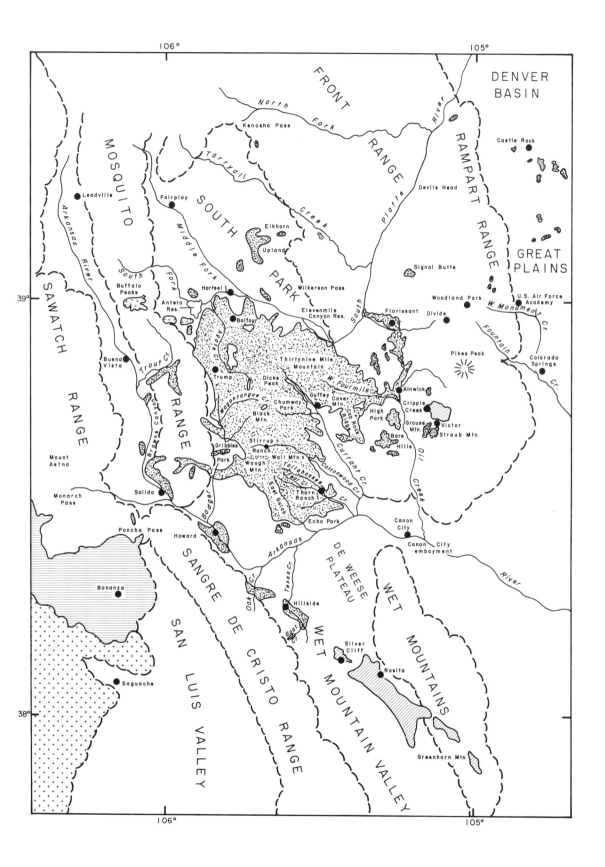

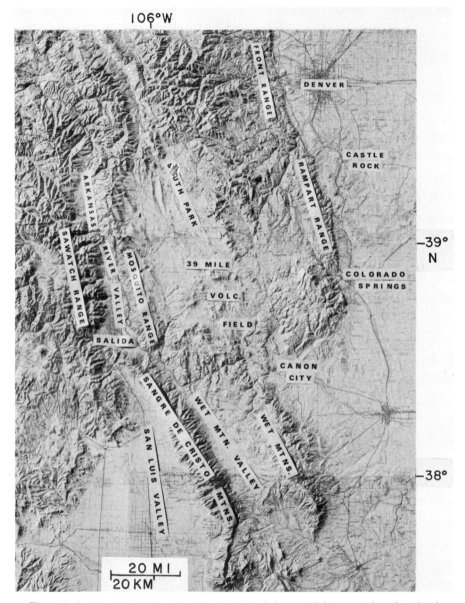

Figure 2. Composite of Army Map Service plastic relief, maps (2-degree quadrangles) showing major geomorphic elements of central Colorado.

determinations from samples at four widely separated localities of the Wall Mountain Tuff indicate an age of 35 to 36 m.y., that is, earliest Oligocene or Chadron (Epis and Chapin, 1974).[1]

Additional K-Ar ages (Fig. 3) for the voluminous and widespread Thirtynine Mile Andesite (34 m.y.), the Florissant Lake Beds (34 m.y.), and the Antero Formation (33 m.y.)—supplemented by vertebrate fossils in the latter two units (Stark and others, 1949; MacGinitie, 1953; Johnson, 1937a, 1937b, 1937c,; DeVoto,

[1] In this paper, we use time subdivisions for the Cenozoic as given by Harland and others (1964).

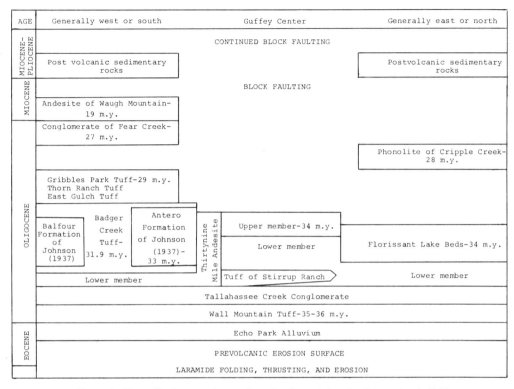

Figure 3. Generalized composite stratigraphy of the Thirtynine Mile volcanic field.

1971)—confirm the early Oligocene age of the Wall Mountain Tuff. It is clear that the prevolcanic surface can be no younger than very earliest Oligocene.

At various localities within the area shown on Figures 1 and 2, the prevolcanic erosion surface truncates Laramide uplifts, their bounding structural features, cores of Precambrian rocks, and upwarped flanking sedimentary units of pre-Oligocene age. The youngest sedimentary rocks involved in Laramide deformation and truncation by the surface establish a lower limit on the age of the surface. In eastern South Park, continental sedimentary and interbedded andesitic rocks, which are termed the Denver Formation (Johnson, 1937b; Stark and others, 1949) or South Park Formation (Sawatzky, 1967; DeVoto, 1971), have been folded and overthrust by Precambrian to Mesozoic rocks of the Front Range along the Elkhorn thrust. Stark and others (1949) and Sawatzky (1964, 1967) showed that the prevolcanic surface ("Elkhorn Upland") truncates the Elkhorn thrust and the folded strata of the South Park Formation. An andesite flow at the base of the South Park Formation yielded a K-Ar age of 56.8 ± 2.6 m.y. (Sawatzky, 1967) which, together with paleontologic data of Johnson (1937a, 1937b) and Brown (1962), indicates that the South Park Formation was deposited in late Paleocene and probably early Eocene time. Scattered outcrops of the Wall Mountain Tuff rest on the beveled upper plate of the Elkhorn thrust; therefore, in this area, the age of the prevolcanic surface is bracketed between late Paleocene (or early Eocene) and earliest Oligocene time.

Similar relations of the prevolcanic surface to Laramide structural features and early Tertiary sedimentary rocks can be documented elsewhere in central Colorado. In the Castle Rock area, outcrops of tuff that was dated at 34.8 ± 1.1 m.y. (Izett and others, 1969; Welsh, 1969) have been correlated with the Wall Mountain Tuff on the basis of similarities in age, stratigraphic position, modal mineral content, 2 V of sanidine, and chemical composition (Epis and Chapin, 1974). The age of the tuff is corroborated by an early Oligocene vertebrate fauna in the overlying Castle Rock Conglomerate (Richardson, 1915; Izett and others, 1969). The tuff rests on an erosion surface carved on the Dawson Formation of Late Cretaceous and Paleocene age (Brown, 1943). The Dawson is involved in folding and thrusting along the east side of the Front Range in a manner analogous to that of the South Park Formation along the west side. Accordingly, the age of the prevolcanic surface in the Denver basin is bracketed between Paleocene and earliest Oligocene. Similar relations in the Canon City embayment exist between the Poison Canyon Formation of Late Cretaceous and Paleocene age and outliers of Oligocene volcanic rocks of the Thirtynine Mile volcanic field (G. R. Scott, 1973, oral commun.).

Beneath the main body of the Thirtynine Mile volcanic field, the prevolcanic surface was developed largely on beveled Precambrian rocks. Within this area, the youngest rocks beneath the surface are reddish, arkosic, conglomeratic mudstone, conglomerate, and minor sandstone units named the Echo Park Alluvium (Epis and Chapin, 1974). These deposits fill a narrow graben that trends north-northwest from the Arkansas River Canyon to the vicinity of Guffey; minor volumes of similar material are found in several paleovalleys on the prevolcanic surface, as for example near Florissant (Niesen, 1969) and at High Park (Tobey, 1969). In such cases, they are directly overlain by the Wall Mountain Tuff. No fossils have been found in the Echo Park Alluvium, but it is similar in lithologic content, color, and stratigraphic position to the Huerfano Formation and Farisita Conglomerate in the Huerfano Park area. The Huerfano and Farisita overlie the Poison Canyon Formation of Late Cretaceous-Paleocene age and interfinger with the Cuchara Formation of probable Eocene age (Johnson and Wood, 1956; Johnson, 1959; Scott and Taylor, 1975). Development of the surface is thus bracketed here between deposition of sediment of probable Eocene age and eruption of volcanic rocks of earliest Oligocene age.

Carving of the erosion surface began with the start of the Laramide orogeny at the close of Campanian time (Tweto, 1975). Thousands of meters of Mesozoic and Paleozoic sedimentary rocks and an unknown thickness of Precambrian rocks were eroded from uplifts during Late Cretaceous, Paleocene, and Eocene time. Erosion apparently kept pace with uplift so that mountains never stood very high above adjacent aggrading basins. When active deformation ceased in late Eocene time, weathering that occurred under a warm, temperate to subtropical climate (MacGinitie, 1953; Leopold and MacGinitie, 1972) and lateral planation by streams produced a surface of very low relief that embraced both beveled uplifts and aggraded basins. Accordingly, the surface as preserved represents a period of post-Laramide tectonic and magmatic quiescence during late Eocene and possibly very earliest Oligocene time. For the sake of convenience, we refer to the surface throughout this paper as the late Eocene surface.

EXTENT OF THE SURFACE

The extent of the late Eocene erosion surface can be documented most clearly in central Colorado by the distribution of the Wall Mountain Tuff and by correlation of exhumed remnants of the surface often strikingly evident in panoramic views of the region (see Figs. 8, 9, 15, and 16). Reconstruction of the surface has been facilitated by an optimum level of erosion that is deep enough to exhume large portions of the surface from beneath its middle Tertiary cover but not so deep as to remove all remnants of the overlying Wall Mountain Tuff or to obscure the nature of the surface.

Because of the importance of the Wall Mountain Tuff to this reconstruction, a brief summary of its lithologic characteristics is given below:

The Wall Mountain Tuff is a multiple-flow, simple cooling unit of moderate to dense welding with conspicuous eutaxitic foliation. It tends to form clifflike outcrops of reddish-brown to buff lithoidal tuff, sometimes with a prominent black vitrophyre zone near the base. Primary and secondary laminar flow structures are common (Lowell and Chapin, 1972). Chemically, the tuff is a potassic calc-alkalic rhyolite but mineralogically, it is a trachyte. Clear, "glassy" sanidine and fresh to intensely argillized andesine are the principal phenocrystic minerals. Biotite and lesser opaque oxides compose 6 to 9 percent of the tuff; traces of pyroxene are frequently present. Quartz is absent as a phenocryst. The sanidine/plagioclase ratio varies from 0.9 to 4.4 and averages about 2.0. The 2V values for sanidine (or anorthoclase) range from 22° to more than 50° with median values for the four ash-flow members of 30.5°, 34°, 38°, and 42° (Chapin, 1965; Lowell, 1969). Tan to pink felsite lithic fragments are moderately abundant in the Browns Canyon area but sparse to absent elsewhere. The tuff has normal magnetic polarity (Graebner, 1967; Graebner and Epis, 1968) and generally occurs in isolated outcrops less than 30 m thick. In major paleovalleys, however, thicknesses greater than 150 m are preserved.

The Wall Mountain Tuff has been correlated across an area of about 10,400 km^2. Most of the outcrops are in paleovalleys in the late Eocene surface. Erosion has stripped most of the tuff from interfluves, but in many paleovalleys, it is protected by overlying sedimentary and (or) volcanic deposits. The tuff most commonly rests on Precambrian rocks, but locally it rests on other pre-Oligocene rocks beveled by the late Eocene surface or on Eocene sedimentary deposits in basins and valleys that were aggrading during carving of the surface (Fig. 4). The source of the Wall Mountain Tuff is not known with certainty; it may have been erupted from vents related to the Mount Princeton batholith in the Sawatch Range. A deeply eroded cauldron complex in the Mount Aetna area north of Monarch Pass (Dings and Robinson, 1957; Lipman and others, 1969; P. J. Toulmin, 1973, oral commun.) is a likely source. From this general area, the ash flows moved eastward down major paleovalleys, such as those at Trout Creek and between Browns Canyon and Waugh Mountain (Fig. 1), and spread widely over the late Eocene surface that sloped gently to the east and southeast.

The late Eocene surface can be traced discontinuously through much of Colorado and New Mexico by applying similar criteria, that is, by examining the distribution of early Oligocene deposits on a surface of low relief which truncates Laramide structures. The original extent of the late Eocene surface is yet to be worked out. However, termination of Laramide deformation in late Eocene time (Tweto, 1975; Coney, 1972) and the existence of a late Eocene hiatus in magmatism over most of the Southern Rocky Mountains and Basin and Range provinces (Damon and Mauger, 1966; Livingston and others, 1968; Cross, 1973) lead us to suggest that the surface documented in this paper may be part of a very widespread surface of regional importance.

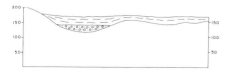

Figure 4. Diagram illustrating typical relations of topography of the late Eocene surface (cut on middle Eocene or older rocks) and overlying Echo Park Alluvium (circles and dots) and Wall Mountain Tuff (wavy lines). Vertical scale is in feet.

GEOMORPHIC CHARACTER OF THE SURFACE

Relief

The late Eocene erosion surface was characterized by relatively low relief and was channeled by shallow, broad stream valleys (Fig. 4). The ash flows that deposited the Wall Mountain Tuff followed these valleys where convenient but had no difficulty in jumping interfluves between them. Except in major valleys near the source of the eruptions, outcrops of the Wall Mountain Tuff are generally less than 30 m thick. Had there been considerably more relief on the surface, the ash flows would have puddled in the valleys and filled them before overriding the interfluves. Instead, the drainage was quickly re-established in the shallow paleovalleys, and the Wall Mountain Tuff was extensively dissected so that in many areas only thin, scattered remnants remained. These remnants were buried by the Tallahassee Creek Conglomerate, mudflows of the Thirtynine Mile Andesite, and shallow lacustrine deposits, such as those in the Antero basin, at Balfour, and in the Florissant basin where lakes formed along the north side of the Thirtynine Mile volcanic field as a result of volcanic damming of drainages.

In his paleontologic studies of the Florissant Lake Beds, MacGinitie (1953, p. 11) stated the following about the late Eocene surface:

This time of erosion probably occupied the greater part of the later Eocene, and, to judge from the low relief of the surface developed on the Pikes Peak granite, and from the comparatively fine sediments associated with the base of the Florissant series, the surface had reached an early old-age stage of the erosion cycle.

MacGinitie further concluded (p. 77) that:

The Oligocene forest occupied streamside and lakeside habitats in a piedmont of low relief and moderate elevation which bordered the Rocky Mountain uplift on the east. The drainage was disorganized and partly ponded by successive volcanic outbursts which covered the area with dust, pumice, and mudflows. The fossil fauna and flora were deposited in the resulting shallow and ephermeral lakes.

MacGinitie's interpretations of the nature of the late Eocene surface in the Florissant area can be applied to the surface in general. Stark and others (1949, p. 138) drew similar conclusions regarding late Eocene topography in eastern South Park: ". . . by the end of the Eocene the region had been reduced to a gently rolling upland, above which rose a few well-rounded monadnocks."

Additional information about the relief of the late Eocene surface can be obtained where the surface is now being exhumed from beneath the Thirtynine Mile volcanic field (Fig. 5). As the edges of the volcanic pile retreat, the relatively flat, gently rolling terrain uncovered usually has shallow paleovalleys and scattered outcrops of the Wall Mountain Tuff. This type of terrain is especially conspicuous in the eastern portion of the Thirtynine Mile volcanic field where the Elkhorn Upland in eastern South Park can be traced eastward and southeastward as a gently rolling terrain underlain by Precambrian rocks on which rest isolated outcrops of the Wall Mountain Tuff (Fig. 6). Similar topographic and geologic relations are present south of West Fourmile Creek between Currant Creek on the west and Oil Creek on the east (Fig. 7). Here, shallow paleovalleys containing remnants of the Wall Mountain Tuff, overlain by boulder alluvium, project from beneath the Thirtynine Mile Andesite. The surface can be traced southward across the highly dissected country near the Arkansas River Canyon to the De Weese Plateau in the northwestern foothills of the Wet Mountains (Fig. 8). Here, again, gently rolling terrain with concordant summits has been developed on Precambrian rocks on which rest scattered outcrops of Oligocene volcanic rocks.

Whereas the late Eocene surface was generally characterized by relatively flat, gently rolling terrain as described above, appreciably greater relief was present locally. Major valleys heading in the Laramide Sawatch uplift were filled with as much as 300 m of Oligocene sedimentary and volcanic rocks (Lowell, 1969, 1971; Chapin and others, 1970). Elsewhere in the Thirtynine Mile volcanic field,

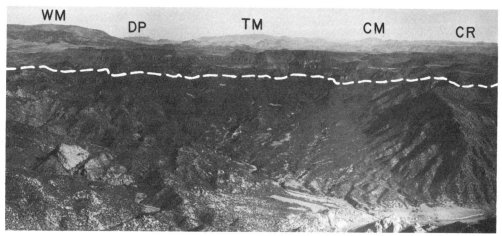

Figure 5. Oblique aerial view northward into the head of Echo Park from above the canyon of the Arkansas River (lower right). The smooth prevolcanic surface is shown by the dashed line. A prevolcanic paleovalley in Precambrian rocks, filled with Echo Park Alluvium, is extensively exhumed (right of center). The Wall Mountain Tuff and the lower member of the Thirtynine Mile Andesite rest on Precambrian rocks and Echo Park Alluvium. Cliffs of the Gribbles Park Tuff and overlying andesite of Waugh Mountain form the rim of Echo Park. Eastern Waugh Mountain (WM), Dicks Peak (DP), Thirtynine Mile Mountain (TM), and Cover Mountain (CM) are volcanic constructional features of the Thirtynine Mile volcanic field. Caprock Ridge (CR) is a thin sheet that consists of Wall Mountain Tuff, Tallahassee Creek Conglomerate, and Thirtynine Mile Andesite resting on the prevolcanic surface. Length of dashed line represents about 11 km.

Figure 6. Oblique aerial view from above High Park northeastward across the extensively exhumed late Eocene surface (middle foreground). Note the low, rolling relief of the surface carved into Precambrian rocks. Miocene-Pliocene gravel units rest on the surface in the vicinity of Divide (d). The Wall Mountain Tuff and Florissant Lake Beds are preserved in a shallow paleovalley in the Florissant basin (f). Southward, at the junction of Fourmile Creek (fm) and Oil Creek (oc) the Tallahassee Creek Conglomerate and lower member of the Thirtynine Mile Andesite rest on the surface in related paleovalleys. Along the east side of Oil Creek and the Florissant basin, the surface has been upfaulted a few hundred meters to the general level of Divide. The smooth late Eocene surface is clearly visible on the skyline along the crest of the Rampart Range where it is upfaulted about 300 m above the general level of Divide. The monadnock, called Devils Head, stands above the surface at the left end of the skyline. Isolated outcrops of gravel units correlative with those at Divide and small outcrops of the Wall Mountain Tuff occur on the surface atop the Rampart Range.

isolated ridges of Precambrian rock stood as high as 240 m above the general level of the surface. In the Tallahassee Creek area, a meandering main-stem drainage carved a narrow, steep-walled valley through such a ridge (Chapin, 1965).

A few larger areas of positive relief existed—because even today they stand as monadnocks above the general level of the surface exhumed in their immediate surroundings. Examples are Devils Head in the northern Rampart Range (Figs. 6 and 15) and the summit portion of Pikes Peak above about 3,800 m (Fig. 20). Many of the present high mountains above the level of the surface, such as Thirtynine Mile and Waugh Mountains and the Buffalo Peaks, are modified volcanic landforms constructed on the surface. Essentially all other mountainous areas that now rise above adjacent segments of the surface are considered to have resulted from late Cenozoic block faulting and ensuing erosion.

In a few cases, the late Eocene erosion surface coincides with stripped surfaces that are pre-Paleozoic and (or) early Mesozoic unconformities. Examples of such areas are found in the vicinity of High Park and the Bare Hills, in the southern portion of the Front Range south of the Cripple Creek volcanic pile, and along the crest of the Wet Mountains south of Greenhorn Mountain. These areas are exceptional in that fortuitous circumstances allowed pre–late Eocene unconformities to be elevated by Laramide orogenic movements to levels that eventually were to coincide with local profiles of the late Eocene surface. Elsewhere, it is clear that thousands of meters of Paleozoic to middle Eocene sedimentary rocks were removed and Precambrian crystalline rocks were deeply eroded in the cores of large truncated Laramide structures.

Drainage Patterns and Regional Slope

The regional eastward and southeastward slope on the late Eocene surface was such that it merged with the western Great Plains. This conclusion is based on mapping and reconstruction of (1) paleovalleys, (2) distribution of early Oligocene volcanic and sedimentary rocks, (3) flow directions deduced from laminar flow

Figure 7. Oblique aerial view across the northeast rim of the valley of Currant Creek (CC). The prevolcanic erosion surface is carved into Precambrian rocks, extends beneath a thin sheet of volcanic rocks at the southeastern end of Caprock Ridge (CR), and emerges in the vicinity of High Park (HP) west of upper Oil Creek. The erosion surface (ES) is upfaulted on the east side of upper Oil Creek and the Florissant basin and extends northward (left) to the vicinity of Divide. (PP) is northern base of Pikes Peak.

structures and other features of the Wall Mountain Tuff, (4) damming of drainages and formation of lakes along the north and northwest sides of the Thirtynine Mile volcanic pile, (5) asymmetry of the Guffey volcano, and (6) considerations of the regional paleogeography during early and middle Tertiary time.

Numerous paleovalleys, well preserved on exhumed portions of the late Eocene surface, enable reconstruction of major portions of the paleodrainage. Figure 9 shows the locations of these paleovalleys and the drainage patterns interpreted from them. Several paleovalleys are exposed along the east side of the upper Arkansas River valley between Salida and Buena Vista where westward-forking tributaries and relatively deeper canyons indicate that the paleovalley headwaters lay to the west on the Laramide Sawatch uplift (Chapin and others, 1970). The two largest and best preserved of these paleovalleys are located between Browns Canyon and Waugh Mountain (Lowell, 1969, 1971) and along Trout Creek near Buena Vista. The Wall Mountain Tuff, unwelded ash-flow tuffs in the Antero Formation, Badger Creek Tuff, and Gribbles Park Tuff all attain maximum thickness and number of member units in one or the other of the two major paleovalleys. These observations suggest that sources of most of the ash-flow tuff units of the Thirtynine Mile volcanic field lay to the west of the field and that such major valleys were channels through which the ash flows moved before spreading over the relatively gentle terrain of the late Eocene surface to the east and southeast.

Detailed studies of laminar flow structures in the Wall Mountain Tuff along the Gribbles Run paleovalley, 17 km northeast of Salida (Lowell and Chapin, 1972), substantiate an eastward flow direction for the first tuff to be erupted onto the late Eocene surface. The Wall Mountain Tuff traveled at least 240 km from its probable vent area above the Mount Princeton batholith in the Sawatch Range to its distal end in the Castle Rock area on the Great Plains. Abundant light-tan felsite lithic fragments occur in this tuff in the paleovalley at Browns Canyon and indicate nearness to the source of the tuff; the lithic fragments are only occasionally seen in outcrops to the east. Near its distal end in the Castle Rock area, the Wall Mountain Tuff is less welded and has an unusually high ratio of matrix to phenocrysts. Yet another indication of eastward flow is the fact that basal black vitrophyre zones are thickest and most continuous in paleovalleys along the western edge of the Thirtynine Mile volcanic field. Vitrophyre zones are present as far east as Florissant, but east of Agate Creek (Fig. 1) in southwestern South Park, such occurrences are small and isolated. The combined effects of higher temperatures of emplacement, greater thicknesses, and higher retention of original volatile constituents allowed more complete compaction and welding along the western edge of the Thirtynine Mile volcanic field. Because movement of most nuées ardentes is influenced by pre-existing topography, eastward transport of the Wall Mountain Tuff is considered a reliable indication of eastward regional slope of the prevolcanic surface.

The late Eocene surface also had a southward component of slope as indicated by the trends of two main-stem drainages in the central and eastern portions of the Thirtynine Mile volcanic field. One of these streams meandered southeastward through the Cottonwood Creek–Tallahassee Creek area (Fig. 1), and the other flowed southward along the present general trend of Oil Creek. Both of these master streams probably headed on higher slopes west and north of South Park and drained into the Canon City embayment. Their courses are partially known from alluvial trains deposited on top of, and in channels which cut through, the Wall Mountain Tuff. These deposits have been named the Tallahassee Creek Conglomerate (Epis and Chapin, 1974); they are as much as 105 m thick and

Figure 8. View of the late Eocene surface south-southeastward from near the head of Echo Park. The surface in the northern part of the De Weese Plateau is immediately south of the Arkansas River Canyon (central part of the photograph) and is cut across Precambrian rocks. Isolated remnants of the Gribbles Park Tuff rest on the surface. Considerable dissection of the surface has been accomplished by the Arkansas River and its tributaries. The rather smooth profile on the left skyline is the late Eocene surface where it forms the crest of the northern and central Wet Mountains and is upfaulted along the Ilse fault more than 460 m above the general level of the De Weese Plateau (Taylor, 1975). The Wet Mountain Valley and the Sangre de Cristo Mountains are visible in the upper right of the photograph.

3 km wide along Tallahassee Creek. Similar alluvial trains are present in southern South Park, but the courses of these streams beneath the main portion of the Thirtynine Mile volcanic pile are unknown (Fig. 9).

One of the obvious questions arising from this reconstruction of the paleodrainage is that of how the Wall Mountain Tuff was able to flow into the Castle Rock area of the Denver basin. The ash flows may have had sufficient momentum to jump a low drainage divide and proceed eastward onto the Great Plains. Partial filling of shallow valleys by early pulses of this multiple-flow unit may have aided in overriding the divide. That such overriding was indeed possible is indicated by diversion of drainage along the north side of the Thirtynine Mile volcanic field across the Rampart Range in early Oligocene time. The diversion is recorded in the Castle Rock Conglomerate, which contains abundant andesitic detritus from the Thirtynine Mile volcanic field. This interpretation is substantiated by the lack of main-stem alluvium within, or immediately above, the Thirtynine Mile Andesite, which completely disrupted surface drainage within the volcanic field (Chapin and Wyckoff, 1969). A major channel across the Rampart Range, between Divide and West Monument Creek, was formed during the diversion (Scott, 1975). Alluvial deposits within the Divide channel contain clasts of the phonolite of Cripple Creek (28 m.y. old) and are therefore younger than the Castle Rock Conglomerate. Repetitive crossing of the Rampart Range by major middle Tertiary streams shows that the imposing barrier this range presents today is a result of late Cenozoic uplift and block faulting.

Additional evidence for the regional slope in early Oligocene time is provided by the asymmetry of the Guffey volcano (Epis and Chapin, 1968; Wyckoff, 1969). The flanks of this structure are well preserved in an elliptical ring of mountains composed of outward-dipping andesitic flows and breccias. The steepest dips are to the north (10° to 15°) on Thirtynine Mile Mountain and to the west (8°) on Black Mountain. To the east, on Castle, McIntyre, and Witcher Mountains, the flows have dips of only 2° to 6°. The gentlest dips are to the southeast where andesitic lavas flowed at least 11 km down a paleovalley and are now preserved as Cap Rock Ridge and the high tablelands that extend northwestward to Cover Mountain (Fig. 7). Before construction of the Guffey volcano, mudflows were deposited southeastward along this same drainage.

Elevation and Climate

The most direct evidence regarding elevation and climatic conditions prevailing during the formation of the late Eocene surface comes from fossils in the Florissant Lake Beds, which occupy a shallow valley on the surface. The lake beds are slightly younger than the Wall Mountain Tuff and were deposited during the main period of volcanism in the Thirtynine Mile volcanic field. Regarding the Florissant flora, MacGinitie (1953, p. 66) concluded that it indicates ". . . a very moderate elevation for the Florissant region in the Oligocene. It appears unlikely that its altitude could have been greater than 3000 feet at the most." MacGinitie (p. 67) compared the Florissant flora with the Green River flora (middle Eocene): "The Green River and Florissant floras both show a peculiar endemism and very much the same general facies. The Florissant flora is a logical derivative of the Green River [flora], a fact which indicates a gradual evolution under similar physical conditions in the same general area, under the influence of a slow climatic change trending toward drier and cooler conditions." From these and other comparisons, MacGinitie (p. 77) concluded that "The climate was subhumid and warm temperate, not unlike the present climate of Monterrey, in the state of Nuevo Leon, Mexico. Warm winters and hot summers prevailed, and abundant sunshine is indicated."

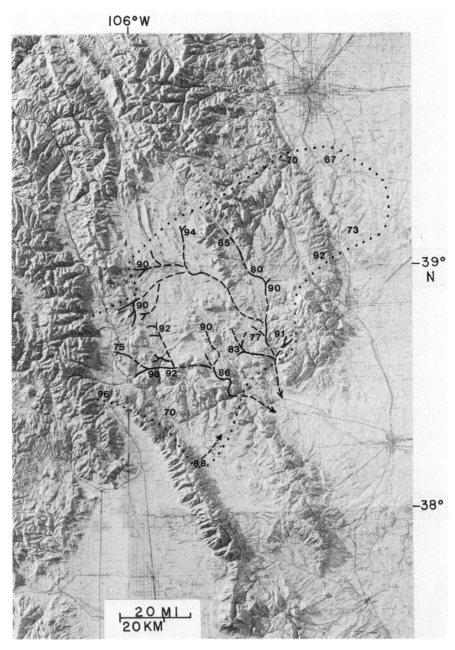

Figure 9. Composite of Army Map Service plastic relief maps (2-degree quadrangles) showing reconstruction of major paleodrainages on the late Eocene erosion surface during deposition of the Wall Mountain Tuff and Tallahassee Creek Conglomerate. Known paleovalleys and alluvial trains are shown with solid lines; interpreted linkage of these is shown with dashed lines. Minimum original extent of the Wall Mountain Tuff (~10,400 km^2) is outlined by the dotted line. Numbers refer to present elevations (in hundreds of feet) of the base of the Wall Mountain Tuff, which lies on the late Eocene surface. Figures 1, 2, 19, and 21 show related information.

He interpreted average annual temperatures as probably never lower than 18°C (65°F), minimum temperatures not less than −7°C (20°F), and annual rainfall not exceeding 640 mm and probably not more than 510 mm (p. 57–58). Leopold and MacGinitie (1972) reaffirmed these conclusions.

Indirect evidence concerning the general elevation and climate at the time the late Eocene surface reached its maximum development can be obtained by tracing regional lithologic and paleontologic trends. Dott and Batten (1971, p. 404–406) summarized these trends as follows:

> About the end of Paleocene time, North America's last epeiric sea apparently retreated to the Gulf of Mexico. . . . During the Eocene Epoch, basin-filling or aggradation continued, but with gradual diminution of coal. Pink, yellow, and red silts and shales are especially characteristic of the Eocene, and typically are eroded into picturesque badlands topography. . . . This widespread sequence represents much of the early basin fill deposited in rivers, lakes, and swamps. . . . On uplands surrounding the Eocene lakes, a lush forest of redwood and other trees thrived. By ecologic analogy with modern counterparts, the fossil flora points to uniformly mild, humid temperate conditions in marked contrast with the present climate. . . . Basin filling was largely completed during late Eocene and early Oligocene time. Oligocene and Miocene strata overlap beyond older ones to rest unconformably upon eroded pre-Cenozoic rocks. . . . The net result of erosion and filling was to smooth the topography and, as uplift and erosion slackened for perhaps 5 or 10 million years, the landscape became relatively stable.

The Florissant data are consistent with regional environmental trends. The late Eocene surface sloped gently southeastward to the Gulf of Mexico, probably at a rate of 0.4 to 0.6 m/km. The elevation estimated by MacGinitie is a maximum elevation; the actual elevation of central Colorado in the late Eocene was probably somewhat less. It appears certain that formation of the surface was completed after Laramide orogenic activity ceased and that it generally did not consist of geomorphic elements controlled by Laramide structures. It clearly must have obliterated such elements; moreover, the climate, altitude and geomorphic character of the surface differed dramatically from those of today.

UPLIFT AND DISRUPTION OF THE SURFACE
BY BLOCK FAULTING

During Miocene and later time, the area was uplifted about 1,500 to 3,000 m. The late Eocene surface, which formed at elevations less than 900 m, now is found at elevations of 2,300 to 3,800 m. Uplift was accompanied by significant block faulting and displacement of the surface. Figure 9 shows the present distribution of the Wall Mountain Tuff at significantly different elevations on the surface. It is clear that the once rather smooth and planar surface was disrupted and differentially raised and lowered in adjacent blocks.

Convincing evidence of major block faulting of the late Eocene surface occurs along the eastern margin of the upper Arkansas River valley between Salida and Trout Creek (Figs. 10 and 19). Here, the Wall Mountain Tuff and associated Oligocene volcanic rocks that fill the paleovalleys discussed above, have been downfaulted onto the floor of the upper Arkansas Valley graben 150 to 600 m below their position on the crest of the southern Mosquito Range (Arkansas Hills) to the east (Van Alstine, 1969; Epis and Chapin, 1968; Lowell, 1969, 1971). The north-trending graben is superimposed on the eastern flank of the Laramide Sawatch uplift and extends northward beyond Leadville. It represents the northernmost

Figure 10. View south-southeastward from Buena Vista illustrating the block-faulted nature of the western margin of the upper Arkansas Valley graben. The late Eocene surface along the crest of the southern Mosquito Range (left skyline) has been progressively step-faulted down about 600 m into the bottom of the valley and modified by later erosion. Thin outcrops of Wall Mountain Tuff occur at the highest and lowest elevations along this profile as well as at intermediate levels. The crest of the Sangre de Cristo Mountains south of Salida is barely visible (right skyline).

Figure 11. Oblique aerial view from above Canon City northward along Oil Creek. The late Eocene surface has been upfaulted from its general level (left skyline) west of the Oil Creek fault zone to the level beneath Mount Pisgah, the conical peak (middle skyline) in the Cripple Creek volcanic field. The surface has been tilted westward by faulting and rises toward Pikes Peak (right skyline). The surface remnant along the high western shoulder of Pikes Peak is interpreted as a segment of the late Eocene surface that was upfaulted to the east above its general level at Cripple Creek.

Figure 12. Oblique aerial view south-ward along Oil Creek showing the late Eocene surface along the crest of the southern Front Range (left skyline). The surface has been downfaulted more than 460 m on the west side of the Oil Creek fault zone to the general level of the white dolomite beds of lower Paleozoic age (lower right corner). The smooth late Eocene surface can be seen again in the distance along the crest of the northern Wet Mountains (right skyline) south of the Canon City embayment.

Figure 13. View northeastward from High Park across the late Eocene surface covered by a thin veneer of Tallahassee Creek Conglomerate. The rather smooth late Eocene surface beneath the Cripple Creek volcanic field can be seen (along the skyline) on either side of the conical phonolite peak of Mount Pisgah. The surface there has been upfaulted about 370 m from the general level of High Park along the Oil Creek fault zone. The course of Oil Creek is along the base of the fault-line scarp below Mount Pisgah.

major graben of the Rio Grande rift that extends southward through southern Colorado and New Mexico (Chapin, 1971). Tweto (1961), Van Alstine (1969), and Tweto and Case (1972) showed that block faulting in the upper Arkansas Valley graben continued during deposition of the graben-fill sediment (called the Dry Union Formation) of Miocene and Pliocene age and that such faulting probably persisted into Holocene time.

Equally convincing examples of significant block faulting of the late Eocene surface and the lower Oligocene rocks on the surface are along the fault zone that extends from the mouth of Oil Creek northward between Cripple Creek and High Park and along the east side of the Florissant basin. The Wall Mountain Tuff, Tallahassee Creek Conglomerate, lower member of the Thirtynine Mile Andesite, and phonolite of Cripple Creek have been displaced about 370 m along this zone from the floor of High Park west of Oil Creek to the base of Straub and Grouse Mountains near Victor (Figs. 11, 12, and 13; Tobey, 1969; Epis and Chapin, 1968).

Figure 14. View northwestward from near Divide of the late Eocene surface, which is overlain by thin Miocene-Pliocene gravel units in the smooth, grassy areas.

Figure 15. View northeastward from the northern flank of Pikes Peak showing Devils Head monadnock (upper left) and the smooth late Eocene surface along the crest of the Rampart Range (on the skyline). The surface has been downfaulted about 300 m across the Ute Pass fault zone to the general level of the tree-covered surface of low relief (middle foreground). Catamount Reservoir is visible on the surface (left foreground).

Figure 16. View westward from Bald Mountain east of Woodland Park (near center of photograph). Grassy area in immediate foreground is covered by thin Miocene-Pliocene gravel units resting on the late Eocene surface atop the Rampart Range. The surface is clearly visible west of Woodland Park where it has been downfaulted about 300 m below the crest of the Rampart Range. U.S. Highway 24 extends from Woodland Park southwestward to Divide along the grassy areas on the surface (left center of the photograph).

Major faults that bound the Rampart Range considerably displace the late Eocene surface. Along the Ute Pass fault, the surface was uplifted more than 300 m above its general level at Divide to where it lies on the crest of the Rampart Range to the east (Figs. 14, 15, and 16). Likewise, displacement along the eastern frontal fault of the Rampart Range is responsible for the lower level of the Wall Mountain Tuff in the vicinity of Castle Rock to the east (Figs. 17 and 18). Scott (1970) demonstrated Quaternary displacement along this fault.

A number of faults with lesser magnitude of displacement but similar style of movement displace the late Eocene surface and overlying Tertiary deposits between the localities described above. Some of these are shown on Figure 19. Many of the faults are known to have been active during Laramide and earlier periods of deformation; this paper emphasizes movement along them in post-Laramide time.

Taylor (1975) and Scott (1975) documented significant Neogene displacement of the late Eocene surface in the region including the San Luis Valley, Sangre de Cristo Mountains, Wet Mountain Valley, Wet Mountains, and the Canon City embayment. Evidence for their interpretation is essentially identical to what has been presented above. It is also clear from their work that the surface has been differentially block faulted several thousands of feet during Miocene and later time.

The high, smooth shoulders on the north, west, and south sides of Pikes Peak at elevations of about 3,800 m are exhumed remnants of a mature erosion surface (Fig. 20). Because nowhere in the region is there evidence of another widespread surface of low relief, we assign these high-level remnants to the late Eocene surface and suggest that they are segments upfaulted from the levels of the surface at Cripple Creek and Divide. That this is the correct interpretation cannot be proved because there are no Tertiary deposits lying on these high shoulders. However, there is abundant evidence of crushing, shearing, and alteration of the Precambrian rocks along zones where faults that could have produced such displacements are located (Fig. 19). This evidence is especially obvious in the roadcuts of Colorado State Highway 67 between Divide and Cripple Creek. The various surfaces of this region (except the high-level ones on Pikes Peak), which were separately named and assigned different ages by Van Tuyl and Lovering (1935), can be proved to be faulted segments of the late Eocene surface described in this paper. The evidence consists of dated and correlated lower Oligocene volcanic and sedimentary deposits that rest on segments of a low-relief surface beveling Laramide structural features and upper Paleocene to middle Eocene rocks.

Figure 21 is a schematic cross section from the vicinity of Castle Rock in the Great Plains southwestward across the mountainous region to beyond the general location of Salida. It is clear that major present-day geomorphic elements of the region postdate the Laramide orogeny and are the direct result of Miocene and younger uplift, basin-and-range style block faulting, and attendant erosion.

REGIONAL SIGNIFICANCE OF THE SURFACE

In addition to the processes of erosion, the late Eocene surface owes its existence to three main factors: (1) cessation of Laramide orogeny in late Eocene time (Tweto, 1975; Coney, 1972, 1973); (2) a 10- to 20-m.y. hiatus in magmatism throughout most of the Southern Rocky Mountains and Basin and Range provinces (Damon and Mauger, 1966; Livingston and others, 1968; Cross, 1973); and (3) the outbreak of volcanism in latest Eocene and earliest Oligocene time throughout the area beveled by the surface (Noble, 1972). Termination of the Laramide orogeny allowed

Figure 17. Oblique aerial view southward along the east frontal fault-line scarp of the Rampart Range showing Pikes Peak (right skyline) and the flat, late Eocene surface on the range upfaulted from the general level of the western Great Plains (left).

erosion to attack uplifts and fill basins without significant tectonic rejuvenation of topographic relief. The hiatus in volcanism allowed destructional geomorphic processes to alter the landscape uninterrupted by added shielding of constructional volcanic piles. Rapid weathering in a warm, temperate to subtropical climate and lateral planation by streams quickly reduced the landscape to a broad and open surface of subdued relief. Laramide uplifts were beveled and adjacent basins filled.

The nearly simultaneous outbreak of volcanism across much of the Southern Rocky Mountains and Basin and Range provinces in earliest Oligocene time preserved the surface beneath a mantle of volcanic and volcaniclastic rocks. Widespread units such as the Wall Mountain Tuff provide stratigraphic and structural datums by which the underlying surface can be reconstructed. Continuation of relatively stable tectonic conditions during middle Tertiary volcanism left the various volcanic fields and the underlying surface almost intact until regional extension and block faulting began in Miocene time. The present level of uplift and erosion in central Colorado is optimum for exposing the late Eocene surface across large areas and makes possible its documentation. The fact that the late Eocene surface is the only widespread low-relief surface that was developed and preserved in the Southern Rocky Mountains during Cenozoic time (Scott, 1975) suggests an unusual combination of events, such as the three listed above.

It is beyond the scope of this paper to attempt to trace the late Eocene surface to its farthest boundaries. It is worth noting, however, a few areas where additional work would probably unravel a story similar to that described here. Steven and Epis (1968) and Steven (1975) have shown that south-central Colorado and adjoining areas were covered by a large composite volcanic field of Oligocene and Miocene age that was disrupted by late Cenozoic block faulting and erosion. The largest and best known remnant of this composite field is the San Juan volcanic field, from which the Thirtynine Mile volcanic field has been isolated by faulting and erosion along the Rio Grande rift. The main portion of the San Juan volcanic field rests on a post-Laramide surface of low relief surrounding a residual Laramide topographic high in the Needle Mountains area (T. A. Steven, 1973, oral commun.). In discussing the age of volcanic activity in the San Juan field, Steven and others (1967, p. D47) stated:

This Tertiary volcanism occurred after a period of erosion had stripped the sedimentary rocks from the more highly upwarped areas. Streams responsible for much of this erosion deposited an apron of gravels, sands, and clays around the margins of the higher areas of exposed Precambrian rocks to form the Telluride and Blanco Basin Formations.

The surface beneath the San Juan field was referred to as the Telluride peneplain by Atwood and Mather (1932, p. 17) and as "the surface at the base of the volcanic rocks" by Larsen and Cross (1956, p. 245). Atwood and Mather (1932, p. 16-18) stated the following about the prevolcanic late Eocene (Telluride) surface:

. . . we may picture the San Juan region at the end of the Telluride epoch as having a maximum relief of only 1,000 or 2,000 feet between the highest land near the center of the San Juan dome, where degradation would still be in progress, and the flat lowland built by aggradation on the flanks of the dome. . . . a region once rugged and mountainous had been reduced to rounded hills and low ridges, standing at only slight altitudes above an undulating plain.

The Blanco Basin Formation was deposited along the southern edge of the San Juan uplift in northern New Mexico and southern Colorado while the Telluride Conglomerate was deposited along the western edge of the uplift. Both formations (1) rest on a surface of low relief which truncates the upturned edges of Cretaceous and older rocks, (2) pinch out against the residual Laramide high near the center of the uplift, (3) contain little or no volcanic detritus, and (4) are overlain in a nearly conformable manner by basal early Oligocene units of the San Juan volcanic field. The Telluride surface therefore is younger than the Late Cretaceous-early Paleocene Animas Formation (Larsen and Cross, 1956; Dane, 1946) and Cimarron Ridge Formation (66 m.y. old; Dickinson and others, 1968) and is older than the early Oligocene San Juan, Picayune, Lake Fork, and Conejos Formations that were erupted beginning about 35 m.y. ago (Lipman and others, 1970, 1973). The age, or ages, of the Telluride and Blanco Basin Formations are as yet uncertain

Figure 18. Oblique aerial view of the late Eocene surface cut into Precambrian Pikes Peak Granite on top of the Rampart Range west of the U.S. Air Force Academy north of Colorado Springs. The steep east front of the range is the fault-line scarp of the Rampart Range fault.

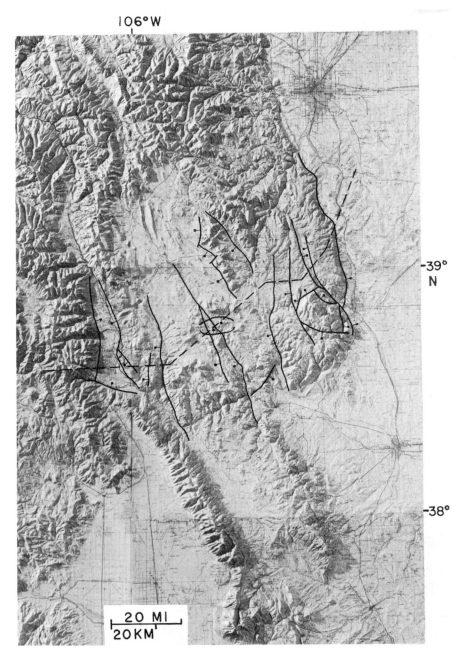

Figure 19. Composite of Army Map Service plastic relief maps (2-degree quadrangles) showing principal normal faults along which the late Eocene surface has been block faulted. Bar and ball symbols indicate downthrown sides of faults. Dashed line is approximate location of schematic structure section shown in Figure 21. See Figures 1 and 2 for geographic references. Figure 9 shows the elevations of the Wall Mountain Tuff on most of the fault blocks.

although several authors (Van Houten, 1957; Simpson, 1948; Dunn, 1964) correlated them on the basis of lithology and stratigraphic position with the San Jose Formation of early Eocene age in the central San Juan basin. The prevolcanic, late Eocene surface must have developed on the upper surfaces of these formations following aggradation of the downwarps in which the formations accumulated. As in the Thirtynine Mile volcanic field and elsewhere, the smoothness of the late Eocene surface is the result of both erosion of uplifts and concomitant aggradation of basins.

In several areas around the eroded periphery of the San Juan volcanic field, large expanses of the prevolcanic surface have been exhumed, and sufficient patches of volcanic and sedimentary rocks remain to encourage dating and reconstruction of the surface. The most promising areas are (1) along the crest of the Tusas Mountains in northern New Mexico (Butler, 1946, 1971; Bingler, 1968; Muehlberger, 1967, 1968), (2) along the crest of the Sangre de Cristo Range in northern New Mexico and southern Colorado (McKinlay, 1956, 1957; Ray and Smith, 1941; Smith and Ray, 1943; Clark and Read, 1972; Pillmore and others, 1973), and (3) between the San Juan and West Elk Mountains where the once continuous volcanic field has been deeply dissected by the Gunnison River and its tributaries (Larsen and Cross, 1956; Olson and others, 1968; Hansen, 1965, 1971; Lipman and others, 1969). Many of the above authors have made reference to erosion surfaces and the occurrence of high-level gravel units beneath, and interbedded with, volcanic rocks correlative with those of the San Juan volcanic field. Interpretations differ, however, as to their age and significance. We suggest that detailed mapping, accompanied by additional dating and correlation of the volcanic rocks and compositional studies of the interbedded gravel units, will demonstrate that the age and geomorphic character of the prevolcanic surface in these areas are similar to those of the late Eocene surface described in this paper.

Scott (1975) suggested that it may be possible to trace the late Eocene surface

Figure 20. View southeastward from Wilkerson Pass across the low, grass-covered portion of the Florissant basin (center) toward Pikes Peak. The late Eocene surface has been upfaulted from the Florissant basin to a level about midway between the basin and timberline, from where the surface extends southward (right) beyond the view of the photograph beneath the Cripple Creek volcanic field. The high smooth surface at about 3,800 m on the north, west, and south shoulders of Pikes Peak is clearly visible above timberline. Even though no confirming Cenozoic deposits have been found there, this surface is interpreted as a remnant of the late Eocene surface faulted even higher above the levels of the Florissant basin and the Cripple Creek volcanic field.

northward from the Thirtynine Mile volcanic field along the Front Range of Colorado into southern Wyoming on the basis of studies by Knight (1953), Soister (1968), Harshman (1968), and Denson and Harshman (1969). Dating and correlation of the surface in this area are handicapped by the scarcity of volcanic rocks, and its reconstruction must depend heavily on studies of the composition and distribution of gravel units and on geomorphic and structural interpretations.

In the Basin and Range province to the southwest, reconstruction of the late Eocene surface is made difficult by complexities of faulting and extensive alluvial cover. Nevertheless, in west-central New Mexico, detailed mapping in the Magdalena area showed that the age and geomorphic character of the prevolcanic surface beneath the Datil-Mogollon volcanic field are similar to those of the late Eocene surface in Colorado. Specifically, the surface is of late Eocene age, truncates Laramide uplifts, and extends across aggraded basins of early Tertiary age. The basal unit of the Datil volcanic rocks in the Magdalena area is the Spears Formation (Tonking, 1957), which consists mainly of a volcaniclastic apron of latitic detritus. The Spears Formation was dated at 37 m.y. B.P. (Burke and others, 1963) and rests on beveled Paleozoic rocks of the Magdalena uplift; only 17 km to the north, the Spears Formation overlies, with gradational contact, Eocene arkose units of the Baca Formation (Snyder, 1970). The Baca, in turn, overlies a thick section of Mesozoic and Paleozoic rocks in a Laramide basin. Additional details of the geomorphic character of this surface are difficult to obtain, but there is no reason to expect that the surface in the Magdalena area differs substantially from the exhumed late Eocene surface of Colorado. Moreover, when the distribution of middle Tertiary volcanic fields in Colorado and New Mexico is superimposed on a map of Laramide basins and uplifts, it is readily apparent that the volcanic

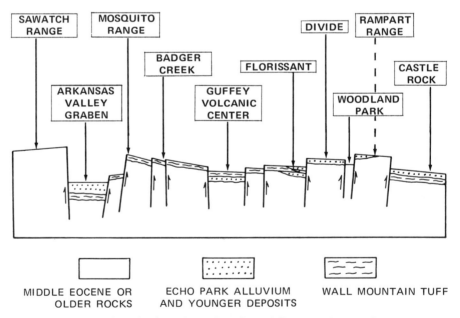

Figure 21. Schematic section from the southern Sawatch Range northeastward approximately 200 km to the vicinity of Castle Rock, illustrating post-Laramide, middle to late Cenozoic block faulting of the late Eocene erosion surface, Wall Mountain Tuff, and associated sedimentary deposits. Figure 19 shows approximate location of line of section.

piles spread across both beveled uplifts and adjacent aggraded basins. If volcanic materials were able to blanket uplifts and basins alike, there must have been a widespread surface of low relief. In contrast, late Cenozoic volcanic rocks are confined largely to basins because pronounced topographic relief generally accompanied the structural relief in late Cenozoic time.

The existence and economic significance of a middle Tertiary erosion surface of regional extent has long been recognized in the porphyry copper province of southern Arizona and southwestern New Mexico. More than 40 years ago, Harrison Schmitt (1933, p. 321–322) wrote:

. . . most of the country was attacked by erosion as soon as it was elevated above the sea. This produced the well known Eocene (and Oligocene) peneplain and erosion surface of continental extent, and recognized from southwestern Colorado southward along the Arizona-New Mexico line to the Mexican border. It is covered by a thick series of intermediate to acid volcanic rocks, believed to be of Miocene-Pliocene age; and by coarse continental conglomerates, lake beds, and bolson deposits dated Miocene, Pliocene, and Pleistocene.

Schmitt's observations were made nearly three decades before K-Ar ages became widely available and before numerous detailed studies of Cenozoic volcanic rocks were accomplished. Yet all that is necessary to update his statement is to delete the word "peneplain," substitute "regional" for "continental," and adjust the age of the Tertiary volcanic rocks to begin in the Oligocene.

P. E. Damon and his associates at the University of Arizona have documented the bimodal distribution with reference to time of Cenozoic magmatism in the Basin and Range province (Damon and Mauger, 1966; Livingston and others, 1968). The 10-m.y. hiatus in magmatism that is responsible for the bimodal distribution coincides with the time of carving of the late Eocene erosion surface. Livingston and others (1968) pointed out that most of the Laramide porphyry copper deposits of Arizona were eroded, leached, and enriched during this quiescent period; were then buried by volcanic and sedimentary rocks of middle and late Cenozoic age; and today have been, or are being, exhumed from beneath this cover. The youngest of these prophyry copper deposits (52 to 55 m.y. old; Livingston, 1973) was emplaced in latest Paleocene-earliest Eocene time. The onset of middle Tertiary volcanism that buried the deposits was nearly contemporaneous with the beginning of volcanism in the Thirtynine Mile, San Juan, and Datil-Mogollon volcanic fields. It seems certain, therefore, that fundamental patterns of tectonism, magmatism, and geomorphic evolution are similar in the Cenozoic history of both the southern Rocky Mountains and Basin and Range provinces. The late Eocene erosion surface may provide a structural datum by which these histories can be more clearly understood.

ACKNOWLEDGMENTS

We wish to acknowledge the many benefits derived from field and office discussions with graduate students who have worked with us at the Colorado School of Mines and the New Mexico Institute of Mining and Technology (see theses listed below) and with P. W. Lipman, G. R. Scott, T. A. Steven, R. B. Taylor, R. E. Van Alstine, and C. T. Wrucke of the U.S. Geological Survey. We are indebted to J. D. Obradovich of the U.S. Geological Survey who has kindly given us permission to publish several new radiometric dates. The work of G. R. Scott and R. B. Taylor has been especially helpful in our analysis of the erosion surface.

BIBLIOGRAPHY

Atwood, W. W., and Mather, K. F., 1932, Physiography and Quaternary geology of the San Juan Mountains, Colorado: U.S. Geol. Survey Prof. Paper 166, 176 p.

Bingler, E. C., 1968, Geology and mineral resources of Rio Arriba County, New Mexico: New Mexico Bur. Mines and Mineral Resources Bull. 91, 158 p.

Brown, R. W., 1943, Cretaceous-Tertiary boundary in the Denver basin, Colorado: Geol. Soc. America Bull., v. 54, p. 65-86.

——1962, Paleocene flora of the Rocky Mountains and Great Plains: U.S. Geol. Survey Prof. Paper 375, 119 p.

Buchanan, P. H., 1967, Volcanic geology of the Guffey area, Park County, Colorado [M. Sc. thesis]: Golden, Colorado School of Mines, 95 p.

Burbank, W. S., and others, compilers, 1935, Geologic map of Colorado: U.S. Geol. Survey, in cooperation with Colorado State Geol. Survey Board and Colorado Metal Mining Fund, scale 1:500,000.

Burke, W. H., Kenny, G. S., Otto, J. B., and Walker, R. D., 1963, Potassium-argon dates, Socorro and Sierra Counties, New Mexico, in New Mexico Geol. Soc. Guidebook 14th Field Conf., Socorro region, 1963: p. 224.

Butler, A. P., Jr., 1946, Tertiary and Quaternary geology of the Tusas-Tres Piedras area, New Mexico [Ph.D. thesis]: Cambridge, Mass., Harvard Univ., 188 p.

——1971, Tertiary volcanic stratigraphy of the eastern Tusas Mountains, southwest of the San Luis Valley, Colorado-New Mexico, in New Mexico Geol. Soc. Guidebook 22nd Field Conf., San Luis Basin, 1971: p. 289-300.

Chapin, C. E., 1965, Geologic and petrologic features of the Thirtynine Mile volcanic field, central Colorado [D. Sc. thesis]: Golden, Colorado School of Mines, 176 p.

——1971, The Rio Grande rift, Pt. I: Modifications and additions, in New Mexico Geol. Soc. Guidebook 22nd Field Conf., San Luis Basin, 1971: p. 191-201.

Chapin, C. E., and Epis, R. C., 1964, Some stratigraphic and structural features of the Thirtynine Mile volcanic field, central Colorado: Mtn. Geologist, v. 1, no. 3, p. 145-160.

Chapin, C. E., and Wyckoff, B. S., 1969, Formation of the 60-cubic-mile andesitic breccia sheet in the Thirtynine Mile volcanic field of central Colorado: Geol. Soc. America, Abs. for 1968, Spec. Paper 121, p. 52.

Chapin, C. E., Epis, R. C., and Lowell, G. R., 1970, Late Eocene paleovalleys and Oligocene volcanic rocks along the upper Arkansas Valley segment of the Rio Grande rift zone in Colorado [abs.], in The Rio Grande depression, New Mexico and Colorado: New Mexico Geol. Soc., Program (24th Ann. Mtg.), p. 6.

Clark, K. F., and Read, C. B., 1972, Geology and ore deposits of Eagle Nest area, New Mexico: New Mexico Bur. Mines and Mineral Resources Bull. 94, 152 p.

Coney, P. J., 1972, Cordilleran tectonics and North American plate motion: Am. Jour. Sci., v. 272, p. 603-628.

——1973, Non-collision tectogenesis in western North American, in Tarling, D. H., and Runcorn, S. K., eds., Implications of continental drift to the earth sciences: New York, Academic Press, p. 713-727.

Cross, T. A., 1973, Implications of igneous activity for the early Cenozoic tectonic evolution of western United States: Geol. Soc. America, Abs. with Programs (Ann. Mtg.), v. 5, no. 7. p. 587.

Cross, Whitman, 1894, Description of Pikes Peak sheet, Colorado: U.S. Geol. Survey Geol. Atlas, Folio 7, 8 p., 5 maps.

Damon, P. E., and Mauger, R. L., 1966, Epeirogeny-orogeny viewed from the Basin and Range province: Soc. Mining Engineers Trans., v. 235, p. 99-112.

Dane, C. H., 1946, Stratigraphic relations of Eocene, Paleocene, and latest Cretaceous formations of eastern side of San Juan basin, New Mexico: U.S. Geol. Survey Oil and Gas Inv. Prelim. Chart 24.

Denson, N. M., and Harshman, E. N., 1969, Map showing areal distribution of Tertiary rocks, Bates Hole-Shirley Basin area, south-central Wyoming: U.S. Geol. Survey Misc. Geol. Inv. Map I-570, scale 1:125,000.

De Voto, R. H., 1961, Geology of southwestern South Park, Park and Chaffee Counties, Colorado [D. Sc. thesis]: Golden, Colorado School of Mines, 323 p.

——1964, Stratigraphy and structure of Tertiary rocks in southwestern South Park: Mtn. Geologist, v. 1, no. 3, p. 117-126.

——1971, Geologic history of South Park and geology of the Antero Reservoir quadrangle, Colorado: Colorado School Mines Quart., v. 66, no. 3, 90 p.

Dickinson, R. G., Leopold, E. B., and Marvin, R. F., 1968, Late Cretaceous uplift and volcanism on the north flank of the San Juan Mountains, Colorado, in Epis, R. C., ed., Cenozoic volcanism in the southern Rocky Mountains: Colorado School Mines Quart., v. 63, no. 3, p. 125-148.

Dings, M. G., and Robinson, C. S., 1957, Geology and ore deposits of the Garfield quadrangle, Colorado: U.S. Geol. Survey Prof. Paper 289, 110 p.

Dott, R. H., Jr., and Batten, R. L., 1971, Evolution of the earth: New York, McGraw-Hill, 649 p.

DuHamel, J. E., 1968, Volcanic geology of the upper Cottonwood Creek area, Thirtynine Mile volcanic field [M. Sc. thesis]: Golden, Colorado School of Mines, 120 p.

Dunn, D. E., 1964, Evolution of the Chama Basin and Archuleta anticlinorium, eastern Archuleta County, Colorado [Ph.D. thesis]: Austin, Univ. Texas, 112 p.

Epis, R. C., and Chapin, C. E., 1968, Geologic history of the Thirtynine Mile volcanic field, central Colorado, in Epis, R. C., ed., Cenozoic volcanism in the southern Rocky Mountains: Colorado School Mines Quart., v. 63, no. 3, p. 51-85.

——1974, Stratigraphic nomenclature of the Thirtynine Mile volcanic field, central Colorado: U.S. Geol. Survey Bull. 1395-C, 23 p.

Graebner, Peter, 1967, Remanent magnetism in major rock units of the Thirtynine Mile volcanic field, central Colorado [M. Sc. thesis]: Golden, Colorado School of Mines, 165 p.

Graebner, Peter, and Epis, R. C., 1968, Remanent magnetism in major rock units of the Thirtynine Mile volcanic field, central Colorado [abs.], in Epis, R. C., ed., Cenozoic volcanism in the southern Rocky Mountains: Colorado School Mines Quart., v. 63, no. 3, p. 87-88.

Hansen, W. R., 1965, The Black Canyon of the Gunnison, today and yesterday: U.S. Geol. Survey Bull. 1191, 76 p.

——1971, Geologic map of the Black Canyon of the Gunnison River and vicinity, western Colorado: U.S. Geol. Survey Misc. Geol. Inv. Map I-584, scale 1:31,680.

Harland, W. B., Smith, A. G., and Wilcock, Bruce, eds., 1964, The Phanerozoic time scale—A symposium dedicated to Professor Arthur Holmes: Geol. Soc. London Quart. Jour. Supp., v. 120s, 458 p.

Harshman, E. N., 1968, Geologic map of the Shirley Basin area, Albany, Carbon, Converse, and Natrona Counties, Wyoming: U.S. Geol. Survey Misc. Geol. Inv. Map I-539, scale 1:48,000.

Hawley, C. C., 1969, Geology and beryllium deposits of the Lake George (or Badger Flats) beryllium area, Park and Jefferson Counties, Colorado: U.S. Geol. Surv. Prof. Paper 608-A, 44 p.

Izett, G. A., Scott, G. R., and Obradovich, J. D., 1969, Oligocene rhyolite in the Denver Basin, Colorado, in Geological Survey research 1969: U.S. Geol. Survey Prof. Paper 650-B, p. B12-B14.

Johnson, J. H., 1935, Stratigraphy of northeastern and east-central parts of South Park, Colorado: Am. Assoc. Petroleum Geologists Bull., v. 19, p. 1339-1356.

——1937a, Algae and algal limestones from the Oligocene of South Park, Colorado: Geol. Soc. America Bull., v. 48, p. 1227-1235.

——1937b, The Tertiary deposits of South Park, Colorado [abs.]: Colorado Univ. Studies, v. 25, no. 1, p. 77.

——1937c, Tertiary deposits of South Park, Colorado, with a description of Oligocene algal limestones [Ph.D. thesis]: Boulder, Colorado Univ., 68 p.

Johnson, R. B., 1959, Geology of the Huerfano Park area, Huerfano and Custer Counties, Colorado: U.S. Geol. Survey Bull. 1071-D, p. 87-119.

Johnson, R. B., and Wood, G. H., 1956, Stratigraphy of Upper Cretaceous and Tertiary rocks of Raton basin, Colorado and New Mexico: Am. Assoc. Petroleum Geologists, v. 40, no. 4, p. 707-721.

Knight, S. H., 1953, Summary of the Cenozoic history of the Medicine Bow Mountains, Wyoming: Wyoming Geol. Assoc. Guidebook 8th Ann. Field Conf., Laramie Basin, Wyoming, and North Park, Colorado, 1953: p. 65-76.

Larsen, E. S., Jr., and Cross, Whitman, 1956, Geology and petrology of the San Juan region, southwestern Colorado: U.S. Geol. Surv. Prof. Paper 258, 303 p.

Leopold, E. B., and MacGinitie, H. D., 1972, Development and affinities of Tertiary floras in the Rocky Mountains: in Aham, A. G., Floristics and paleofloristics of Asia and eastern North America: Amsterdam, Elsevier Pub. Co., chap. 12, p. 147-200.

Lipman, P. W., Mutschler, F. E., Bryant, Bruce, and Steven, T. A., 1969, Similarity of Cenozoic igneous activity in the San Juan and Elk Mountains, Colorado, and its regional significance, in Geological survey research 1969: U.S. Geol. Survey Prof. Paper 650-D, p. D33-D42.

Lipman, P. W., Steven, T. A., and Mehnert, H. H., 1970, Volcanic history of the San Juan Mountains, Colorado, as indicated by potassium-argon dating: Geol. Soc. America Bull, v. 81, p. 2329-2352.

Lipman, P. W., Steven, T. A., Luedke, R. G., and Burbank, W. S., 1973, Revised volcanic history of the San Juan, Uncompahgre, Silverton, and Lake City calderas in the western San Juan Mountains, Colorado: U.S. Geol. Survey Jour. Research, v. 1, no. 6, p. 627-642.

Livingston, D. E., 1973, A plate tectonic hypothesis for the genesis of prophyry copper deposits of the southern Basin and Range province: Earth and Planetary Sci. Letters, v. 20, p. 171-179.

Livingston, D. E., Mauger, R. L., and Damon, P. E., 1968, Geochronology of the emplacement, enrichment, and preservation of Arizona prophyry copper deposits: Econ. Geology, v. 63, p. 30-36.

Lowell, G. R., 1969, Geologic relationships of the Salida area to the Thirtynine Mile volcanic field of central Colorado [D. Sc. thesis]: Socorro, New Mexico Inst. Mining and Technology, 113 p.

——1971, Cenozoic geology of the Arkansas Hills region of the southern Mosquito Range, central Colorado, in New Mexico Geol. Soc. Guidebook 22nd Field Conf., San Luis Basin, 1971: p. 209-217

Lowell, G. R., and Chapin, C. E., 1972, Primary compaction and flow foliation in ash-flow tuffs of the Gribbles Run paleovalley, central Colorado: Geol. Soc. America, Abs. with Program (Ann. Mtg.), v. 4, no. 7, p. 725-726.

Lozano, Efraim, 1965, Geology of the southwestern Garo area, South Park, Park County, Colorado [M. Sc. thesis]: Golden, Colorado School of Mines, 115 p.

MacGinitie, H. D., 1953, Fossil plants of the Florissant beds, Colorado: Carnegie Inst. Washington Pub. 599, Contr. Paleontology, 198 p.

McKinlay, P. F., 1956, Geology of Costilla and Latir Peak quadrangles, Taos County, New Mexico: New Mexico Bur. Mines and Mineral Resources Bull. 42, 32 p.

——1957, Geology of Questa quadrangle, Taos County, New Mexico: New Mexico Bur. Mines and Mineral Resources Bull. 53, 23 p.

Morris, Gary R., 1969, Geology of the Dicks Peak area, Park County, Colorado [M. Sc. thesis]: Golden, Colorado School of Mines, 69 p.

Muehlberger, W. R., 1967, Geology of the Chama quadrangle, New Mexico: New Mexico Bur. Mines and Mineral Resources Bull. 89, 114 p.

——1968, Geology of Brazos Peak quadrangle, New Mexico: New Mexico Bur. Mines and Mineral Resources Geol. Map 22, scale 1:48,000.

Niesen, P. L., 1969, Stratigraphic relationships of the Florissant Lake Beds to the Thirtynine Mile volcanic field of central Colorado [M. Sc. thesis]: Socorro, New Mexico Inst. Mining and Technology, 65 p.

Noble, D. C., 1972, Some observations on the Cenozoic volcano-tectonic evolution of the Great Basin, western United States: Earth and Planetary Sci. Letters, v. 17, p. 142-150.

Olson, J. C., Hedlund, D. C., and Hansen, W. R., 1968, Tertiary volcanic stratigraphy in the Powderhorn-Black Canyon region, Gunnison and Montrose Counties, Colorado: U.S. Geol. Survey Bull. 1251-C, p. C1-C29.

Pillmore, C. L., Obradovich, J. D., Landreth, J. O., and Pugh, L. E., 1973, Mid-Tertiary volcanism in the Sangre de Cristo Mountains of northern New Mexico: Geol. Soc. America, Abs. with Programs (Rocky Mtn. Sec.), v. 5, no. 6, p. 502.

Ray, L. L., and Smith, J. F., Jr., 1941, Geology of the Moreno Valley, New Mexico: Geol. Soc. America Bull., v. 52, p. 177-210.

Richardson, G. B., 1915, Description of the Castle Rock quadrangle, Colorado: U.S. Geol. Survey Geol. Atlas, Folio 198, 13 p.

Sawatzky, D. L., 1964, Structural geology of southeastern South Park, Park County, Colorado: Mtn. Geologist, v. 1, no. 3, p. 133-139.

——1967, Tectonic style along the Elkhorn thrust, eastern South Park and western Front Range, Park County, Colorado [D. Sc. thesis]: Golden, Colorado School of Mines, 206 p.

Schmitt, H., 1933, Summary of the geological and metallogenetic history of Arizona and New Mexico, in Ore Deposits of the Western States: New York, Am. Inst. Mining and Metall. Engineers, p. 316-326.

Scott, G. R., 1970, Quaternary faulting and potential earthquakes in east-central Colorado: U.S. Geol. Survey Prof. Paper 700-C, p. C11-C18.

——1975, Cenozoic surfaces and deposits in the Southern Rocky Mountains, in Curtis, Bruce, ed., Cenozoic history of the Southern Rocky Mountains: Geol. Soc. America Mem. 144, p. 227-248.

Scott, G. R., and Taylor, R. B., (1975), Post-Paleocene Tertiary rocks and Quaternary volcanic ash of the Wet Mountain Valley, Colorado: U.S. Geol. Survey Prof. Paper (in press).

Simpson, G. G., 1948, The Eocene of the San Juan basin, New Mexico: Am. Jour. Sci., v. 246, no. 5, p. 247-282; v. 246, no. 6, p. 363-385.

Smith, J. F., Jr., and Ray, L. L., 1943, Geology of the Cimarron Range, New Mexico: Geol. Soc. America Bull., v. 54, p. 891-924.

Snyder, D. O., 1970, Fossil evidence of Eocene age of Baca Formation, New Mexico: New Mexico Geol. Soc. Guidebook 21st Field Conf., Tyrone-Big Hatchet Mountains-Florida Mountains region, 1970: p. 65-67.

Soister, P. E., 1968, Stratigraphy of the Wind River Formation in south-central Wind River Basin, Wyoming: U.S. Geol. Survey Prof. Paper 594-A, p. A1-A50.

Stark, J. T., and others, 1949, Geology and origin of South Park, Colorado: Geol. Soc. America Mem. 33, 188 p.

Steven, T. A., 1975, Middle Tertiary volcanic field in the Southern Rocky Mountains, in Curtis, Bruce, ed., Cenozoic history of the Southern Rocky Mountains: Geol. Soc. America Mem. 144, p. 75-94.

Steven, T. A., and Epis, R. C., 1968, Oligocene volcanism in south-central Colorado, in Epis, R. C., ed., Cenozoic volcanism in the southern Rocky Mountains: Colorado School Mines Quart., v. 63, no. 3, p. 241-258.

Steven, T. A., Mehnert, H. H., and Obradovich, J. D., 1967, Age of volcanic activity in the San Juan Mountains, Colorado, in Geological Survey research 1967: U.S. Geol. Survey Prof. Paper 575-D, p. D47-D55.

Taylor, R. B., 1975, Neogene tectonism in south-central Colorado in Curtis, Bruce, ed., Cenozoic history of the Southern Rocky Mountains: Geol. Soc. America Mem. 144, p. 211-226.

Tobey, E. F., 1969, Geologic and petrologic relationships between the Thirtynine Mile volcanic field and the Cripple Creek volcanic center [M. Sc. thesis]: Socorro, New Mexico Inst. Mining and Technology, 61 p.

Tonking, W. H., 1957, Geology of the Puertocito quadrangle, Socorro County, New Mexico: New Mexico Bur. Mines and Mineral Resources Bull. 41, 67 p.

Tweto, Ogden, 1961, Late Cenozoic events of the Leadville district and upper Arkansas Valley, Colorado, in Short papers in the geologic and hydrologic sciences: U.S. Geol. Survey Prof. Paper 424-B, p. B133-B135.

——1975, Laramide (Late Cretaceous-early Tertiary) orogeny in the Southern Rocky Mountains, *in* Curtis, Bruce, ed., Cenozoic history of the Southern Rocky Mountains: Geol. Soc. America Mem 144, p. 1-44.

Tweto, Ogden, and Case, J. E., 1972, Gravity and magnetic features as related to geology in the Leadville 30-minute quadrangle, Colorado: U.S. Geol. Survey Prof. Paper 726-C, 31 p.

Van Alstine, R. E., 1969, Geology and mineral deposits of the Poncha Springs NE quadrangle, Chaffee County, Colorado: U.S. Geol. Survey Prof. Paper 626, 52 p.

Van Houten, F. B., 1957, Appraisal of Ridgway and Gunnison "tillites," southwestern Colorado: Geol. Soc. America Bull., v. 68, p. 383-388.

Van Tuyl, F. M., and Lovering, T. S., 1935, Physiographic development of the Front Range: Geol. Soc. America Bull., v. 46, p. 1291-1350.

Welsh, Fred, 1969, The geology of the Castle Rock area, Douglas County, Colorado [M. Sc. thesis]: Golden, Colorado School Mines, 93 p.

Wyckoff, B., 1969, Geology of the east side of the Guffey volcanic center, Park County, Colorado [M. Sc. thesis]: Socorro, New Mexico Inst. Mining and Technology, 64 p.

MANUSCRIPT RECEIVED BY THE SOCIETY MAY 8, 1974

Geological Society of America
Memoir 144
© 1975

Middle Tertiary Volcanic Field in the Southern Rocky Mountains

Thomas A. Steven

U.S. Geological Survey
Federal Center
Denver, Colorado 80225

ABSTRACT

A widespread volcanic field covered most of the Southern Rocky Mountains in middle Tertiary time, 40 to 25 m.y. ago (approximately Oligocene time). This field covered an erosion surface that beveled structures formed during the Laramide orogeny in Late Cretaceous and early Tertiary time. The source vents from which the volcanic rocks were derived were largely restricted to the deformed area. Recognized volcanic centers lie mostly within a broad triangular area bounded on the east by the Rocky Mountain front, on the northwest by the northeast-trending Colorado mineral belt lineament, and on the southwest by the southern margin of the recurrently active Uncompahgre-San Luis uplift. Local volcanic centers existed also in the Never Summer Mountains and Rabbit Ears Range north of the mineral belt lineament. The resulting volcanic field thus consisted of a major southern segment covering all of south-central Colorado and adjacent New Mexico and a northern segment extending into the mountain areas of north-central Colorado. The two segments were linked along the trend of the Colorado mineral belt.

Most of the volcanic field consisted of volcanic rock of intermediate composition derived from many widely scattered volcanoes. Extensive aprons of volcaniclastic rocks around these volcanoes coalesced to provide a virtually continuous cover of volcanic material. In places, the volcanic activity became more silicic with time, great ash-flow tuff sheets were erupted, and the source areas subsided to form calderas. The largest ash-flow field and most of the calderas formed in the San Juan Mountains. Other ash flows apparently were derived from areas in the Sawatch Range in central Colorado and the Never Summer Mountains in northern Colorado.

Large near-surface batholiths were emplaced beneath the San Juan Mountains and beneath the region that extends northeastward from the Elk Mountains and Sawatch Range along the trend of the Colorado mineral belt as far as the Rocky Mountain front. These batholiths are manifested by two large gravity lows and by many exposed epizonal plutons that represent cupolas on the larger underlying bodies. The batholith in central Colorado probably consists of plutons of both Laramide (70 to 55 m.y.) and middle Tertiary (40 to 25 m.y.) ages.

INTRODUCTION

In a symposium a few years ago on volcanism in the Southern Rocky Mountains, Steven and Epis (1968, p. 241-258) suggested that south-central Colorado was covered by a widespread volcanic field in middle Tertiary time and that this field was fragmented by later Cenozoic block faulting and erosion. Other papers in that symposium (Corbett, 1968, p. 1-28; Taylor and others, 1968, p. 39-50; Epis and Chapin, 1968, p. 51-88; Siems, 1968, p. 89-124; Luedke and Burbank, 1968, p. 175-208) not only verified this conclusion but made it obvious that the middle Tertiary volcanic field also covered most of the remainder of the Southern Rocky Mountains. The purpose of this paper is to explore the nature and extent of this volcanic field more fully and to consider its relation to other geologic terranes in the Southern Rocky Mountain region.

The volcanic field formed largely in Oligocene time,[1] but evidence exists that related igneous activity began in latest Eocene time (about 40 m.y. ago) and persisted into early Miocene time (25 to 20 m.y. ago). The late Cenozoic basalts and associated high-silica rhyolites that were erupted widely throughout the Southern Rocky Mountains from early Miocene to Holocene time are believed to have evolved under a different tectonic regime (Lipman and others, 1970, p. 219; Christiansen and Lipman, 1972) and are not included in the assemblage of rocks that is the primary focus of this paper.

For the purpose of the present discussion, the volcanic field is limited to near-source lava and breccia and to the adjacent volcaniclastic mudflow and fluviatile outwash apron. A vast quantity of volcanic ash undoubtedly blanketed much of the surrounding area, particularly downwind to the east and southeast. Some of this material probably is represented by the large amounts of air-fall and reworked ash in the early Oligocene White River Group and other middle Tertiary sedimentary formations in the plains area to the east, but most of the blanketing ash was eroded and transported out of the region. It is beyond the scope of this paper to correlate any of the outlying ash-derived deposits with activity in the volcanic field itself.

GENERAL GEOLOGY

The middle Tertiary volcanic field (Figs. 1 and 2) consisted largely of intermediate-composition volcanic rocks that are typical of the continental interior of the western United States. Most are calc-alkalic or alkali-calcic andesitic, rhyodacitic, and quartz latitic lavas and breccias that formed many widespread central-vent volcanoes with coalescing aprons of volcaniclastic rocks. Rocks of more silicic composition, mostly somewhat younger ash-flow tuffs and associated quartz latitic and rhyolitic lavas, were erupted in large volume in parts of the field. A significant variant exists along the eastern fringe of the field where the more alkalic syenodiorites of the Spanish Peaks area, trachytes of the Rosita and Silver Cliff area, and phonolites of the Cripple Creek center occur. A fairly regular eastward increase in the K_2O/SiO_2 ratio at a given silica content across the field has been indicated by Lipman and others (1972, p. 236).

---

[1]The limits of the Oligocene used in this report, 38 to 26 m.y. ago, are from Harland and others (1964).

The middle Tertiary field developed in postorogenic time in an area that was strongly deformed during the Laramide orogeny in Late Cretaceous and early Tertiary time (Tweto, 1975) and during an earlier orogeny in late Paleozoic time that formed the Ancestral Rocky Mountains (Mallory, 1960). Structural trends resulting from both periods of prevolcanic deformation exercised major control over the distribution of middle Tertiary igneous activity, but individual volcanic centers generally show little relation to specific earlier structures. The volcanic field covered the late Eocene surface described by Epis and Chapin (1975) and is capped by widespread late Cenozoic basalt flows whose genesis and tectonic environment are interpreted in the accompanying papers by Lipman and Mehnert (1975) and Larson and others (1975).

Source Areas

The source areas from which the rocks of the middle Tertiary volcanic field were derived are largely confined within a large triangular-shaped area that covers most of the Southern Rocky Mountains in Colorado and northern New Mexico

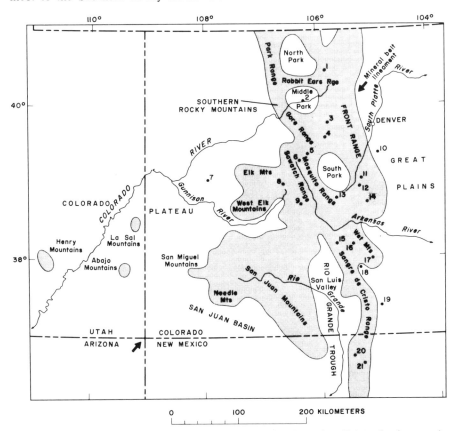

Figure 1. **Geographic reference map of Southern Rocky Mountains. Points of reference: 1, Never Summer Mountains, Specimen Mountain, Lulu Mountain, Mount Richthofen; 2, Hot Sulphur Springs; 3, Red Mountain stock; 4, Montezuma batholith; 5, Climax, Chalk Mountain stock; 6, Leadville; 7, Grand Mesa; 8, Grizzly Peak caldera, Twin Lakes stock; 9, Mount Princeton batholith; 10, Castle Rock; 11, Signal Butte; 12, Florissant; 13, Thirtynine Mile volcanic field; 14, Cripple Creek; 15, Cottonwood (Rito Alto) stock; 16, Westcliffe, Silver Cliff, Rosita Hills; 17, Deer Peak; 18, Huerfano Park, Devils Hole; 19, Spanish Peaks; 20, Red River; 21, Eagles Nest.**

(Figs. 1 and 2). Igneous centers extend south of this triangle along the trend of the Southern Rocky Mountains to join with areas of major middle Tertiary volcanism in central New Mexico. Scattered laccolithic centers were formed in the Colorado Plateau west of the composite middle Tertiary field, and some of these may have vented to form local volcanic accumulations.

In the southern segment of the composite volcanic field, the northwest margin of the source area is broadly limited by the well-known Colorado mineral belt lineament (Fig. 2). This lineament follows the trend of a major shear zone of Precambrian age (Tweto and Sims, 1963) and is marked by a belt of Laramide intrusions that extends from northeast Arizona to the eastern front of the Southern Rocky Mountains northwest of Denver (Steven and others, 1972). Middle Tertiary plutons in the Elk and West Elk Mountains extend at least 50 km northwest of the mineral belt lineament in west-central Colorado, and intrusions in the core of the San Miguel Mountains extend 20 to 22 km west of the Laramide intrusions at Rico in southwest Colorado. However, on a regional scale (Fig. 2), the mineral belt lineament appears to have had a definite limiting effect on middle Tertiary igneous activity.

The southwest margin of the source area, in the southern segment of the composite volcanic field, extends from the San Miguel Mountains southeastward along the margin of the San Juan basin into New Mexico, where it crosses younger fault-block mountains and basins to terminate finally at the eastern front of the Southern Rocky Mountains. The southwest boundary closely parallels the southwest margin of the recurrently active Uncompahgre–San Luis highland, a structural element

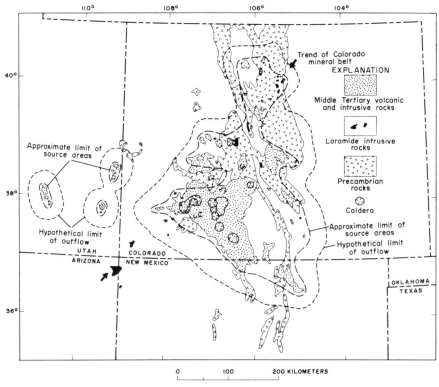

Figure 2. Map of Southern Rocky Mountains showing extent of middle Tertiary volcanic field (geology modified from Steven and others, 1972).

first uplifted in late Paleozoic time as part of the so-called Ancestral Rockies (Tweto, 1975; Mallory, 1960, p. 27, Fig. 2). This uplift, extending from eastern Utah across southwestern Colorado to northern New Mexico, was reactivated in Laramide time; several subsidiary folds, such as the Needle Mountains uplift, then formed along its borders.

The eastern boundary of the source area in the southern segment of the volcanic field virtually coincides with the eastern front of the Southern Rocky Mountains and, thus, with the edge of major Laramide deformation. Neglecting the minor Signal Butte center in the southern Front Range, the eastern margin extends from a southern laccolithic intrusive center near Eagles Nest on the east side of the Sangre de Cristo Range northward to the Cripple Creek center near the south end of the Front Range. From here, the limit of the source area swings northwestward in a broad arc to the Colorado mineral belt in the vicinity of Climax in central Colorado.

The northern segment of the composite volcanic field is linked to the southern segment along the trend of the Colorado mineral belt. The linkage area contains most of the intrusions with well-established middle Tertiary ages in central Colorado. Few Tertiary plutons of any kind have been mapped in the central Front Range southeast of the Colorado mineral belt, and therefore this zone almost certainly marks the southern limit of sources in the northern segment of the middle Tertiary volcanic field. Other igneous centers of middle Tertiary age are known north of the mineral belt as far as the Never Summer Mountains and Rabbit Ears Range in north-central Colorado (Fig. 2).

Tectonic Setting

The source vents for the southern segment of the composite field are not only limited by pre-existing trends but also appear to be confined largely to an area of recurrent tectonic activity. The San Juan and West Elk Mountains were built largely on top of the recurrently active Uncompahgre-San Luis uplift. To the northeast, the plutons in the Elk Mountains and Sawatch and Sangre de Cristo Ranges and the volcanic rocks in the Thirtynine Mile volcanic field and Wet Mountains were emplaced largely in the Central Colorado trough, a feature that formed in late Paleozoic time when it was filled with a thick accumulation of sediments. Parts of the trough were extensively compressed and thrust faulted during the Laramide orogeny (Burbank and Goddard, 1937; Tweto, 1975).

The volcanic centers of middle Tertiary age in the Leadville-Climax area in central Colorado were emplaced near the eastern margin of the Central Colorado trough and along the eastern flank of the Sawatch anticline of Laramide age. To the northeast, the eruptive vents follow the mineral belt lineament onto the top of the Front Range highland element of the late Paleozoic Ancestral Rockies (Tweto, 1968b, p. 561). This ancestral structural element trends northwest and embraces the area that now includes the igneous centers in the Never Summer Mountains and Rabbit Ears Range. The middle Tertiary igneous centers in this segment of the field show no obvious control by Laramide structural features, however, because the Never Summer centers are along the west flank of the Front Range anticline, and the Rabbit Ears centers extend west across the trend of the North Park-Middle Park synclinal basin.

Despite the earlier history of recurrent tectonic activity, the area covered by the composite volcanic field appears to have been structurally stable when the field was emplaced. The Laramide structures were widely beveled by Eocene erosion. A regional surface of low relief was developed in the southern part of Colorado (Epis and Chapin, 1975), but the Rabbit Ears Volcanics in north-central Colorado accumulated on irregular topography (Izett, 1968, p. 36). Most of the structural features developed during Oligocene volcanism were of volcano-tectonic origin, were of limited extent, and were related to eruptions or movements of magmas at local centers. Regional block faulting that began in early Miocene time (Lipman and Mehnert, 1975; Izett, 1975; Taylor, 1975) heralded the end of the middle Tertiary volcanism.

San Juan Mountains

The San Juan Mountains in the southwestern part of Colorado consist largely of a dissected volcanic plateau and form the largest remnant of the middle Tertiary volcanic field. Volcanic eruptions in this portion of the field took place largely between 35 and 26 m.y. ago, although locally they persisted for another several million years (Lipman and others, 1970). Numerous igneous centers are associated with the volcanic rocks throughout the San Juan Mountains, but they are notably sparse in adjacent areas to the west and southwest (Steven and others, 1974), except for the intrusions in the core of the San Miguel Mountains that extend nearly 30 km west of the San Juan volcanic pile. Erosion evidently has removed nearly all of the outflow apron of volcanic and volcaniclastic rocks in the area west and southwest of the San Juans; the thick near-source accumulations associated with the core volcanoes are all that remain.

Early intermediate-composition lavas and breccias constitute about two-thirds of the volcanic pile of the San Juan Mountains (Lipman and others, 1970, p. 2331). These rocks are largely andesite, rhyodacite, and mafic quartz latite, and they formed numerous widely scattered central-vent volcanoes whose volcaniclastic aprons coalesced to form an unbroken volcanic field much wider than the present area of exposure. Volcanic activity in the San Juans culminated in more silicic, commonly explosive eruptions, and great ash-flow tuff sheets were emplaced. The thick sequence of welded ash-flow tuffs that covers a large part of the San Juan Mountains was erupted from numerous vents in the central and western San Juans, and major calderas have been identified at the sources of all major individual sheets (Fig. 2).

The full extent of volcanic cover that once existed over the flat-lying sedimentary rocks south and west of the San Juan Mountains cannot be established. The Needle Mountains (Figs. 1 and 2), which lie along the south side of the western San Juan Mountains, are the remains of a Laramide uplift. This uplift had an eroded core of Precambrian rocks and stood as a knot of hills (30 to 35 km across) that may never have been completely covered by the middle Tertiary volcanic accumulations. No comparable barrier existed east or west of the Needle Mountains, however, and a coalescing outflow apron of volcaniclastic rocks from the San Juan Mountains probably spread around this local uplift and extended many kilometers to the south, southwest, and west.

The San Juan volcanic plateau is bounded on the east by the San Luis Valley segment of the Rio Grande trough, a late Cenozoic rift zone that extends northward from southern New Mexico along the trend of the Southern Rocky Mountains at least as far as central Colorado. Along the eastern edge of the San Juan Mountains, the volcanic formations dip eastward beneath alluvial fill of the San Luis Valley, and no fault is evident along the margin. However, gravity studies (Gaca and Karig, 1965) have indicated that rough fault-block topography beneath the alluvial fill in the Rio Grande trough cuts off the San Juan volcanic rocks. In southernmost Colorado and northern New Mexico, the Rio Grande trough is filled with 3.5- to 4.5-m.y.-old basalt flows (Ozima and others, 1967) that obscure the bounding faults. Intermediate-composition volcanic rocks that probably represent the eastern flank of the San Juan volcanic accumulation are exposed in the San Luis Hills in the southern part of the San Luis Valley (Burroughs, 1971, p. 280) and are exposed in places on the Sangre de Cristo mountain block east of the Rio Grande trough. These patches extend from a point northwest of the Spanish Peaks in Colorado (Johnson, 1969) to the Red River in New Mexico. Some of these rocks just south of the Colorado–New Mexico border have been dated as 35.6 m.y. old (Pillmore and others, 1973), and Burroughs (1971, p. 280) reported ages of 27.4 and 27.9 m.y. for intrusive rock cutting the volcanic rocks in the San Luis Hills.

Spanish Peaks and Wet Mountains

The Spanish Peaks intrusive center (Johnson, 1969) just east of the Sangre de Cristo Range in southern Colorado is believed to represent the roots of former volcanoes of middle Tertiary age (J. D. Vine, 1972, written commun.). K-Ar ages reported by Stormer (1972, p. 2445) indicate that igneous activity at this center was largely in early Miocene time, 26 to 22 m.y. ago, and fission-track ages reported by Smith (1973, p. 513–514) range from 28.5 to 19.8 m.y. in confirmation. Any volcanic edifice that existed above this center has been removed by erosion, and its former extent can only be conjectured. On the other hand, one of the patches of volcanic rock that has been mapped along the southern Sangre de Cristo Range (Johnson, 1969) is only 19 km west of the Spanish Peaks center, and it seems likely that volcanic rocks covered the intervening area. If so, a virtually continuous cover of volcanic rocks may have existed between the San Juan Mountains and the eastern edge of the volcanic pile that accumulated above the Spanish Peaks intrusive center.

Middle Tertiary volcanic rocks form a discontinuous cap on the Wet Mountains from the vicinity of Westcliffe southward. The Rosita and Silver Cliff volcanic centers (Siems, 1968) near the northern part of this assemblage of volcanic rocks are the remnants of former central-vent volcanoes that erupted lavas and breccias ranging from andesite to rhyolite in composition. Pyroclastic rocks form only a minor part of either volcano. An age of 31.7 m.y. has been reported for pumice from the Rosita center (Scott and Taylor, 1974). Equivalent mudflow and conglomeratic outflow debris is found in the Devils Hole area in Huerfano Park to the south (Scott and Taylor, 1974). The Deer Peak center in the southern Wet Mountains (Scott and Taylor, 1974) exposes the root of another volcano of intermediate composition; related volcanic rocks are still preserved in adjacent parts of the Wet Mountains, and outflow debris has been recognized in the Devils Hole area to the southwest.

The eastern flank of the Wet Mountains volcanic pile has been eroded, and the original position of the eastern margin of the pile cannot be established. Some of the volcanic rocks pass beneath younger alluvial fill in the Wet Mountain Valley west of the Silver Cliff and Rosita Hills volcanic centers. Similar rocks that reappear on the west side of the valley and volcanic rocks that belong to the Thirtynine Mile volcanic field farther north are in fault contact with deformed sedimentary rocks in the Sangre de Cristo mountain block (Scott and Taylor, 1974). Evidently the volcanic rocks of the Wet Mountains extended much farther west prior to late Cenozoic block faulting, and they may have been continuous with those in the San Juan Mountains.

Dikes, sills, plugs, and stocks of intermediate to silicic composition have been mapped in the Sangre de Cristo Range and adjacent parts of Huerfano Park northwest of the Spanish Peaks. These intrusive bodies extend beyond Cottonwood (Rito Alto) Peak (Toulmin, 1953) in the northern part of the Sangre de Cristo Range.

The age of most of these intrusions has not been established, but some of the more southerly ones cut or deform Eocene sedimentary rocks and probably were emplaced in middle Tertiary time. Preliminary fission-track ages on the Cottonwood (Rito Alto) stock indicate a middle Oligocene age (R. B. Taylor and C. W. Naeser, 1973, oral commun.). Some of these intrusions may have vented at the surface and contributed to the volcanic cover between the San Juan and Wet Mountains.

Thirtynine Mile Volcanic Field and Southern Front Range

The Thirtynine Mile volcanic field (Epis and Chapin, 1968) consists largely of andesitic lavas and breccias derived from a cluster of central-vent volcanoes. The field was built largely in Oligocene time in the southern part of the South Park intermontane basin. The Wall Mountain Tuff at the base of the succession has been dated as 35 to 36 m.y. old (Van Alstine, 1969, p. 15). K-Ar ages obtained on the overlying Antero Formation of Johnson (1937) and the Gribbles Park Tuff are, respectively, 33 m.y. and 29 m.y. Most of the outflow apron of volcanic and volcaniclastic rocks around the margins of this field has been removed by erosion, and a complex of near-source lavas and breccias that accumulated around clustered volcanic centers in the heart of the field are all that remain. Rock units that can be correlated with some of those in the Thirtynine Mile field have been recognized (1) along the east side of the Sangre de Cristo Range south of the Arkansas River (Scott and Taylor, 1974) within 10 to 20 km of the volcanic centers at Silver Cliff and Rosita (Fig. 1), and (2) near the northern end of the San Luis Valley associated with locally derived volcanic rocks in the northeast part of the San Juan volcanic field (Lowell, 1971, p. 216). It thus seems a certainty that the Thirtynine Mile volcanic field was once connected with the accumulation of volcanic rocks in the Wet Mountains and San Juan Mountains.

R. C. Epis and C. E. Chapin (1973, written commun.) have mapped a series of ash-flow units in the Thirtynine Mile volcanic field that appear to have had sources to the west of the upper Arkansas River valley, possibly in the Sawatch Range. Two of these units, the Wall Mountain Tuff and the Gribbles Park Tuff, are sheets of sufficient volume that caldera subsidence probably took place at their sources. These sources are being actively sought by the combined efforts of several workers, and speculations here as to their locations would be premature.

The Cripple Creek mining district (Loughlin and Koschmann, 1935; Koschmann, 1949), 15 to 18 km east of the Thirtynine Mile volcanic field, is largely within a deeply eroded remnant of a middle Tertiary volcano. Most of the volcanic rocks at Cripple Creek are within a subsided block enclosed within walls of Precambrian granite, but Tobey (1969) has mapped patches of volcanic rock that are related to the rocks at Cripple Creek and overlie Precambrian rocks in adjoining areas. Rocks derived from the Cripple Creek center overlie one of the ash-flow tuff sheets from the Sawatch Range, and phonolite typical of that at Cripple Creek forms plugs that cut andesitic breccias derived from centers in the Thirtynine Mile field. The phonolite at Cripple Creek has been dated as 28 m.y. old (R. C. Epis and C. E. Chapin 1973, written commun.).

Other localities in the southern Front Range where there are small volcanic accumulations of Oligocene age are Florissant and Signal Butte. At Florissant (MacGinitie, 1953), tuffaceous lake sediments containing Oligocene plant fossils were deposited during the period of volcanism that produced the Thirtynine Mile volcanic field (Niesen, 1969). The lake at Florissant formed behind a barrier of andesitic volcanic rocks derived from the Thirtynine Mile volcanic field, and the tuffaceous sediments deposited in the lake were subsequently covered by additional volcanic rocks from the same source. At Signal Butte, about 13 km north of Florissant, a deeply eroded remnant of a volcanic dome of intermediate composition rests on the late Eocene erosion surface. The vent that fed the dome is probably beneath the existing remnant. The volcanic materials at Signal Butte appear to be locally derived, and no continuous cover of volcanic materials between this locality and the Thirtynine Mile field is required by the evidence.

Izett and others (1969) dated a rhyolite ash-flow tuff near Castle Rock, Colorado, as early Oligocene (34.8 m.y.) by the K-Ar method. This tuff overlies the Dawson Arkose of Late Cretaceous and Paleocene age and underlies the Castle Rock Conglomerate of early Oligocene age. The source of the tuff was postulated to lie in the mountain area to the southwest, and the unit may correlate with one of the ash-flow tuff sheets mapped in the Thirtynine Mile volcanic field (Epis and Chapin, 1968). Subsequent microprobe analytical work on minerals from the tuff at Castle Rock by George Desborough of the U.S. Geological Survey (1973, written commun.) suggests that the tuff may be the distal end of a major sheet at the base of the Thirtynine Mile volcanic succession that has been mapped widely in and adjacent to the southern part of South Park. If this correlation proves valid, Oligocene volcanic rocks must have been widely distributed over the southern Front Range, although perhaps largely confined to topographically lower portions of the area.

Elk and West Elk Mountains

The West Elk Mountains consist largely of a great mass of volcanic breccia called the West Elk Breccia. This rock probably was derived from volcanoes that may have existed above granodiorite plutons exposed in the northern West Elk Mountains (Olson and others, 1968; Lipman and others, 1969), where stocks and related dikes, laccoliths, and sills cut both the West Elk Breccia of Oligocene age and lower Eocene sedimentary rocks of the Wasatch Formation (Godwin and Gaskill, 1964). In addition, Gaskill and others (1973) identified a volcano source for some of the West Elk Breccia in this area. To the south, near the Gunnison River, the West Elk Breccia coalesces with similar volcanic breccias (35 to 30 m.y. old) derived from volcanoes in the northern and western San Juan Mountains, and great ash-flow sheets (29 to 27 m.y. old) from caldera sources in the San Juans overlie breccias from both areas (Olson and others, 1968; Hansen, 1965; Lipman and others, 1970).

The core of the Elk Mountains (which extend northwest from the Sawatch Range in central Colorado) contains many epizonal plutons of Oligocene age that are between 34 and 29 m.y. old (Obradovich and others, 1969, p. 1749-1756). These rocks are largely granodiorite and are closely similar to the intrusive rocks in the West Elk Mountains to the south. No volcanic rocks are preserved in the Elk Mountains proper, but the nearby Grizzly Peak caldera in the vicinity of Independence Pass in the central Sawatch Range (Cruson, 1972; Obradovich and others, 1969, p. 1854) contains volcanic rocks, including ash-flow tuff. These rocks are more than a kilometer thick and are preserved in a downdropped block about 8.5 km across. Some of the Elk Mountains plutons may have vented to form significant volcanic accumulations similar to those in the West Elk Mountains. Lipman and others (1969) called attention to the close similarity in age and composition of the intrusive rocks in the Elk Mountains and the intrusive and extrusive rocks in the West Elk and San Juan Mountains. However, the original extent of any volcanic pile that might have existed above the Elk Mountains is unknown.

Mount Princeton Batholith

The Mount Princeton batholith in the southern Sawatch Range was largely mapped by Dings and Robinson (1957) and has been studied intensively in recent years by P. Toulmin III of the U.S. Geological Survey. According to Toulmin (1973, written commun.), K-Ar age determinations by C. Hedge of the U.S. Geological Survey indicate that intrusion of the batholith took place in latest Eocene and early Oligocene time. Toulmin (in U.S. Geol. Survey, 1963, p. A88) suggested that the roots of a Tertiary volcano may be preserved in the Mount Aetna area in the southern part of the batholith and that the deeply eroded remnant of a caldera may be represented. A similar conclusion was reached by Lipman and others (1969, p. D38) by interpreting the geological map of Dings and Robinson (1957, p. 1). Brock and Barker (1965, p. 320) described a dike of welded tuff more than 1.5 km long and 9 to 17 m thick near the north margin of the Mount Princeton batholith. Chapin and others (1970) and Lowell (1971, p. 216) described paleovalleys filled with volcanic rocks derived in part from the Sawatch Range along the east side of the upper Arkansas River valley across from the Mount Princeton batholith. These relations all suggest that a major volcano or cluster of volcanoes was present above the Mount Princeton batholith in Oligocene time.

Colorado Mineral Belt

Intrusive igneous rocks in north-central Colorado are concentrated along the northeast-trending Colorado mineral belt (Fig. 2). Two distinct episodes of igneous activity are represented: (1) a Late Cretaceous–early Tertiary (Laramide) episode in which abundant intrusions were emplaced between 70 and 55 m.y. ago from the west side of the Sawatch Range to the northeast end of the mineral belt, and (2) an Oligocene episode in which intrusions were emplaced 39 to 26 m.y. ago in the Climax-Chalk Mountain area in the Tenmile and southern Gore Ranges, on the west side of the Front Range (Montezuma batholith), and in the Red Mountain area. Reworked volcanic rocks formed during the first episode were deposited in sedimentary basins that flank the mountains on either side of the mineral belt (Tweto, 1975). Volcanic rocks produced during the second episode occur in Middle Park (Izett and others, 1969; Naeser and others, 1973).

Wallace and others (1968) described the complex sequence of intrusive and hydrothermal events leading to the development of the highly mineralized igneous center at Climax, Colorado. The last thermal event in that sequence, probably related to hydrothermal activity, has a K-Ar age of about 30 m.y. The Chalk Mountain stock, 3 to 5 km west of Climax, and a rhyolite plug a few kilometers north of Climax were reported by V. E. Surface (Tweto and Case, 1972, p. C7) to be about 27 and 35 m.y. old, respectively. Tweto (1968a, p. 693-694) and Tweto and Case (1972, p. C7) pointed out that other intrusive bodies in the Leadville-Climax area also might have been emplaced in post-Laramide (middle Tertiary?) time. No volcanic rocks related to the Leadville-Climax intrusive rocks have been identified, and no evidence has been cited to indicate that volcanic venting took place. However, the intrusives clearly indicate that magmatism occurred in the mineral belt of central Colorado in Oligocene time, and the possibility of concurrent volcanic activity cannot be ruled out.

The Montezuma batholith is a major intrusive body on the west side of the Front Range along the trend of the Colorado mineral belt. The Montezuma pluton has been described as a stock, but a recent study of underground exposures in the H. D. Roberts water diversion tunnel and geophysical evidence indicate that the body is of batholithic proportions. Before the advent of isotopic dating techniques, this pluton was believed to belong to the Laramide sequence of igneous intrusions in the Colorado mineral belt (Lovering, 1935, p. 32; Lovering and Goddard, 1938, 1950, p. 44-47) and was placed near the middle of that sequence. K-Ar dates by McDowell (1966) and Hedge and others (in U.S. Geol. Survey, 1970, p. A34) show, however, that the batholith was emplaced about 39 m.y. ago, roughly at the end of Eocene time, and that it is decidedly younger than the numerous bodies of Laramide age (summarized by Tweto, 1968b, p. 565) that exist throughout the mineral belt. Geological maps and sections of the H. D. Roberts water diversion tunnel that traverses the batholith (Wahlstrom and Hornback, 1962, Pl. 7; Warner and Robinson, 1967, Pl. 1) show that the west margin of the batholith approximately underlies the edge of outcrop, but that the east margin is nearly 3 km east of the exposed edge of the batholith. Gravity studies by Brinkworth (1970) indicate that the pronounced gravity low produced by the batholith extends at least 19 km north of its surface outcrop; the gravity low includes the Red Mountain igneous center that has been dated as 27 to 26 m.y. and is therefore of latest Oligocene age (Taylor and others, 1968, p. 42; Naeser and others, 1973).

It is unknown if any volcanic accumulation existed above the Montezuma batholith. However, Taylor and others (1968, p. 46) believed that at Red Mountain, intrusion breccias, the shattering of wall rock, and the appearance of pumice fragments and glass shards in the upper part of the intrusion indicate volcanic venting. They suggested (p. 43) that the rhyolitic pyroclastic rocks of approximately the same age as the Red Mountain intrusion (29 m.y.) in the Fraser basin a few kilometers to the north are probably derived from nearby centers. Other rhyolite intrusions at Cabin Creek and Leavenworth Creek, 8 km northeast of the exposed part of the Montezuma batholith and within the gravity low shown by Brinkworth (1970), have been assigned a middle Tertiary age by R. B. Taylor and R. U. King on the basis of geologic evidence (in U.S. Geol. Survey, 1968, p. A30).

Never Summer Mountains

Wahlstrom (1944) described the eroded roots of a silicic volcano at Specimen Mountain in the northwestern part of Rocky Mountain National Park, and Corbett (1968) mapped contiguous remnants of a related volcanic field northwest of Specimen Mountain in the Never Summer Mountains. Corbett separated the volcanic rocks into two units: (1) an older assemblage of relatively mafic volcanic rocks, the Cameron Pass Volcanic Group, that probably forms part of a volcano centered above an andesite porphyry plug on Mount Richthofen (Corbett, 1968, p. 2-3), and (2) a younger assemblage of silicic volcanic rocks, the Specimen Mountain Group of Corbett (1968), derived from central vents at Specimen and Lulu Mountains. The older sequence was postulated to be Eocene in age, based on its compositional similarity to andesitic rocks interlayered with sedimentary rocks of Cretaceous to Eocene age in the nearby North Park basin. It seems equally plausible that some of the older rocks mapped by Corbett are generally equivalent to andesitic volcanic rocks in the Rabbit Ears Range to the west, which Izett (1966) has shown to be Oligocene in age. The younger silicic volcanic rocks in the Never Summer Mountains are largely pyroclastic rocks and local lava flows. A remnant of one formerly widespread sheet of ash-flow tuff has been preserved. The younger rocks were dated by Corbett (1968, p. 8) as 27 to 28 m.y. old.

The Never Summer stock of granodiorite and quartz monzonite cuts the older volcanic units along the west side of the area mapped by Corbett (1968, p. 28, Pl. 1). A K-Ar age of 28 m.y. obtained on the stock accords closely with the 27- to 28-m.y. ages obtained on the silicic volcanic rocks of the Specimen Mountain Group of Corbett.

The remnant of volcanic rocks preserved in the Never Summer Mountains appears to be only a small, near-source part of the original accumulation. Little detailed geological mapping is available on adjacent areas to the north, east, or south, and there is little evidence to indicate the former extent of the volcanic field in these directions. To the northwest, however, a wedge of volcanic rocks in the southeast part of North Park is considered by Kinney (1970a) as probably contemporaneous with the Rabbit Ears Volcanics that lie to the south and southwest. The volcanic rocks in southeast North Park are only 8 to 20 km northwest of volcanic centers in the Specimen Mountain area and probably were derived from them.

Intrusive plugs cut older sedimentary rocks in several places southwest of the Never Summer Mountains in the eastern part of the Rabbit Ears Range (Kinney, 1970b). Some of these intrusions probably mark the roots of volcanoes that may have produced a coalescing pile of volcanic rocks extending from the Never Summer Mountains to the large mass of volcanic rocks in the western part of the Rabbit Ears Range.

Rabbit Ears Range

The Rabbit Ears Range between North and Middle Parks, Colorado, is capped by an extensive sequence of mafic, intermediate, and silicic volcanic rocks that are cut by a series of volcanic necks and pluglike intrusive bodies that mark the roots of ancient volcanoes (Kinney and others, 1968, p. 3-6). Most of these volcanic rocks are included in the Rabbit Ears Volcanics (Izett, 1966, p. B42; 1968, p. 29-40) of Oligocene and Miocene(?) age. A sample of rhyolite obtained near the middle of this volcanic sequence was dated by K-Ar methods as about 33 m.y. old (Izett, 1966, p. B45), and a silicic ash-flow tuff at Troublesome Creek (on the south slope of the range) was dated by the fission-track method as 30 m.y. old (Naeser and others, 1973). Other fission-track ages from rocks in the Rabbit Ears Range (Naeser and others, 1973) are about 28 m.y. for a stock at Haystack Mountain and about 23 m.y. for a stock at Poison Ridge. The Rabbit Ears Volcanics rest unconformably on Upper Cretaceous and lower Tertiary sedimentary rocks and are overlain by, and perhaps interfinger with, Miocene basin-fill sediments of the Troublesome Formation (Izett, 1968, p. 36).

Volcanic breccia of the Rabbit Ears Volcanics has been mapped as far west as the crest of the Park Range near Rabbit Ears Peak (Hail, 1968, Pl. 2) and north into the southern part of North Park (Hail, 1968, Pls. 2 and 3; Kinney and Hail, 1970). Izett (1966, Fig. 1; 1968, p. 29) traced the same unit south to the Colorado River near Hot Sulphur Springs.

Tweto (1957, p. 24) described an intrusive center surrounded by patches of volcanic rocks at Green Mountain in the Blue River valley 28 km southwest of Hot Sulphur Springs. On the basis of geologic evidence, Tweto favored a late Tertiary age for the igneous activity of this locality, but a recent fission-track age on a sill at Green Mountain (Naeser and others, 1973) indicates emplacement about 30 m.y. ago. Other intrusive bodies and a rhyolite flow that occur along the west side of the Front Range 20 to 30 km northeast of Hot Sulphur Springs and south of the Never Summer Mountains have fission-track ages (Naeser and others, 1973) of generally 25 to 27.5 m.y.

Colorado Plateau

Numerous laccolithic intrusive centers have been emplaced in the flat-lying sedimentary rocks of the Colorado Plateau, and as summarized by Witkind (1964, p. 79), all these centers were originally believed to have been intruded at about the same time. Hunt (1956, p. 42) postulated that this was during the late Tertiary, but Witkind (1964, p. 79-81) concluded that the preponderance of available evidence favored Late Cretaceous–early Tertiary intrusion. More recently, isotopic age determinations (Armstrong, 1969; Dickinson and others, 1968; Damon, 1968) have shown clearly that two general episodes of intrusion were involved: (1) an older Laramide episode during which laccoliths were emplaced in the linear belt extending from the western San Juan Mountains southwestward to northeast Arizona, and (2) a middle Tertiary episode during which more widely scattered laccoliths were emplaced in southwest Colorado and southeast Utah (Armstrong, 1969, Fig. 1; Steven and others, 1972, Fig. 1). The Laramide intrusions are limited to a regional northeast-trending belt that includes numerous intrusions in the Colorado mineral belt of central Colorado, whereas the middle Tertiary laccoliths are widely dispersed, as are most other middle Tertiary igneous centers.

The three largest middle Tertiary laccolithic centers, the Henry, La Sal, and Abajo Mountains, have all been studied in detail. In the Henry Mountains, Hunt and others (1953, p. 147-148) concluded that the overburden at the time of intrusion was sufficiently thin so that volcanic venting probably took place, although no definite supporting evidence was found. The La Sal Mountains (Hunt, 1958, p. 319) contain pipelike masses of intrusion breccia that very likely indicate concurrent volcanic venting. Witkind (1964) cited no comparable evidence in the Abajo Mountains, although the logic applied to the Henry Mountains (Hunt and others, 1953, p. 147) would seemingly apply here as well. Although unproven, it thus is possible that scattered middle Tertiary volcanoes were active in the Colorado Plateau area west of the main composite volcanic field in the Southern Rocky Mountains.

RELATIONS TO BATHOLITHS

The area occupied by the composite volcanic field of middle Tertiary age is in part coincident with two large gravity lows (Fig. 3) that have been interpreted as reflecting the presence of shallow batholiths (Plouff and Pakiser, 1972, p. B186; Case, 1965; Tweto and Case, 1972, p. C18). The outlines of batholiths shown on Figure 3 were drawn along the trends of steepest gravity gradients bordering the gravity lows shown by Behrent and Bajwa (1972). The southern gravity low covers much of the San Juan Mountains and is nearly coincident with the distribution of calderas that mark the sources of major ash-flow deposits. The northern gravity low extends northeastward from the Elk Mountains and southern Sawatch Range along the trend of the Colorado mineral belt, encompasses the Montezuma batholith, and terminates at the east side of the central Front Range. This gravity low embraces the source areas of igneous rocks of both Laramide and middle Tertiary age.

Whereas the gravity low in the San Juan Mountains is almost coincident with the area of ash-flow–related calderas, it shows little relation to the distribution of early intermediate-composition central-vent volcanoes. This can be interpreted to indicate that the early central-vent volcanoes were fed from deep-seated, relatively mafic magma chambers, and that the high-level silicic batholith that fed the ash-flow eruptions and produced the gravity low did not rise to shallow depths until the middle of the volcanic episode. The silicic ash flows that were erupted indicate that the upper parts of the batholith were significantly more differentiated than the deeper magma chambers that fed the early intermediate-composition volcanoes. The less dense silicic rocks of the batholith have sufficient density contrast with the Precambrian host rocks to produce a clear-cut gravity low (Tweto and Case, 1972, p. C12-C13). The compositional zoning displayed by many major ash-flow sheets, which range from crystal-poor rhyolites at the base to crystal-rich quartz latites at the top, indicates a layered aspect of the differentiated magma chamber. This conclusion is further substantiated by the eruption of andesitic to rhyodacitic lavas from local centers during the general period of ash-flow eruptions; these centers presumably tapped the batholith at lower, less differentiated levels (Lipman and others, 1970, p. 2347).

The gravity low underlying central Colorado (Behrent and Bajwa, 1972; Tweto and Case, 1972, p. C17; Brinkworth, 1970) is a composite of several individual gravity lows produced by intrusions of two distinct ages. A gravity low underlying the southern Sawatch Range is centered on the Mount Princeton batholith and seems clearly related to this major middle Tertiary pluton. This gravity low is separated from a broad gravity low that extends west from the Sawatch Range

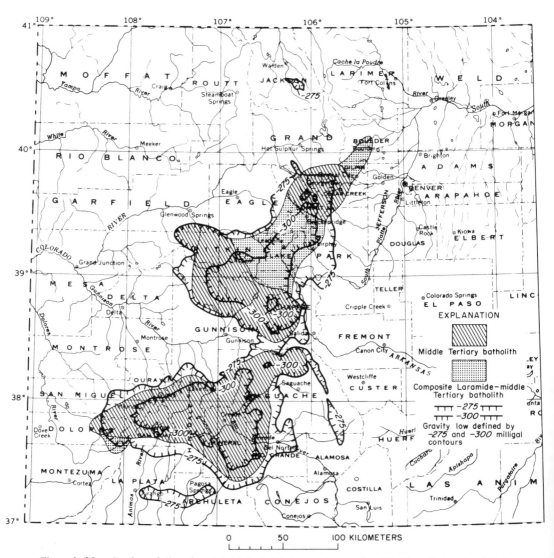

Figure 3. Map showing relation of postulated batholith to gravity lows (modified from Behrent and Bajwa, 1972; Tweto and Case, 1972; and Plouff and Pakiser, 1972).

under the Elk Mountains and the northern part of the West Elk Mountains by a low gravity divide. The broad gravity low coincides with an area of abundant middle Tertiary epizonal plutons (Obradovich and others, 1969) and probably reflects the presence of a large underlying batholith. The east end of this broad gravity low merges into a narrower northeast-trending low that extends from the Sawatch Range to the east side of the Front Range. In the Sawatch Range, the Grizzly Peak caldera of middle Tertiary age (Cruson, 1972; Obradovich and others, 1969) cuts the Twin Lakes stock of probable Laramide age. To the northeast—across the Sawatch Range, the upper Arkansas River valley, and the Gore and Mosquito Ranges—most of the hypabyssal intrusions are Laramide in age, but middle Tertiary intrusive bodies have been documented in places, and these may be more numerous than are presently known. The postulated batholith underlying this segment of the gravity low in central Colorado is very likely a composite of numerous plutons of both Laramide and middle Tertiary age. To the northeast, the deeper parts of the gravity low in central Colorado embrace the Montezuma batholith and seem largely to correspond to this major middle Tertiary intrusion. A shallower gravity trough extends northeast from the Montezuma batholith along the trend of the mineral belt to the eastern Rocky Mountain front. Numerous shallow intrusions are exposed along this trend. Most of those dated are of Laramide age, but some may have been emplaced in middle Tertiary time.

The parts of the composite middle Tertiary volcanic fields in the Rabbit Ears Range and Never Summer Mountains, in the Thirtynine Mile volcanic field-Cripple Creek area, in the Wet Mountains-Spanish Peaks area, and in the West Elk Mountains lack a clear-cut gravity expression. Like the andesitic rocks in the San Juan Mountains, the rocks of intermediate composition in these areas were probably derived from deep and relatively undifferentiated bodies of magma that did not contrast markedly in density with the adjacent wall rocks.

RELATION TO EARLIER IGNEOUS ROCKS

Most Laramide and middle Tertiary igneous rocks of equivalent silica content in the Southern Rocky Mountains are virtually identical in general composition and lithology. Both are predominantly intermediate-composition calc-alkalic rocks or somewhat more silicic differentiates. Some middle Tertiary rocks along the eastern fringe of the volcanic field are distinctly more alkalic, in common with many igneous areas along the cordillera of western North and South America. Although the distributions of the igneous rocks of the two ages differ greatly, both are strongly limited in one way or another by the Colorado mineral belt lineament. The batholith that probably underlies the mineral belt in central Colorado seems likely to be a composite of plutons of both ages. It thus seems likely that the two periods of igneous activity were related magmatic pulses generated by recurrent geologic processes.

Lipman and others (1972, p. 234–240) have proposed a genetic scheme that fits the broad requirements concerning composition, distribution, and tectonic setting of the rocks under discussion. In general, they proposed that an imbricate subduction system underlay western North America in Laramide and middle Tertiary time, with one zone dipping eastward under the Great Basin area of Nevada and western Utah and a parallel one descending under the eastern Colorado Plateau and Southern Rocky Mountains. These zones apparently were joined along a zone of horizontal shear within the low-velocity zone of the mantle. Using the method devised by Dickinson and Hatherton (1967) and Hatherton and Dickinson (1969), which related the K_2O/SiO_2 ratios of andesitic rocks to the depth of the underlying Benioff zone, they determined that these imbricate subduction zones dipped 15° to 20° E., a dip similar to those of modern continental-margin Benioff zones, although considerably more gentle than most intraoceanic Benioff zones.

Whereas this scheme provides a general mechanism for generating magmas of appropriate compositions in the proper gross tectonic settings, many detailed questions remain to be answered. Especially, why should the mineral belt lineament, which trends diagonally northeastward across the more north-trending Laramide uplifts and basins, have had such a profound influence in localizing igneous activity during both Laramide and middle Tertiary time? Except for deflecting the eastern margin of the Front Range somewhat, this lineament had little apparent effect on major Laramide structural blocks, and detailed accounts of the geology of many parts of the belt give no hint of localized stress fields or regional dislocations that might have influenced the recurrent generation and upward migration of magma. The lineament broadly follows the trend of a Precambrian shear zone (Tweto and Sims, 1963), but this shear zone is only one of several in the Southern Rocky Mountains and seems subordinate to the major Mullens Creek-Nash Fork shear zone in southern Wyoming that separates Precambrian terranes of greatly differing ages (Houston and others, 1968, p. 140-141). An adequate answer to this question would contribute greatly to an understanding of both the igneous rocks and the related ore deposits in the Southern Rocky Mountains.

ACKNOWLEDGMENTS

This review cites new data on areas under study by numerous colleagues who shared their information and ideas with me. I thank R. C. Epis, G. A. Izett, P. W. Lipman, G. R. Scott, R. B. Taylor, and Ogden Tweto for many stimulating discussions over the years while the ideas expressed in this paper were evolving.

REFERENCES CITED

Armstrong, R. L., 1969, K-Ar dating of laccolithic centers of Colorado Plateau and vicinity: Geol. Soc. America Bull., v. 80, p. 2081-2086.

Behrent, J. C., and Bajwa, L. Y., 1972, Bouguer gravity map of Colorado: U.S. Geol. Survey Open-File Rept.

Brinkworth, G. L., 1970, Complete Bouguer anomaly gravity map and generalized geology of an area west of Denver, Colorado: U.S. Geol. Survey Open-File Map.

Brock, M. R., and Barker, Fred, 1965, Intrusive welded tuff in the Sawatch Range, Colorado: Geol. Soc. America, Abs. for 1964, Spec. Paper 82, p. 320-321.

Burbank, W. S., and Goddard, E. N., 1937, Thrusting in Huerfano Park, Colorado, and related problems of orogeny in the Sangre de Cristo Mountains: Geol. Soc. America Bull., v. 48, p. 931-976.

Burroughs, R. L., 1971, Geology of the San Luis Hills, south-central Colorado, in James, H. L., ed., Guidebook of the San Luis Basin, Colorado: New Mexico Geol. Soc., Ann. Field Conf., no. 22, p. 277-287.

Case, J. E., 1965, Gravitational evidence for a batholithic mass of low density along a segment of the Colorado mineral belt: Geol. Soc. America, Abs. for 1964, Spec. Paper 82, p. 26.

Chapin, C. E., Epis, R. C., and Lowell, G. R., 1970, Late Eocene paleovalleys and Oligocene volcanic rocks along the upper Arkansas Valley segment of the Rio Grande rift zone in Colorado [abs.]: New Mexico Geol. Soc., Program, 24th ann. mtg., p. 8.

Christiansen, R. L., and Lipman, P. W., 1972, Cenozoic volcanism and plate tectonic evolution of the western United States; Pt. II, Late Cenozoic, in A discussion on volcanism and the structure of the Earth: Royal Soc. London Philos. Trans., ser. A, v. 271, p. 249-284.

Corbett, M. K., 1968, Tertiary volcanism of the Specimen-Lulu-Iron Mountain area, north-central Colorado, in Epis, R. C., ed., Cenozoic volcanism in the Southern Rocky Mountains: Colorado School Mines Quart., v. 63, no. 3, p. 1-37.

Cruson, M. G., 1972, Grizzly Peak cauldron complex, Sawatch Range, Colorado: Geol. Soc. America, Abs. with Programs (Cordilleran Sec.), v. 4, no. 3, p. 142.

Damon, P. E., 1968, Application of the potassium-argon method to the dating of igneous and metamorphic rock within the Basin Ranges of the southwest, in Titley, S. R., ed., Southern Arizona guidebook: Tucson, Arizona Geol. Soc., p. 7-20.

Dickinson, R. G., Leopold, E. B., and Marvin, R. F., 1968, Late Cretaceous uplift and volcanism on the north flank of the San Juan Mountains, Colorado, in Epis, R. C., ed., Cenozoic volcanism in the Southern Rocky Mountains: Colorado School Mines Quart., v. 63, no. 3, p. 125-148.

Dickinson, W. R., and Hatherton, T., 1967, Andesitic volcanism and seismicity around the Pacific: Science, v. 157, p. 801-803.

Dings, M. G., and Robinson, C. S., 1957, Geology and ore deposits of the Garfield quadrangle, Colorado: U.S. Geol. Survey Prof. Paper 289, 110 p.

Epis, R. C., and Chapin, C. E., 1968, Geologic history of the Thirtynine Mile volcanic field, central Colorado, in Epis, R. C., ed., Cenozoic volcanism in the Southern Rocky Mountains: Colorado School Mines Quart., v. 63, no. 3, p. 51-86.

——1975, Geomorphic and tectonic implications of the post-Laramide, late Eocene erosion surface in the Southern Rocky Mountains, in Cenozoic history of the Southern Rocky Mountains: Geol. Soc. America Mem. 144, p. 45-74.

Gaca, J. R., and Karig, D. E., 1965, Gravity survey in the San Luis Valley area, Colorado: U.S. Geol. Survey Open-File Rept.

Gaskill, D. L., Bartleson, B. L., and Larson, E. E., 1973, West Elk volcanic center, Gunnison County, Colorado—A preliminary report: Geol. Soc. America, Abs. with Programs (Rocky Mtn. Sec.), v. 5, no. 6, p. 481.

Godwin, L. H., and Gaskill, D. L., 1964, Post-Paleocene West Elk laccolithic cluster, west-central Colorado: U.S. Geol. Survey Prof. Paper 501-C, p. C66-C68.

Hail, W. J., Jr., 1968, Geology of southwestern North Park and vicinity, Colorado: U.S. Geol. Survey Bull. 1257, 119 p.

Hansen, W. R., 1965, The Black Canyon of the Gunnison, today and yesterday: U.S. Geol. Survey Bull. 1191, 76 p.

Harland, W. B., Smith, A. G., and Wilcock, B., eds., 1964, The Phanerozoic time-scale—A symposium: Geol. Soc. London Quart. Jour., v. 120s, 458 p.

Hatherton, T., and Dickinson, W. R., 1969, The relationship between andesitic volcanism and seismicity in Indonesia, the Lesser Antilles, and other island arcs: Jour. Geophys. Research, v. 74, p. 5301-5310.

Houston, R. S., and others, 1968, A regional study of rocks of Precambrian age in that part of the Medicine Bow Mountains lying in southeastern Wyoming—With a chapter on the relationship between Precambrian and Laramide structure: Wyoming Geol. Survey Mem. 1, 167 p.

Hunt, C. B., 1956, Cenozoic geology of the Colorado Plateau: U.S. Geol. Survey Prof. Paper 279, 99 p.

——1958, Structural and igneous geology of the La Sal Mountains, Utah: U.S. Geol. Survey Prof. Paper 294I, p. 305-364.

Hunt, C. B., Averitt, Paul, and Miller, R. L., 1953, Geology and geography of the Henry Mountains region, Utah: U.S. Geol. Survey Prof. Paper 228, 234 p.

Izett, G. A., 1966, Tertiary extrusive volcanic rocks in Middle Park, Grand County, Colorado, in Geological Survey Research 1966: U.S. Geol. Survey Prof. Paper 550-B, p. B42-B46.

——1968, Geology of the Hot Sulphur Springs quadrangle, Grand County, Colorado: U.S. Geol. Survey Prof. Paper 586, 79 p.

——1975, Late Cenozoic sedimentation and deformation in northern Colorado and adjoining areas, in Cenozoic history of the Southern Rocky Mountains: Geol. Soc. America Mem. 144, p. 179-210.

Izett, G. A., Scott, G. R., and Obradovich, J. D., 1969, Oligocene rhyolite in the Denver basin, Colorado, in Geological Survey Research 1969: U.S. Geol. Survey Prof. Paper 650-B, p. B12-B14.

Johnson, J. H., 1937, The Tertiary deposits of South Park, Colorado, with a description of the Oligocene algal limestones [abs.]: Colorado Univ. Studies, v. 25, no. 1, p. 77.

Johnson, R. B., 1969, Geologic map of the Trinidad quadrangle, south-central Colorado: U.S. Geol. Survey Misc. Geol. Inv. Map I-558.

Kinney, D. M., 1970a, Preliminary geologic map of the Gould quadrangle, North Park, Jackson County, Colorado: U.S. Geol. Survey Open-File Map.

———1970b, Preliminary geologic map of the Rand quadrangle, North and Middle Parks, Jackson and Grand Counties, Colorado: U.S. Geol. Survey Open-File Map.

Kinney, D. M., and Hail, W. J., Jr., 1970, Preliminary geologic map of the Hyannis Peak quadrangle, North and Middle Parks, Jackson and Grand Counties, Colorado: U.S. Geol. Survey Open-File Map.

Kinney, D. M., Izett, G. A., King, R. U., and Taylor, R. B., 1968, The Poison Ridge volcanic center and related mineralization, Grand and Jackson Counties, Colorado: U.S. Geol. Survey Circ. 594, 8 p.

Koschmann, A. H., 1949, Structural control of the gold deposits of the Cripple Creek district, Teller County, Colorado: U.S. Geol. Survey Bull. 955-B, p. 19-60.

Larson, E. E., Ozima, Minoru, and Bradley, W. C., 1975, Late Cenozoic basic volcanism in northwestern Colorado and its implications concerning tectonism and the origin of the Colorado River system, in Cenozoic history of the Southern Rocky Mountains: Geol. Soc. America Mem. 144, p. 155-178.

Lipman, P. W., and Mehnert, H. H., 1975, Late Cenozoic basaltic volcanism and development of the Rio Grande depression in the Southern Rocky Mountains, in Cenozoic history of the Southern Rocky Mountains: Geol. Soc. America Mem. 144, p. 119-154.

Lipman, P. W., Mutschler, F. E., Bryant, Bruce, and Steven, T. A., 1969, Similarity of Cenozoic igneous activity in the San Juan and Elk Mountains, Colorado, and its regional significance, in Geological Survey Research 1969: U.S. Geol. Survey Prof. Paper 650-D, p. D33-D42.

Lipman, P. W., Steven, T. A., and Mehnert, H. H., 1970, Volcanic history of the San Juan Mountains, Colorado, as indicated by potassium-argon dating: Geol. Soc. America Bull., v. 81, p. 2329-2352.

Lipman, P. W., Prostka, H. J., and Christiansen, R. L., 1972, Cenozoic volcanism and plate tectonic evolution of the western United States; Pt. I; Early and middle Cenozoic, in A discussion on volcanism and the structure of the Earth: Royal Soc. London Philos. Trans., ser. A, v. 271, p. 217-248.

Loughlin, G. F., and Koschmann, A. H., 1935, Geology and ore deposits of the Cripple Creek district, Colorado: Colorado Sci. Soc. Proc., v. 13, no. 6, p. 217-435.

Lovering, T. S., 1935, Geology and ore deposits of the Montezuma quadrangle, Colorado: U.S. Geol. Survey Prof. Paper 178, 119 p.

Lovering, T. S., and Goddard, E. N., 1938, Laramide igneous sequence and differentiation in the Front Range, Colorado: Geol. Soc. America Bull., v. 49, p. 35-68.

———1950, Geology and ore deposits of the Front Range, Colorado: U.S. Geol. Survey Prof. Paper 223, 319 p.

Lowell, G. R., 1971, Cenozoic geology of the Arkansas Hills region of the southern Mosquito Range, central Colorado, in James, H. L., ed., Guidebook of the San Luis Basin, Colorado: New Mexico Geol. Soc., Ann. Field Conf., no. 22, p. 209-217.

Luedke, R. G., and Burbank, W. S., 1968, Volcanism and cauldron development in the western San Juan Mountains, Colorado, in Epis, R. C., ed., Cenozoic volcanism in the Southern Rocky Mountains: Colorado School Mines Quart., v. 63, no. 3, p. 175-208.

MacGinitie, H. D., 1953, Fossil plants of the Florissant beds, Colorado: Carnegie Inst. Washington Pub. 599, 198 p.

Mallory, W. W., 1960, Outline of Pennsylvanian stratigraphy of Colorado, in Weimer, R. J., and Haun, J. D., eds., Guide to the geology of Colorado: Rocky Mtn. Assoc. Geologists, p. 23-33.

McDowell, F. W., 1966, K-Ar dating of Cordilleran intrusives [Ph.D. dissert.]: New York, Columbia Univ., 242 p.

Naeser, C. W., Izett, G. A., and White, W. H., 1973, Zircon fission-track ages from some

Tertiary igneous rocks in northwestern Colorado: Geol. Soc. America, Abs. with Programs (Rocky Mtn. Sec.), v. 5, no. 6, p. 498.

Niesen, P. L., 1969, Stratigraphic relationships of the Florissant Lake Beds to the Thirtynine Mile volcanic field of central Colorado [M.S. thesis]: Socorro, New Mexico Inst. Mining and Technology, 65 p.

Obradovich, J. D., Mutschler, F. E., and Bryant, Bruce, 1969, Potassium-argon ages bearing on the igneous and tectonic history of the Elk Mountains and vicinity, Colorado: A preliminary report: Geol. Soc. America Bull., v. 80, p. 1749-1756.

Olson, J. C., Hedlund, D. C., and Hansen, W. R., 1968, Tertiary volcanic stratigraphy in the Powderhorn-Black Canyon region, Gunnison and Montrose Counties, Colorado: U.S. Geol. Survey Bull. 1251-C, 29 p.

Ozima, M., Kuno, M., Kaneoka, I., Kinoshita, H., Kobuyashi, Kazuo, and Nagata, T., 1967, Paleomagnetism and potassium-argon ages of some volcanic rocks from the Rio Grande gorge, New Mexico: Jour. Geophys. Research, v. 72, p. 2615-2622.

Pillmore, C. L., Obradovich, J. D., Landreth, J. O., and Pugh, L. E., 1973, Mid-Tertiary volcanism in the Sangre de Cristo Mountains of northern New Mexico: Geol. Soc. America, Abs. with Programs (Rocky Mtn. Sec.), v. 5, no. 6, p. 502.

Plouff, Donald, and Pakiser, L. C., 1972, Gravity study of the San Juan Mountains, Colorado: U.S. Geol. Survey Prof. Paper 800-B, p. B183-B190.

Scott, G. R., and Taylor, R. B., 1974, Post-Paleocene Tertiary rocks and Quaternary volcanic ash of the Wet Mountain Valley, Colorado: U.S. Geol. Survey Prof. Paper 868, 15 p.

Siems, P. L., 1968, Volcanic geology of the Rosita Hills and Silver Cliff districts, Custer County, Colorado, in Epis, R. C., ed., Cenozoic volcanism in the Southern Rocky Mountains: Colorado School Mines Quart., v. 63, no. 3, p. 89-124.

Smith, R. P., 1973, Age and emplacement structures of Spanish Peak dikes, south-central Colorado: Geol. Soc. America, Abs. with Programs (Rocky Mtn. Sec.), v. 5, no. 6, p. 513-514.

Steven, T. A., and Epis, R. C., 1968, Oligocene volcanism in south-central Colorado, in Epis, R. C., ed., Cenozoic volcanism in the Southern Rocky Mountains: Colorado School Mines Quart., v. 63, no. 3, p. 241-258.

Steven, T. A., Smedes, H. W., Prostka, H. J., Lipman, P. W., and Christiansen, R. L., 1972, Upper Cretaceous and Cenozoic igneous rocks, in Geologic atlas of the Rocky Mountain region: Rocky Mtn. Assoc. Geologists, p. 229-232.

Steven, T. A., Lipman, P. W., Hail, W. J., Jr., Barker, Fred, and Luedke, R. G., 1974, Geologic map of the Durango quadrangle, southwestern Colorado: U.S. Geol. Survey Misc. Geol. Inv. Map I-764.

Stormer, J. D., Jr., 1972, Ages and nature of volcanic activity on the southern High Plains, New Mexico and Colorado: Geol. Soc. America Bull., v. 83, p. 2443-2448.

Taylor, R. B., 1975, Neogene tectonism in south-central Colorado, in Cenozoic history of the Southern Rocky Mountains: Geol. Soc. America Mem. 144, p. 211-226.

Taylor, R. B., Theobald, P. K., and Izett, G. A., 1968, Mid-Tertiary volcanism in the central Front Range, Colorado, in Epis, R. C., ed., Cenzoic volcanism in the Southern Rocky Mountains: Colorado School Mines Quart., v. 63, no. 3, p. 39-50.

Tobey, E. F., 1969, Geologic and petrologic relationships between the Thirtynine Mile volcanic field and the Cripple Creek volcanic center [M.S. thesis]: Socorro, New Mexico Inst. Mining and Technology, 61 p.

Toulmin, Priestley, III, 1953, Petrography and petrology of Rito Alto stock, Custer and Saguache Counties, Colorado [M.S. thesis]: Boulder, Colorado Univ., 49 p.

Tweto, Ogden, 1957, Geologic sketch of southern Middle Park, Colorado, in Finch, W. C., ed., Guidebook to the geology of North and Middle Parks basin, Colorado: Rocky Mtn. Assoc. Geologists, p. 18-31.

——1968a, Leadville district, Colorado, in Ridge, J. D., ed., Ore deposits of the United States, 1933-1967 (Graton-Sales volume): Am. Inst. Mining Metallurgy Petroleum Engineers, v. 1, p. 681-705.

——1968b, Geologic setting and interrelationships of mineral deposits in the mountain province of Colorado and south-central Wyoming, *in* Ridge, J. D., ed., Ore deposits of the United States, 1933-1967 (Graton-Sales volume): Am. Inst. Mining Metallurgy Petroleum Engineers, v. 1, p. 551-588.

Tweto, Ogden, 1975, Laramide (Late Cretaceous-early Tertiary) orogeny in the Southern Rocky Mountains, *in* Cenozoic history of the Southern Rocky Mountains: Geol. Soc. America Mem. 144, p. 1-44.

Tweto, Ogden, and Case, J. E., 1972, Gravity and magnetic features as related to geology in the Leadville 30-minute quadrangle, Colorado: U.S. Geol. Survey Prof. Paper 726-C, 31 p.

Tweto, Ogden, and Sims, P. K., 1963, Precambrian ancestry of the Colorado mineral belt: Geol. Soc. America Bull., v. 74, p. 991-1014.

U.S. Geological Survey, 1963, Geological Survey research 1963: U.S. Geol. Survey Prof. Paper 475-A, 300 p.

——1968, Geological Survey research 1968: U.S. Geol. Survey Prof. Paper 600-A, 371 p.

——1970, Geological Survey Research 1970: U.S. Geol. Survey Prof. Paper 700-A, 426 p.

Van Alstine, R. E., 1969, Geology and mineral deposits of the Poncha Spring NE quadrangle, Chaffee County, Colorado: U.S. Geol. Survey Prof. Paper 626, 52 p.

Wahlstrom, E. E., 1944, Structural petrology of Specimen Mountain, Colorado: Geol. Soc. America Bull., v. 55, p. 77-90.

Wahlstrom, E. E., and Hornback, V. Q., 1962, Geology of the H. D. Roberts Tunnel, Colorado: West portal to station 468 + 49: Geol. Soc. America Bull., v. 73, p. 1477-1498.

Wallace, S. R., Muncaster, N. K., Johnson, D. C., Mackenzie, W. B., Bokstrom, A. A., and Surface, V. E., 1968, Multiple intrusion and mineralization at Climax, Colorado, *in* Ridge, J. D., ed., Ore deposits of the United States, 1933-1967 (Graton-Sales volume): Am. Inst. Mining Metallurgy Petroleum Engineers, v. 1, p. 603-640.

Warner, L. A., and Robinson, C. S., 1967, Geology of the H. D. Roberts Tunnel, Colorado: Station 468 + 49 to east portal: Geol. Soc. America Bull., v. 78, p. 87-119.

Witkind, I. J., 1964, Geology of the Abajo Mountains area, San Juan County, Utah: U.S. Geol. Survey Prof. Paper 453, 110 p.

MANUSCRIPT RECEIVED BY THE SOCIETY MAY 8, 1974

Geological Society of America
Memoir 144
© 1975

Controls of Sedimentation and Provenance of Sediments in the Oligocene of the Central Rocky Mountains

JOHN CLARK

Field Museum of Natural History
Roosevelt Road and Lake Shore Drive
Chicago, Illinois 60605

ABSTRACT

Oligocene rocks on the eastern side of the central Rocky Mountains are part of a broad expanse of such deposits extending from central Colorado to Saskatchewan. Resting unconformably on older rocks, they record renewal of sedimentation following an erosional episode. Only in western South Dakota and Nebraska and southeastern Wyoming do the succeeding Miocene beds lie conformably on the Oligocene; however, with few exceptions, the Oligocene strata have not been disturbed from their original attitude. Heavy mineral suites of the early Oligocene beds indicate derivation from local sedimentary rocks, from volcanic sources, and from the Black Hills, the Laramie Range, the Front Range, and perhaps other mountain uplifts. Late Oligocene strata in mapped channels were derived from about the same sources, whereas finer sediments of this age are mostly of pyroclastic origin.

Consideration of these beds and others previously mapped in central and western Wyoming permits reconstruction of the major elements of a probable Oligocene drainage system. Most of the streams flowed generally eastward, but local exceptions are indicated, and the drainage pattern differed in details from the present-day network.

The development of an aggrading fluvial regime, extending from central Colorado to Saskatchewan in Oligocene time, can scarcely have resulted from simple transgressive overlap, from basin downwarping near source mountains, from downwarp along a regional hinge, nor solely from overloading of the streams by volcanic ash. Instead, the evidence points to progressively drying climate with increased erosion as the major factor causing widespread fluvial deposition.

INTRODUCTION

The Oligocene Epoch was a time of change. Eocene climates differed little from Cretaceous ones; progressive cooling and drying during Oligocene time marked the beginning of the climatic deterioration that culminated in Pleistocene time. The principal locus of continental sedimentation shifted from a restricted group of intermontane basins to a broad area, reaching from Saskatchewan southeastward to central eastern Colorado, east of the Continental Divide. The final vestiges of Laramide tectonism ended, apparently, before Oligocene time; the central and northern Rocky Mountains area entered an episode of quiescence, which endured into Miocene time. Even the color of sediments changed, from predominantly rich browns, greens, purples, and reds of upper Eocene rocks to pale tans, greenish grays, and shades of light cream to white.

Biologically, the changes were equally striking. Eocene faunas would have appeared strange to an observer accustomed to modern animals; by latest Oligocene time, horses, rhinoceroses, camels, canids, and felids had become recognizable, and most of the archaic types of animals had disappeared.

In terms of human history, the Oligocene Epoch might best be compared with the Renaissance—a time when archaic patterns disappeared and modern systems first clearly established themselves.

This paper constitutes a progress report rather than a completed study. It discusses only provenances of sediment and controls of sedimentation, because those factors are most pertinent to this symposium. A more complete study of all aspects of Oligocene paleogeography is in progress. It is hoped, therefore, that the reader will regard the data as demonstrable but incomplete and the interpretations as reasonable but tentative.

Only superficial personal studies have been made on the Norwood Formation of Utah and the White River Formation of Bates Hole, Wyoming. I have not observed the Wiggins Formation of Wyoming or the Castle Rock Formation of Colorado. Statements and interpretations regarding these four are, therefore, derived from the literature.

Numbers, such as G 5071, refer to specimens in the Field Museum collection of sedimentary specimens.

DEFINITION OF OLIGOCENE TIME AND STRATA

The white ash bed that occurs near the top of Sheep Mountain (T. 43 N., R. 44 W., Shannon County) and elsewhere in the Big Badlands of South Dakota has long been used as a convenient boundary marker between Oligocene and accordant overlying Miocene strata (Matthew, 1907). Although the mammalian faunas above and below it do not differ sharply, adequate determination is possible.

Unfortunately, the base of the Oligocene is not so easily defined. Widely scattered strata of small areal extent in Utah, California, Wyoming, Texas, and South Dakota yield confusing mixtures of Eocene and Oligocene faunal elements. These formations have been generally assigned to the Duchesnean time interval (Duchesne River equivalents), which is variously regarded as Eocene or Oligocene in age, or as overlapping both epochs.

Of the Duchesnean strata, the Slim Buttes Formation and its probable correlatives in South Dakota, and the possibly Duchesnean Big Sand Draw Lentil of central Wyoming, are related in their depositional history to the general Oligocene regimen. The La Point Member of the Duchesne River Formation of Utah is clearly a closing phase of deposition that had continued through much of Eocene time.

Nevertheless, its fauna plus a recent K-Ar age determination suggest that it may be Oligocene. These rock units, therefore, are discussed in this paper, with the clear understanding that their positions in the general time scale are not yet certain.

Pertinent Oligocene K-Ar age determinations are listed in Figure 1. Four notable facts emerge from it:

1. The majority of rocks suitable for dating are of early Oligocene age. This is partly due to the restricted extent of middle and upper Oligocene rocks (compare Figs. 2 and 3) and partly to the relative preservation of datable mineral grains within them.

2. The Oligocene Epoch lasted for approximately 9 to 15 m.y., depending on boundary interpretation.

3. Early Oligocene, embracing Duchesnean and Chadronian stages, comprises more than half of total Oligocene time.

4. Berggren's placement of the Oligocene-Miocene boundary would include within the Oligocene all American continental formations presently regarded as lower Miocene. It is not followed in this paper.

GENERAL STRUCTURAL RELATIONS

Oligocene sediments rest unconformably on older rocks throughout the area. The unconformity ranges in time value from simple erosional disconformity with the underlying Uintan Wagon Bed Formation along the Beaver Rim of Wyoming, to overlap of the Precambrian rocks of the Black Hills and the Laramie Range. Everywhere, the basal Oligocene sediments represent incipience or renewal of sedimentation in sites of previously erosional regimens.

The age of the basal Oligocene sediments is not everywhere the same, due in part to transgressive onlap of the mountains. Apparently, Oligocene sedimentation

FORMATION AND LOCATION	DUCHESNE RIVER UTAH	NORWOOD TUFF UTAH	CHADRONIAN RATTLESNAKE HILLS WYOMING	VIEJA GROUP TEXAS	CLARNO OREGON	CHADRON S.DAKOTA	EUROPA	STRATIGRAPHIC BOUNDARIES
22							22.5 MOULIN DE BERNACRON FRANCE	M̅ / O 1
25								M / O 3,4
28							28.8 - 31.2 CHATTIAN, GERMANY	
31			31.6 - ASH J 32.6 ASH G					
			33.7 ASH F (BIOTITE)	33.9 ±1.1 BRITE				
34			33.3 ASH B (BIOTITE) 35.2 SANIDINE	32.4-34.2 PANTERA				
37				36.5±1.2 BRACKS	36.5 ± 0.9	36.3 ± 0.7		O / E 1
		37.5 (BIOTITE)			37.5		37.5 ± 3 NEERREPEN, BELGIUM	
				38.6 ±1.2 BUCKSHOT				
40	39.3 ± 0.8			40.0±2.0 GILL BRECCIA			O	O / E 2 E 4
SOURCE	Mc DOWELL, WILSON, CLARK, IN PRESS	EVERNDEN ET AL 1964[3]	EVERNDEN ET AL 1964	WILSON ET AL 1968[2]	EVERNDEN ET AL 1964	Mc DOWELL, WILSON, CLARK, IN PRESS	BERGGREN 1972[1]	HOLMES 1959[4]

Figure 1. Potassium-argon dates of Oligocene rocks from various localities. Numbers in column on stratigraphic boundaries are key to authorities given as sources.

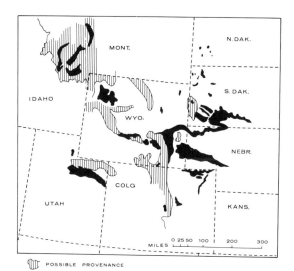

Figure 2. Outcrop areas of Oligocene sediments. Wiggins Formation of northwestern Wyoming is mostly Eocene; it is mapped here as Oligocene, because channel fills of Oligocene lie within it (data not yet published). Map adapted and modified from Geologic Atlas of the Rocky Mountain region (Robinson, 1972).

started almost contemporaneously in the drainages of South Dakota, but the earliest (Duchesnean) sediments were limited both in thickness and in extent. Many of the preserved sediments were partially eroded before the succeeding Chadronian sedimentary episode occurred. The combination of nondeposition, partial intraformational erosion, later erosion, and progressive onlap creates a false impression of differing beginnings in the various Oligocene drainages.

The only exception to the above description is the Duchesne River Formation of Utah (Peterson, 1932; Kay, 1934; Andersen and Picard, 1972), which rests with complete conformity on the underlying Uinta Formation. However, it also onlaps with angular unconformity on older formations.

In the central Great Plains (Big Badlands, western Nebraska, southeastern Wyoming, but not northeastern Colorado), the Oligocene is conformably overlain

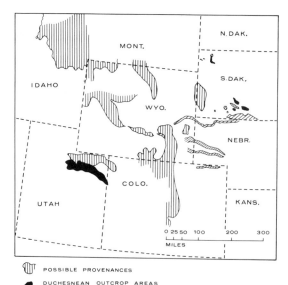

Figure 3. Outcrop areas of earliest (Duchesnean) and latest (Whitneyan) Oligocene sediments, for comparison with Figure 1. Note that widest expanse of Oligocene sediment is lower (Chadronian) to middle (Orellan) Oligocene in age.

by lowermost Miocene sediments. A hiatus separates the locally uppermost Oligocene beds from whatever Miocene or later sediments may overlie them in all other places. The hiatus usually includes at least middle and late Oligocene time, plus earliest Miocene.

In the Slim Buttes area (Lillegraven, 1970), a series of Oligocene fault blocks are overlain by flat-lying Miocene strata with angular unconformity. Middle and upper Oligocene sediments are included in at least two of the blocks, which demonstrates that tectonic movement and later erosion occurred during Miocene time. This is the only locality that gives evidence of movement between earliest Oligocene and latest Miocene time.

Rejuvenation of basement faulting in post-early–Miocene time has produced a series of drape structures in the Oligocene of the Big Badlands and northern Nebraska (Clark and others, 1967). The depositional dips of this area have been subtly but definitely warped, so that probably no place is undisturbed, without notably disrupting the primary southeastward depositional slope. Post-Oligocene movement in southeastern Wyoming, west of the Lance Creek oil field, has tilted the Chadron to dips of 8° to 12° SE., which rest on Cretaceous strata that dip 10° NW. (see Fig. 4). Active post-Oligocene faulting has also affected the Oligocene of the Granite Mountains (Love, 1960).

With these exceptions, however, Oligocene strata lie essentially undisturbed. The major Holocene streams run at gradients not too different in slope or direction from those of their Oligocene progenitors, and probably the present elevations roughly equal the Oligocene ones (Sharp and others, 1964).

PROVENANCES OF SEDIMENTS IN VARIOUS LOCALITIES
(FIGS. 5 AND 6)

Big Badlands of South Dakota

The Oligocene section in the Big Badlands is composed of the following formations:

Miocene	Rosebud Formation	
Oligocene	Brule Formation 120 m	⎡ Poleslide Member; 80 m ⎣ Scenic Member; 24 to 40 m
	Chadron Formation 40 to 43 m	⎡ Peanut Peak Member; 6 to 9 m ⎢ Crazy Johnson Member; 6 to 12 m ⎣ Ahearn Member; 0 to 24 m
	Slim Buttes Formation 0 to 6 m	

These rest on an erosional surface of moderate relief, weathered 18 to 21 m deep, that developed upon the Cretaceous Pierre and Niobrara Formations (Clark and others, 1967).

Locally derived clays plus clastics from the Black Hills can be recognized in the scattered lenses that make up the Slim Buttes Formation (see nos. s1 and s2, Fig. 5; also Fig. 6). The purely siliceous basal gravels may be remanié gravels weathered from sediments of a yet older depositional episode. In the southwestern Badlands, the gravels include rose quartz from the Precambrian and Fairburn-type

Figure 4. Angular unconformity of Chadron Formation dipping 13° southeast at right of picture, on Cretaceous dipping 10° west at left. Twenty miles west of Lance Creek, Wyoming.

agates (no. 3, Fig. 5) from the Mississippian Pahasapa Formation. Farther to the northeast and east, rose quartz and Fairburn agates are absent.

The heavy minerals consist of magnetite and ilmenite; shards of pink garnet; small, colorless, pink and yellow zircons; and smoky to olive schorlite; actinolite, red rutile, indicolite, staurolite, and yellow-brown allanite occur uncommonly to rarely. The provenance of this assemblage is unknown; it is believed to be a mixture of materials, both local and from the Black Hills.

The Chadron Formation offers definite evidence of derivation from three sources plus a possible fourth: (1) local, (2) volcanic, (3) southern Black Hills, and (4) northern Black Hills.

Abundant clay, usually red or brown, with a higher percentage of kaolinite and

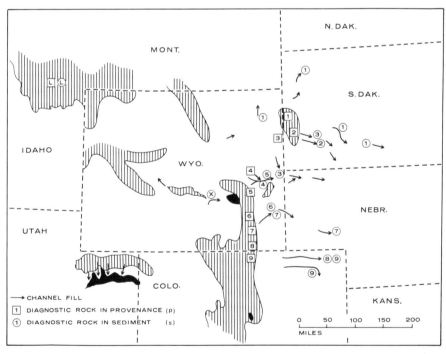

Figure 5. Provenances of Oligocene sediments. Numbers refer to rock and mineral suites described in text. Arrows indicate directional flow as recorded by sedimentary structures in channel fills. Provenance references in text are indicated by numbers preceded by ''p''; sediment references are indicated by numbers preceded by ''s.'' Dark areas, lacustrine.

illite than is generally typical of Oligocene mudstones, was apparently derived by slopewash from local hills of Cretaceous shale. The colors and composition resemble those of the "Interior" weathered zone developed upon the Cretaceous beneath the Oligocene beds (see Wanless, 1923).

The volcanic materials are represented first by very high percentages of montmorillonite in the mudstones. Nowhere in the Big Badlands, save in one lacustrine siltstone near Oelrichs, South Dakota (G 4485), have I found any unaltered glass shards. Two lenses of pure, bentonitized ash replete with unweathered biotite (Clark and others, 1967, p. 22; this paper, Fig. 7) confirm the pyroclastic origin. Andesine and sanidine grains are also clearly of pyroclastic derivation.

The assemblage of heavy minerals characteristic of the southern Black Hills (no. s2, Fig. 5) includes black to olive tourmaline, pink and gray garnet, brown staurolite, glauconite, muscovite, and rounded chips of fossil bone.

Channel fills in the eastern part of the Badlands display a different assemblage. The commonest heavy minerals are magnetite, ilmenite, greenish-yellow epidote, actinolite, hornblende, and pink garnet. Less common are lemon-yellow sphene, staurolite, black tourmaline, colorless zircon, rutile, and muscovite.

This assemblage (no. s1, Fig. 5), particularly the abundant magnetite-ilmenite, hornblende, epidote, and actinolite, associated with lemon-yellow sphene, characterizes all sediments from a northern Black Hills provenance. The sphene and magnetite probably derive from the Laramide intrusives of the northern hills, in some of which they have been found.

Sandstone channel fills of the Brule Formation contain two similar assemblages of clastic heavy minerals. The individual channel fills have been mapped (Clark and others, 1967). It is probable that more detailed work will reveal precisely which Oligocene channel fill is the ancestor of which Holocene stream.

The fine sediments of the Brule Formation consist of fine sand- to coarse silt-sized quartz and feldspar fragments, euhedral biotite crystals, tiny zircons of various proportions, and occasional magnetite and hematite grains set in a matrix of montmorillonite and more-or-less cemented with zeolites and microgranular calcite. The smallest grains of feldspar average about 1 microcline to 20 or 30 andesine; the largest are all albite and microcline. Zircons, all colorless, range from sharply

Figure 6. Basal Oligocene strata north of Weta, South Dakota, SW¹/₄ sec. 2, T. 3 S., R. 20 E., Jackson County. In foreground is "Interior soil zone" developed on Cretaceous Pierre Shale (KP). Dark- and white-banded sandy sites are Slim Buttes Formation (OSB), overlain disconformably by dark, bentonitic Chadron mudstone (OC), which caps the left butte and partially cuts out Slim Buttes on the right butte. Off picture to right, Chadron Formation rests upon the Pierre Shale.

Figure 7. Lens of volcanic ash in the Chadron Formation. Four-foot lens of volcanic ash near top of Ahearn Member, in the SE¼ of sec. 14, T. 4 S., R. 10 E., Custer County, South Dakota. Overlying sandstones form base of Crazy Johnson Member.

euhedral to some that are small cylinders with almost hemispherical termini. This lithofacies must be attributed mostly to a pyroclastic origin, with a small addition of waterborne clastics.

This interpretation differs notably from that of Sato and Denson (1967). The difference probably arises from the fact that their Big Badlands sample was taken in sec. 5, T. 4 S., R. 13 E., which happens to lie between a major Oligocene channelway and an overflow channel. The rocks at that place do contain a preponderance of hydroclastics, but they are not typical of the bulk of sediments in the Big Badlands. The Scenic Member of the Brule Formation also includes several lithotopes composed chiefly of hydroclastics (Fig. 8; Clark and others, 1967); a sample from one of these could cause Sato and Denson to misinterpret the whole mass.

Lance Creek and Hat Creek, Northwestern Nebraska

Consistent with the situation elsewhere in the Great Plains, the basal Chadron sediments include much reddish and brown clay derived from the Interior soil zone on Cretaceous uplands.

The pyroclastic constituents are more easily recognized than in the Big Badlands. The mudstones generally are a little coarser grained, and several contain unaltered glass fragments. Characteristically, the glass shows flow structures and elongated bubbles. Typical shards from between vesicles are rare to absent.

A major set of channel fills trending south-southeast, SE¼ sec. 12, T. 34 N., R. 61 W., Niobrara County, Wyoming (no. s3, Fig. 5), has yielded one Fairburn type (Pahasapa Formation) agate (G 4884), which defines the provenance for this stream as the southwestern Black Hills (p3, Fig. 5). A large channel fill northwest of Lost Spring, Wyoming, is composed almost entirely of chert and limonite chips from Cretaceous rocks that crop out in the vicinity (p4–s4, Fig. 5).

Granite cobbles and pebbles in the Oligocene fill of La Bonte Creek, and those in the channel fills southeast of the town of Douglas, are obviously derived from the Precambrian of the Laramie Range (p5–s5, Fig. 5). The channel fills have clear directional trends, and they outcrop in or near the debouchures of the present streams from their canyons.

No evidence has been reported of any sediments from the Granite Mountains or any others farther west. The drainage pattern in Figure 11, therefore, shows the Black Hills, the northern Laramie Range, and local areas of Cretaceous outcrops as sources of tributaries that joined to produce a single major, eastward-flowing stream through northwestern Nebraska. This stream was first recognized by Schultz and Stout (1955), but the provenances of its sediments were not established.

Goshen Hole, Central Nebraska

Two excellent, easily identifiable provenances in the central Laramie Range provide evidence of a Goshen Hole drainageway. Anorthosite and ilmenite-magnetite exposed along the course of Sybille Creek yielded cobbles and pebbles that have been recovered from Oligocene sediments as far east as Mitchell, Nebraska (p7–s7, Fig. 5; specimens G 4581, G 4701).

A series of hornblende-amphibole-garnet metamorphics immediately north of the anorthosite (Condie, 1969) yielded abundant hornblende, actinolite, garnet, and mica (p6–s6, Fig. 5), which mixed with the mafic waste to produce a highly characteristic mixture (G 4392–4395, G 4882–4883, G 4585, and others). The mixture plus a variety of igneous and metamorphic clasts and colored chert pebbles of unknown origin makes up the intricate series of channel fills of Goshen Hole.

The channel fills are interbedded with flood-plain silt and tuff of mixed pyroclastic and hydroclastic origin. They overlie a thin, discontinuous basal member (the Yoder Member), which consists of reddish and green silty sandstone with weathered granite pebbles. The Yoder Member shows the mixed local and mountain provenance that characterizes basal Oligocene sediments all the way from northwesternmost South Dakota to northeastern Colorado.

Northeastern Colorado

At least one volcanic and three clastic provenances contributed to the 140- to 170-m blanket of Oligocene sediment in Weld and Logan Counties of northeastern Colorado. Erosion has removed the middle Tertiary beds from the immediate proximity of the Rocky Mountain front, producing a broad valley on Cretaceous rocks between the mountains to the west and the low Oligocene scarp to the east. The Oligocene section (Galbreath, 1953) consists of

White River Formation	Vista Member; 0 to 30 m
	Cedar Creek Member; 91 to 110 m
	Horsetail Creek Member; 24 to 57 m

These correlate approximately with the Chadron Formation, Scenic Member, and Poleslide Member of South Dakota.

The basal portion of the Horsetail Creek Member is composed of red and greenish silty sandstones with irregular lenses of basal conglomerate. Like the basal Oligocene section of Nebraska and South Dakota, it consists of muds and sand washed from the underlying Cretaceous rocks, mixed with granitic waste from the Front Range.

This grades upward into massive white and pale-tan tuff. Thin strata and laminae of reddish mudstone interbed with the tuff, fingering-out southward; commonly the tuff is transected by small, discontinuous dikes of red mudstone (G 5130).

The tuff consists of fresh fragments of glass, showing strong flow structures. Heavy minerals (G 5129) in general order of frequency are (1) euhedral biotite; (2) prisms and needles of hornblende; (3) subhedral magnetite; (4) colorless, euhedral topaz; and (5) pale-yellow, anhedral epidote.

The exact source of the ash is unknown, but its geographic position and moderately coarse grain size suggest that it came from one of the volcanic centers in Colorado.

A set of channel fills trends straight eastward about through the middle of T.

Figure 8. Brule Formation in Big Badlands. Alternating mudstones (dark) and sandy siltstones (light) of Scenic Member, grading upward into mudstones of Poleslide Member. Mudstones are predominantly altered volcanics; siltstones are predominantly clastics; sec. 22, T. 2 S., R. 15 E., Pennington County, South Dakota.

11 N. (p8–s8 and p9–s9, Fig. 5; Peavey Butte, Fig. 12). Red Sherman Granite (p8) forms a high percentage of the sand-to-cobble size fractions. Reddish mud, probably from the Triassic beds of the Front Range, makes up the finer material within the channel fills as well as the mudstones associated with them. Other igneous and metamorphic clasts (p9–s9, Fig. 5) apparently came from the plutonic rocks of the Front Range south of the Sherman Granite; they constitute usually less than half of the bulk of any one outcrop. The entire set represents the anastomosing channels of one river.

A second river, approximately 40 km south of the first, followed a parallel eastward course until it turned southeastward a few kilometers north of the town of Sterling (p9–s9, Fig. 5). It received no sediment from the Sherman Granite and no red mud, but otherwise its sediments greatly resemble those of its neighbor to the north.

Careful study of the clastics of these two Oligocene streams would almost certainly delineate their headwaters. Such a study would, however, require more knowledge of Front Range plutonic rocks than I possess.

Slim Buttes, Harding County, Northwestern South Dakota

The Slim Buttes Formation consists of red, purple, and brown mud derived from the underlying Cretaceous Hell Creek Formation, fine- to medium-grained sand, also derived from Hell Creek, brilliant whitish sand of mixed derivation, greenish-white arkosic sand, and conglomerates of chert and quartz pebbles (Fig. 9). Determination of plutonic provenance is complicated by the fact that the underlying Cretaceous sands contain garnets, hornblende, magnetite, tourmaline, epidote, and feldspars.

However, the red- and green-banded silty sandstone of the type locality (G 5084) shows a highly characteristic assemblage: magnetite and ilmenite (glossy, euhedral to subhedral); schorlite tourmaline; dravite tourmaline; greenish-yellow glassy epidote; zircon (colorless, euhedral, columnar; pink, euhedral, columnar; pink, rounded subhedral; gray, rounded subhedral); rutile (very dark brown, glassy, usually fractured); sphene, lemon-yellow; sphene, euhedral, almost colorless; staurolite, yellow to orange; cassiterite, yellow-brown, uncommon; biotite, very rare, euhedral to subhedral; and muscovite, uncommon. Garnet and hornblende are notably absent.

Some of the magnetite, ilmenite, zircon, schorlite, and cassiterite grains are very well rounded with a high sphericity. Other grains of the same species, and all of the epidote and sphene, show fresh fractures of crystal faces and shiny, fresh surfaces. These are grains inherited from an unknown older formation (very possibly underlying Cretaceous or neighboring Paleocene beds), plus an addition that suggests derivation from parts of the northern Black Hills. The presence of cassiterite strongly suggests origin in the western Black Hills, but the absence of hornblende, anthophyllite, actinolite, and tremolite indicates otherwise.

The upper strata of the Slim Buttes Formation, and the sand of the Chadron Formation, contain a northern Black Hills clastic assemblage. Granite pebbles have been found in a conglomerate at the base of the Slim Buttes Formation. In addition, outcrops of presumed Chadron channel fills at the Twin Buttes (T. 4-5 E., R. 12 N.; Butte County, South Dakota) contain abundant granite pebbles (shown on Fig. 5 as arrow between north edge of Black Hills and the northernmost s1, which marks the Slim Buttes).

The total evidence seems to indicate a Black Hills provenance for the clastics of the Slim Buttes area, with a second source in some older, local sandstones.

The pyroclastics that occur abundantly in both the Chadron and Brule Formations,

Figure 9. Slim Buttes Formation (Duchesnean). Type locality, Antelope Creek, Harding County, South Dakota. Alternating variegated silty sands, locally derived, and white sands of mixed local and distant provenance.

but not in the Slim Buttes Formation, have not been studied in detail. Superficially, they resemble pyroclastics of the Big Badlands. The total Oligocene section (Lillegraven, 1970, p. 834–836) is slightly thicker than that of the Big Badlands, but the difference is not as great as are the variations within the two areas.

Missouri Buttes

Small patches of post-Eocene sedimentary beds lie scattered from the northwest flank of the Black Hills westward to the Missouri Buttes. The lowermost, topographically and stratigraphically, consist of greenish to gray sandstones and conglomerates, not more than 9 to 12 m thick. Higher strata are tan to pale-gray conglomerates, siltstones, and mudstones with considerable, variably altered volcanic ash, not more than 46 m thick and usually much less.

Most of these outliers have not yielded fossils. It is, therefore, impossible to determine whether they are Oligocene, Miocene, or Pliocene in age. Deposits of any of these ages might be present (Robinson and others, 1964). The basal sandstones have yielded one titanothere tooth, which dates them as Chadronian.

The extensive outlier west-northwest of the Missouri Buttes (secs. 28 and 33, T. 54 N., R. 66 W.; Crook County, Wyoming) has been studied in detail. The basal sandstone extends as a sheet 1 to 2.5 m thick, approximately 0.4 km in east-west extent and 2.4 km north-south. A thicker zone approximately 46 m wide trends directly north-south; channel ways and minor cross-beds within this zone indicate current motion from south to north (see Fig. 5, s1, west of the Black Hills).

The rock consists of small pebbles, grit, and sand, well cemented by opal and chalcedony. Pebbles above 1-cm size (G 4071) are all of fine-grained sandstone; many in the 1-mm to 1-cm size are black or brown chert. Subrounded white to greenish quartz makes up about one-third of the coarse fraction. In the 1.0- to 2.0-mm-size fraction, the quartz-feldspar ratio is roughly 4:1. The feldspar is fresh, pale-greenish perthitic microcline. A few granite pebbles reach sizes of 1 cm.

The heavy minerals offer important evidence of provenance; commonest are ilmenite-magnetite in ratio 3:1; hornblende; anthophyllite; epidote, glassy yellow-green; garnet, both pink and orange; and actinolite. Less common are staurolite; zircon, pink to plum colored; tremolite; sphene, pale lemon-yellow; and cassiterite, yellow, clear. Rare are zircon, clear; biotite, euhedral, black to red-brown; schorlite tourmaline; monazite; rutile; and apatite.

The association of magnetite–yellow sphene–hornblende–actinolite is characteristic of the northern Black Hills provenance. The addition of cassiterite confirms the provenance. Careful comparison of the optical properties of cassiterite from mines of the Hill City district with the optical properties of these grains reveals no differences.

The Chadron Formation of the Missouri Buttes area, therefore, was deposited by streams that flowed west and then north from the Black Hills, somewhat as the tributaries of Belle Fourche River do today.

Pumpkin Buttes

The Oligocene series of the Pumpkin Buttes (Campbell County, Wyoming) was first recognized by Darton (1905) and first adequately described by Love (1952). More details were added by Sharp and others (1964). My own studies have added little to those of the previous authors.

The outcrops cap five buttes, which lie in a northwest-trending line. Fallen blocks between the buttes leave no doubt that the outcrops were once coextensive, but they did not extend far beyond the buttes in any direction. The usual thickness is 9 to 15 m.

The Pumpkin Buttes Oligocene sediment consists of clean sandstone and conglomerate (G 4104, G 4106, G 4107, and Fig. 10). Abundant cross-beds in sets usually 15 to 76 cm thick dip in either planar or concave fashion to the east-northeast, almost at right angles to the elongation of the deposit. Over-steepened dips are common in the more massive sets, especially in the better-sorted ones. Cut-and-fill structures do not commonly extend more than 30 cm vertically, but one 1.8 m cut face was observed. Minor conglomerates occur at the base of local members, and two major channelways with cobbles as much as 20 cm in diameter trend east-northeast across the upper part of the outcrops. Longitudinal bars, local channelways, internal cross-bedding, and trends of entire units indicate that the flow in these channels was northeastward.

All beds are free of clay, most are well sorted, and all are moderately to very well cemented. These outcrops are unique among Oligocene deposits of the entire area. They appear to represent a subaqueous delta along the southwest shore of a lake where there was sufficient winnowing to remove fines. The channel and distributary mouth bar subenviroments as mapped by Coleman and Gagliano (1965, p. 144) apparently are represented. The problem is that no other traces of a lake

Figure 10. North Pumpkin Butte, Campbell County, Wyoming. Well-sorted, well-cemented, slabby sandstones of a topset deltaic phase.

remain nor is there any known structural reason for an Oligocene lake in the Powder River basin.

Cobbles of limestone, quartzite, chert, granite, and hornblende-to-biotite gneiss could have been derived from the Bighorn Range to the west or from any other range of central Wyoming.

One small placer from the South Butte (G 4104), however, contains mostly ilmenite in subhedral to euhedral grains 0.25 to 1.0 mm in diameter; plus subordinate quantities of brown, fresh garnet as much as 0.5 mm; rounded grains of brown, cloudy monazite and some pale yellow monazite; and clear to pink, glassy, anhedral fragments of zircon as much as 0.35 mm in diameter. I do not know the provenance of this assemblage, but it cannot have travelled far. The ilmenite grains are larger than those reported from various sandstones in central Wyoming (Osterwald and others, 1966) and are probably first-cycle rather than inherited. Anhedral chips of zircon in the sizes found must have come from a recognizable group of pegmatites or a placer in the Cambrian sandstone.

Love (1952, p. 5) described pebbles of Absaroka volcanics as much as 7.6 cm in diameter in Pumpkin Buttes conglomerates. This certainly indicates a provenance far to the west. Although the pebbles probably came directly from the Absaroka Mountains, they could have been inherited from the late Eocene Wagonbed Formation of Beaver Rim. Because of this, they are not indicated on Figure 5, although they must be taken into account.

Rattlesnake Hills–Clarkson Hill, Wyoming (G 5168–5177), and others (sx, Fig. 5)

Love (1970) has published a thorough study of the Cenozoic geology of central Wyoming.

Approximately 245 m of mudstones and sandstones make up the Chadronian section. A basal conglomerate is succeeded by <30 m of red, greenish, and brown sandstones, silts, and mudstones, all poorly sorted and apparently derived from the underlying Wind River Formation (Eocene). Above these sandstones lies a 45-m-thick series of bentonitic mudstones that resemble those of the Chadron Formation in South Dakota. The remainder of the formation is composed of tuffaceous sandy mudstones and sandstones with a few thin layers of pebble conglomerate and other layers of almost pure volcanic ash.

Pebble counts reveal a preponderance of white and pink granite, sandstone, chert, and quartz; the proportion of granite to other materials increases upward, accompanied by notable decreases in sandstone and chert. The only pebbles with a presumably well-defined provenance are those of aventurine quartz (sx, Fig. 5) in the basal conglomerate. Aventurine is known in pebble- to boulder-size clasts in every Cenozoic age formation of this area from the Eocene (Wind River Formation) to the Pliocene, but its specific source has not been located (D. L. Blackstone, 1971, written commun.). Studies of the sand of central Wyoming have not progressed far enough to discover whether or not mineral grains diagnostic of provenance occur.

The curved arrow beside the sx (Fig. 5) represents the course and direction of a major channel fill at Lone Tree Gulch, 11 km northwest of Alcova, Wyoming. The channel fill enters the area from the south-southwest, then turns east in the direction of the Bates Hole (mapped as lacustrine on Fig. 5). Love (1970, p. C62–C71) ascribed the tuffaceous pyroclastics to sources in the Absaroka Range. He regards the Granite Mountains as the probable source of the hydroclastic sediments. The evidence for this is convincing. The channel fill mentioned above could also have brought in clastics from any of the mountains to the south or southwest, but there is no evidence that this occurred.

Bates Hole, Wyoming

The area under consideration (immediately west of the Laramie Mountains) is mapped with considerable oversimplification as lacustrine (Fig. 5). I have studied the Bates Hole only hastily in 1955. Toots (1965) made pertinent observations; Harshman (1968) and Denson and Harshman (1969) mapped the area.

Generally, the Bates Hole section includes both lower and middle Oligocene deposits. Lacustrine beds constitute a much higher proportion of the sedimentary mass than is usual, especially in the southeastern part of the area. Flood-plain and lacustrine-border deposits also occur. The sediments are tuffaceous and generally finer grained than those of Clarkson Hill.

It is proposed as a working hypothesis that the Bates Hole area represents a blockaded drainage system, with at least one major tributary entering via Clarkson Hill from the west-southwest. Toots (1965) suggested that the Bates Hole area drained northward and, ultimately, out toward the region of Douglas, Wyoming, and thence east to Nebraska. My own information is not sufficient to justify an opinion.

Granite Mountains, Wyoming

Love (1970) follows Van Houten (1964) in identifying two sources of clastics for this area: (1) the Granite Mountains; and (2) the Yellowstone-Absaroka volcanic center. My own studies have corroborated theirs on all essential points. The basal Oligocene Big Sand Draw Lentil (Van Houten, 1964) was deposited apparently by a stream that flowed northwest out of the Granite Mountains. It is represented by the northwest-directed arrow in Figure 5. Abundant volcanic clasts as much as 244 cm in diameter are reported in some of the conglomerates; these apparently came by stream or mudflow transport from the Absaroka Mountains.

Both Love and Van Houten suggest that erosion of previously erupted volcanic materials in the Rattlesnake Hills volcanic field yielded some hydroclastic cobbles of lava to the Oligocene sediments, but they agree that the Oligocene tuff represents ash, windborne from the Absaroka field and water-deposited in their present sites (igneoaqueous beds).

All Oligocene strata in the area have yielded fossils of either positively or possibly early Oligocene age. The Big Sand Draw Lentil is probably Duchesnean; the remainder of the section is probably all Chadronian.

The total picture is one of streams carrying sediment outward from the Granite Mountains, with at least one stream coming from the Absaroka area of northwestern Wyoming. No channel fills that might be considered master streams have been preserved. It is, therefore, impossible to connect this area with the general drainage pattern to the east. Figure 11 indicates the situation by question marks appropriately placed.

Absaroka Mountains

Love (1939) described the Wiggins Formation, a series of volcanic breccias, tuffs, agglomerates, and conglomerates with a maximum thickness of 900 m. The strata rest at elevations usually above 3,170 m. Originally the entire mass was believed to be Chadronian (early Oligocene) on the basis of fossil mammals from two localities. Recent work (Smedes and Prostka, 1972) has shown, however, that the fossils come from Oligocene fills in channelways cut into much older Eocene rocks. The Oligocene sequence of the Absaroka Range, therefore, occupies pre-existing mountain valleys, as does that on the east flank of the Big Horn Mountains and on the north and east flanks of the Laramie Range.

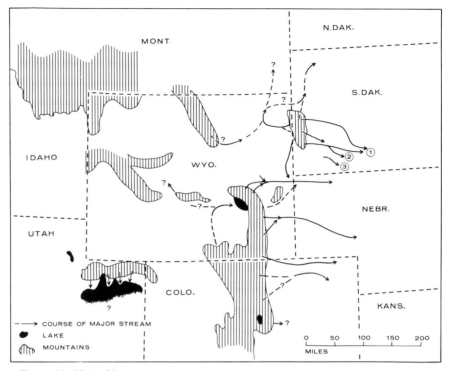

Figure 11. Major Oligocene stream pattern. Figure summarizes and interprets data presented in Figure 5. Numbered courses in South Dakota represent successively younger channelways of what may have been the same stream.

Duchesne River of Utah (Andersen and Picard, 1972)

The Duchesne River Formation, whose uppermost members (LaPoint and Starr Flat) are probably Oligocene in age, consists of fluvial, flood-plain, deltaic, and lacustrine sediments that are derived from the Uinta Mountains and were washed into the northern shore of Uinta Lake. A few bentonitized ash beds with a heavy content of fresh biotite occur in the LaPoint member; their provenance is unknown.

The formation constitutes the closing phase of deposition in Uinta Lake, whose outlet, if it had one, lay somewhere to the west. These beds are not geologically or paleogeographically related to the Oligocene series of the central Rocky Mountains area and are mentioned here only because of their age and proximity.

Major Oligocene Drainage Pattern

All of the major ranges had been eroded to their cores by Oligocene time, and a thick soil zone had developed over the Great Plains westward at least as far as central Wyoming and western Montana. It is not surprising, therefore, that synthesis of the data presented above reveals an integrated major drainage system (Fig. 11).

The generally eastward direction of the drainage probably reflects the initial regional slope following Laramide movements. However, the drainage pattern differs in important respects from that of Holocene streams.

The Cheyenne River, and possibly the Belle Fourche River, captured the centrifugal drainage of the Black Hills durng Pleistocene time. The entire eastern portion of the Black Hills, north to Rapid City, drained to the southeast during Oligocene time as it does now. These streams probably joined the eastward-flowing river of northern Nebraska somewhere in the swampy plains east of the present Chadron, Nebraska.

The modern North Platte River drainage was divided into three segments. First, a drainage from the southwest entered Bates Hole near the present Alcova, Wyoming. Whether or not this ultimately passed through the present gap or east of it is not known. The second segment, composed of headwaters in the area of Douglas, Wyoming, drained eastward into Nebraska. The third portion had the ancestral Laramie River as its headwaters and followed a course similar to the present stream to its junction with the North Platte River. There the ancient stream turned and ran southeastward past Scotts Bluff. Whether the two Colorado tributaries entered the ancestral North Platte River or turned southeastward to an unrecorded ancestral South Platte River is not known.

Several problems beset interpretation of the Oligocene drainage in the Powder River basin. There can be no reasonable doubt of a drainage westward from the northern Black Hills and thence northward past the Missouri Buttes. Whether this stream continued northward, or turned east through the Twin Buttes and then north to the Slim Buttes of South Dakota, is not known. In either case, did these north-flowing streams continue northeastward to the Hudson's Bay drainage, as seems probable, or did they turn due east to the Mississippi? A point of interest is the drainage of the presumed lake at the Pumpkin Buttes. The Pumpkin Buttes stand presently at 1,830 m elevation, 430 m higher than the Oligocene beds at the Missouri Buttes, 130 km to the northeast. Numerous minor structural features intervene, but none are known to have produced or altered a regional slope. The present slope is about 2.8 m per km, roughly equivalent to the Oligocene slopes eastward from the Black Hills and to those of many Holocene streams in the area. What would cause a local blockade; was the lake permanent or temporary, large or small, and did it ultimately drain northeastward or southeastward? I have at present no answer.

CONTROLS OF OLIGOCENE SEDIMENTATION

The development of an Oligocene depositional regime over an area more than 1,600 km long from north to south and 450 to 650 km wide from east to west is one of the major phenomena of Cenozoic paleogeography. Apparently, every drainage from Saskatchewan to central Colorado began to aggrade near its headwaters, at about the same time. Furthermore, every deposit of enough areal extent to record the fact shows maximum thicknesses within 46 km or less of its source mountain range, thinning both downstream and, by transgressive overlap, upstream from its thickest point. This cannot, therefore, be a simple case of transgressive overlap by aging streams in an area undergoing peneplanation. Three independent lines of evidence also militate against such an interpretation.

First, nowhere in the Great Plains do any Uintan fluvial deposits occur, as they should if the Oligocene beds were deposited by simple upstream trangressive overlap.

Second, most of the Oligocene streams concerned drained into the Gulf of Mexico. Yet Oligocene units of the Mississippi Embayment are more restricted in both areal extent and thickness than are the Eocene and Miocene units, which would be impossible if the Oligocene beds of the Great Plains were but a headward extension of a broader downstream sheet.

Third, the eastern moieties of the several Oligocene units of the Dakotas and Nebraska thin rapidly, show rapid decreases in maximum grain size, and reveal a notable increase in proportion of lacustrine, pond, and paludal over fluvial sediments. These changes demonstrate that Oligocene deposition never extended far beyond its present limits, and that excess floodwater ponded but deposited very little sediment beyond the known areas of deposition. Obviously, the Oligocene streams were aggrading only in those segments contiguous to their respective mountain ranges.

Only four other possible hypotheses might adequately explain the facts: (1) downwarp of the various ranges relative to their adjacent basins; (2) downwarp of the entire area along a hinge that would have run approximately north-south through central South Dakota, Nebraska, and Kansas; (3) overloading of all streams by enormous volumes of volcanic ash; and (4) climatic change, which caused a change in weathering of provenances from a predominantly chemical to a predominantly physical mode.

Downwarp of Individual Ranges

This hypothesis can be briefly and definitely disposed of. There is no structural evidence of any Oligocene movement of ranges relative to their neighboring basins or troughs.

Chadron sandstones lap without break into the Battle Creek watergap in the hogback of Cretaceous rocks in the southeastern Black Hills. Brule sediments, south of there and in the northern Black Hills (Darton, 1909), overlap onto Precambrian rocks. It can be demonstrated that no post-Laramide faults transect the area from the Precambrian outcrops eastward for at least 80 to 160 km, and therefore, the Black Hills have not moved relative to the Badlands. A similar situation exists west of Cheyenne, Wyoming, where lower Oligocene sediments lap onto Precambrian rocks.

Oligocene Downwarp of the Entire Area

It has been suggested that subsidence of the entire area constituted the fundamental mechanism that caused deposition by lowering the gradients of all streams.

The first evidence against this hypothesis is structural. No appropriate hinge line is known to exist; indeed, such a hinge would cross at an angle such well-known structural features as the rim of the Williston Basin, and it would certainly have been detected if present.

A 900-m difference in altitude separates lower Oligocene units on the two sides of the Laramie Range (Toots, 1965). Such a discrepancy could not develop in an integrated drainage area subject to epeirogenic downwarp, nor do adequate intervening structural features exist in the Laramie Range.

Oligocene rocks were nowhere deposited on the western side of the Oligocene continental divide (except in locally blockaded drainages in northern Utah). Lower Oligocene deposits now occupy the Absaroka Range, almost abutting the present Continental Divide and also the Beaver Rim, and the Rattlesnake Hills, both within 140 km of the divide. Therefore, any downwarp must have been accompanied by large-scale downfaulting along the Continental Divide from southwestern Saskatchewan to central Colorado. There is no such fault system.

The second line of evidence is sedimentary. Progressive downwarp should bring successively later deposition progressively farther west. That is, it should create a regional onlap. Instead, early Oligocene sedimentation occurred over the whole area (Figs. 2 and 3); middle Oligocene sediments were not, apparently, deposited west of southeastern Wyoming (Robinson, 1972). Lower and middle Oligocene sediments show normal, transgressive overlap against the east flanks of the Black Hills and possibly the Laramie Range and northern Front Range, but nowhere else. Structural movement necessary to produce such a pattern through time would be complex indeed and should have left some evidence of its existence.

The pattern of Oligocene sedimentation does not fit any reasonable theory of structural downwarp. Furthermore, the structures required to produce such a downwarp are absent.

Overload of All Drainages by Enormous Volumes of Volcanic Ash

This theory was advanced by Love (1970) and has been supported by others (McKenna and Love, 1972). The rationale follows: (1) The Yellowstone, central Colorado, and Nevada volcanic fields are known to have undergone intense activity during Oligocene time. (2) Oligocene sediments are generally thickest near these centers and progressively thinner eastward. (3) The Oligocene deposits lie along a gently sloping line that can be drawn from the Absaroka Range eastward through the Big Horn Mountains and the Pumpkin Buttes to the Big Badlands. This suggests that, by the close of Oligocene time, a graded sheet of ash extended from the Absaroka Range eastward to the Great Plains. (4) The bulk of Oligocene sediment consists of ash, either fresh or altered; clastics comprise a minor portion of it. Therefore, accelerated erosion of headwaters was not a cause of sedimentation.

Opposed to these data are several others at least equally cogent:

1. By the beginning of Oligocene time, a coordinated drainage system had been established (Fig. 4). Mature streams of about the same size as Holocene ones were following roughly the same courses as present ones.

2. The total thickness of Oligocene sediment in the Big Badlands is 165 m. The duration of Oligocene time was about 12 to 15 m.y. Sedimentary incrementation in this area was in 20 to 90 cm units.

3. Any stream, flowing at grade and subjected to a sudden influx of fine sediment, will, after initial choking, remove that sediment. The following evidence of this is offered:

A. The Pleistocene loess of China has been cut by canyons as much as 370 m deep, in less than 100,000 yr.

B. Cheyenne River in the Big Badlands established its present course in post-Sangamon (Riss) time. Its small tributaries have had no difficulty cutting canyons 120 m deep through semiconsolidated Oligocene sediments, keeping pace with the downcutting major stream, in less than 200,000 yr.

C. Studies have shown (Lane, 1955; Mackin, 1948) that a modern small stream that is temporarily overloaded will remove as much as 90 m of sediment in a few years, once the overloading is discontinued. Lane and Mackin both mention the case of the Yuba River in California, which achieved a graded deposit as much as 24 m thick in the foothills, due to overloading by mining debris. Mining discharge stopped in 1890, and by 1920 the stream had eroded enough to re-establish its original grade.

D. Nowhere in the Big Badlands, or in northwestern Nebraska, is there any evidence of Oligocene intrastratal erosion.

Therefore, overloading by volcanic ash cannot have been the causative mechanism of deposition. The streams involved would have ripped canyons into the soft ash within months of any one incrementation. They would have been competent to remove several times the volumes concerned during the time available. They would also have been fully competent to remove all the Oligocene ash of Wyoming and Colorado, but the evidence here is more difficult to demonstrate.

4. In the Big Badlands, Nebraska, eastern Colorado, southeastern Wyoming, and the Rattlesnake Hills, the basal 6 to 9 m of Chadron sediments include little or no volcanic ash. They are composed of granitic debris plus mud from the

underlying local sediments. Sedimentation was initiated without benefit of pyroclastics; it cannot therefore be attributed to them.

5. The Oligocene of the Great Plains was deposited in four episodes: Brule, middle-upper Chadron, lower Chadron, and Slim Buttes.

Each of these shows a general trend of coarse clastics in greater quantity near the base, grading toward finer sediments above. This normal gradation would not be the case if deposition were caused by incrementation of fine ash.

6. The situation at Peavey Butte, Logan County, Colorado (Fig. 12), offers evidence that a phenomenon other than ash falls caused deposition. At Peavey Butte, white Chadronian tuff has been eroded to a depth of 24 m. A series of granitic sands and red clastic muds fills the resultant trough; the red sequence is also of Chadronian age, the upper part of the Horsetail Creek Formation.

It may be argued that deposition of the tuff engorged the local stream, and that during a temporary cessation of volcanism, the stream then trenched to a depth of 24 m. Deposition, however, then resumed without the aid of volcanic materials and the entire trough filled again. Obviously, this stream was aggrading toward a graded profile. It used whatever materials were at hand. The reason for the erosional episode is not known, but it probably was a temporary interruption in a predominantly aggradational regimen.

7. A profile line drawn from the Absaroka Range through the Oligocene unit of the Big Horn Mountains, then southeast to the Pumpkin Buttes, then to the Big Badlands, may produce a smooth curve. However, a line drawn from the Absaroka Range to the Oligocene of the Big Horn Mountains, then directly east to the Missouri Buttes, then to the Big Badlands, will show a 430-m notch rather than a smooth curve. Furthermore, because the Pumpkin Buttes Oligocene sediments are clastic rather than volcanic and show a northeastern-flow direction, they do not fit Love's (1970) general pattern. The Missouri Buttes Oligocene sediments, with a straight northward-flow direction, are also anomalous to any generally eastward-sloping, ash-controlled surface.

For all of these reasons, it seems best to consider a hypothesis other than incrementation of airborne ash as the causative agent of Oligocene sedimentation.

Change in Climate Causing a Change in Stream Regimen

Climatic change, starting presumably during Duchesnean time and operating progressively throughout Oligocene time, has been documented (Russell, 1973; Clark and others, 1967; Leopold and MacGinitie, 1972; Dorf, 1969).

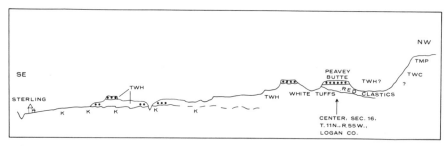

Figure 12. Cross section about 55 km northwest from town of Sterling, Colorado. TMP, Miocene and Pliocene; TWC, Cedar Creek Formation (middle Oligocene); TWH, Horsetail Creek Formation (lower Oligocene); K, Cretaceous. Shows irregular, basal Horsetail Creek sand (the southern channel fill), and at Peavey Butte the intraformational erosion into white tuff, followed by 25 m of red, clastic fill; and whole Horsetail Creek section topped by a massive channel fill.

The general consensus is that climates throughout temperate North America changed from very warm temperate or semitropical, with seasonal rainfall of monsoonal type, to much drier and somewhat cooler.

The hypothesis already proposed (Clark and others, 1967) is that extensive chemical weathering plus a heavy vegetational cover during Eocene time produced a landscape of sharply emergent mountains with closely adjacent lowlands developed on the more readily weathered sediments of the basins. The landscape might have resembled that present in southeastern China or Cuba. The streams probably flowed coffee-colored with organic matter and solutes but almost empty of clastics due to lack of supply.

Change of climate to cooler and drier greatly reduced the effectiveness of the vegetative cover in the mountains. Rainfall certainly became more sporadic; probably the total annual supply of water was decreased at the same time that its eroding ability was increased by lessened vegetative cover in stream headwaters.

Addition of load and decrease in volume would cause each stream to aggrade in the segment adjacent to its headwater range (Mackin, 1948; Lane, 1955). A lens of clastics from the headwaters plus material from neighboring interfluve hills would accumulate at a gradient change near the foot of each range.

This first lens of soft material would then act as a sponge, by abstracting water from succeeding floods, and would help to initiate deposition farther upstream. Eventually, the entire mass would build to a smooth curve, representing the position of grade for that stream under that climatic regimen, with that particular bed and subbed porosity. Leopold and others (1964, p. 441) stated that within recent times "changes in rainfall-runoff relations associated with climatic variations cause even tributaries near the watershed divide to aggrade and degrade, and thus [deposition and erosion] need not depend on changes in base level." My hypothesis applies to Oligocene sedimentation, on a regional scale, principles which they have found operative locally.

The presence of abundant volcanic ash undeniably hastened the aggradation in two ways. First, it provided a considerable supply of clastic particles in sizes that any stream could transport. Second, once deposited, ash formed (except where bentonitized) a highly porous mass, which would help to cause further deposition by abstraction of water. Ash itself, however, could not have been the primary cause of deposition, as already indicated.

Furthermore, the assumption (McKenna and Love, 1972), that occurrence of Oligocene deposits at high elevations requires that the neighboring basins be filled to those levels, is opposed by several lines of evidence.

1. The Chadron Formation southeast of the Black Hills presently lies at slopes of 3 m/km from the outcrops at Fairburn to the Badlands.

2. At Little Pipestone Creek, Jefferson County, Montana, demonstrably initial dips of Chadronian tuff reach 13° to 15°. Demonstrably initial dips in fine tuffaceous sediments are 5° in many areas of Montana. All that is necessary for such a phenomenon is a porous substrate.

3. All of the Oligocene units of the various ranges lie within pre-existing mountain valleys and canyons, which certainly did lead downward to the basins. The climatic hypothesis suggests that these valley fills lie at significant initial dips, not much less than the slopes of the valley bottoms themselves. This is the case in Montana, South Dakota, and the east flank of the Laramie Range.

The climatic hypothesis, in summary, views Oligocene deposition as a response to an aggrading condition caused in the mountain-border segments of all streams over a wide area by a climatic change. Volcanic ash influenced the mode but did not determine the fact of deposition. The drainage was thoroughly integrated at the beginning of Oligocene time and remained so thoughout. This hypothesis fits all facts known at present. Whether it fits better than the major alternatives must be a matter of opinion and hopefully an inspiration for further testing.

REFERENCES CITED

Andersen, D. W., and Picard, M. D., 1972, Stratigraphy of the Duchesne River Formation (Eocene-Oligocene?), northern Uinta Basin, northeastern Utah: Utah Geol. and Mineralog. Survey Bull., v. 97, p. 1-29.

Berggren, W. A., 1972, A Cenozoic time-scale—Some implications for regional geology and paleobiogeography: Lethaia, v. 5, no. 2, p. 195-215.

Clark, J., Beerbower, J. R., and Kietzke, K. K., 1967, Oligocene sedimentation, stratigraphy, paleogeography, and paleoclimatology in the Big Badlands of South Dakota: Fieldiana—Geology Mem., v. 5, p. 1-158.

Coleman, J. M., and Gagliano, S. M., 1965, Sedimentary structures: Mississippi delta plain, in, Middleton, G. V., ed., Primary sedimentary structure: Soc. Econ. Paleontologists and Mineralogists Spec. Pub. 12, p. 1-265.

Condie, K. C., 1969, Petrology and geochemistry of the Laramie Batholith and related metamorphic rocks of Precambrian age—Eastern Wyoming: Geol. Soc. America Bull., v. 80, p. 57-82.

Darton, N. H., 1905, Preliminary report on the geology and underground water resources of the central Great Plains: U.S. Geol. Survey Prof. Paper 32, p. 1-243.

——1909, Geology and water resources of the northern portion of the Black Hills and adjoining regions of South Dakota and Wyoming: U.S. Geol. Survey Prof. Paper 65, p. 1-105.

Denson, N. M., and Harshman, E. N., 1969, Map showing areal distribution of Tertiary rocks Bates Hole-Shirley Basin area, south-central Wyoming: U.S. Geol. Survey Misc. Geol. Inv. Map I-570.

Dorf, E., 1969, Paleobotanical evidence of Mesozoic and Cenozoic climatic changes, in Paleoclimatology: North American Paleont. Conv., Proc., pt. D, p. 323-346.

Galbreath, E. C., 1953, A contribution to the Tertiary geology and paleontology of northeastern Colorado: Kansas Univ. Paleont. Contr., no. 4, p. 1-120.

Harshman, E. N., 1968, Geologic map of the Shirley basin area, Albany, Carbon, Converse, and Natrona Counties, Wyoming: U.S. Geol. Survey Misc. Geol. Inv. Map I-539, scale 1:48,000.

Holmes, A., 1959, A revised geological time-scale: Edinburgh Geol. Soc. Trans., v. 17, p. 183-216.

Kay, J. L., 1934, The Tertiary formations of the Uinta Basin, Utah: Carnegie Mus. Annals, v. 23, p. 357-371.

Lane, E. W., 1955, The importance of fluvial morphology in hydraulic engineering: Am. Soc. Civil Engineers Proc., v. 81, paper 745, p. 1-17; and in Schumm, S. A., ed., River morphology: Stroudsburg, Pa., Dowden, Hutchinson & Ross, p. 181-197, 1972.

Leopold, E. B., and MacGinitie, H. D., 1972, Development and affinities of Tertiary floras in the Rocky Mountains, Chap. 12, in Graham, A., ed., Floristics and paleofloristics of Asia and eastern North America: Amsterdam, Elsevier Pub. Co., 258 p.

Leopold, L. B., Wolman, M. G., and Miller, J. P., 1964, Fluvial processes in geomorphology: San Francisco, W. H. Freeman and Co., p. 1-522.

Lillegraven, J. A., 1970, Stratigraphy, structure, and vertebrate fossils of the Oligocene Brule Formation, Slim Buttes, northwestern South Dakota: Geol. Soc. America Bull., v. 81, p. 831-850.

Love, J. D., 1939, Geology along the southern margin of the Absaroka Range, Wyoming: Geol. Soc. America Spec. Paper 20, p. 1-134.

——1952, Preliminary report on uranium deposits in the Pumpkin Buttes area, Powder River Basin, Wyoming: U.S. Geol. Survey Circ. 176, p. 1-37.

——1960, Cenozoic sedimentation and crustal movement in Wyoming (Bradley Volume): Am. Jour. Sci., v. 258A, p. 204-214.

——1970, Cenozoic geology of the Granite Mountains area, central Wyoming: U.S. Geol. Survey Prof. Paper 495-C, p. 1-154.

Mackin, J. H., 1948, Concept of the graded river: Geol. Soc. America Bull., v. 59, p. 463-512.

Matthew, W. D., 1907, A lower Miocene fauna from South Dakota: Am. Mus. Nat. History Bull., v. 23, p. 169-219.

McKenna, M. C., and Love, J. D., 1972, High-level strata containing early Miocene mammals on the Bighorn Mountains, Wyoming: Am. Mus. Novitates, no. 2490, p. 1-31.

Osterwald, F. W., Osterwald, D. B., Long, J. S., Jr., and Wilson, W. H., 1966, Mineral resources of Wyoming: Wyoming Geol. Survey Bull., v. 50, p. 1-287.

Peterson, O. A., 1932, New species from the Oligocene of the Uinta: Carnegie Mus. Annals, v. 21, p. 61-78.

Robinson, C. S., Mapel, W. J., and Bergendahl, M. H., 1964, Stratigraphy and structure of the northern and western flanks of the Black Hills uplift: U.S. Geol. Survey Prof. Paper 404, p. 1-134.

Robinson, P., 1972, Tertiary history, in Geologic atlas of the Rocky Mountain region: Denver, Rocky Mt. Assoc. Geol., p. 233-242.

Russell, L. S., 1973, Geological evidence on the extinction of some large terrestrial vertebrates: Canadian Jour. Earth Sci., v. 10, no. 2, p. 140-145.

Sato, Y., and Denson, N. M., 1967, Volcanism and tectonism as reflected by the distribution of nonopaque heavy minerals in some Tertiary rocks of Wyoming and adjacent states: U.S. Geol. Survey Prof. Paper 575-C, p. C42-C54.

Schultz, C. B., and Stout, T. M., 1955, Classification of Oligocene sediments in Nebraska: Nebraska Univ. State Mus. Bull., v. 4, no. 2, p. 17-52.

Sharp, W. N., McKay, E. J., McKeown, F. A., and White, A. M., 1964, Geology and uranium deposits of the Pumpkin Buttes area of the Powder River Basin, Wyoming: U.S. Geol. Survey Bull. 1107-H, p. 541-638.

Smedes, H. W., and Prostka, H. J., 1972, Stratigraphic framework of the Absaroka volcanic supergroup in the Yellowstone National Park region: U.S. Geol. Survey Prof. Paper 729-C, p. C1-C33.

Toots, H., 1965, Reconstruction of continental environments: The Oligocene of Wyoming [Ph.D. thesis]: Laramie, Univ. Wyoming.

Van Houten, F. B., 1964, Tertiary geology of the Beaver Rim area, Fremont and Natrona Counties, Wyoming: U.S. Geol. Survey Bull. 1164, p. 1-99.

Wanless, H. R., 1923, The stratigraphy of the White River beds of South Dakota: Am. Philos. Soc. Proc., v. 62, no. 4, p. 190-269.

Wilson, J. A., Twiss, P. C., DeFord, R. K., and Clabaugh, S. E., 1968, Stratigraphic succession, potassium-argon dates, and vertebrate faunas, Vieja Group, Rim Rock Country, Trans-Pecos Texas: Am. Jour. Sci., v. 266, p. 590-604.

MANUSCRIPT RECEIVED BY THE SOCIETY MAY 3, 1974

Printed in U.S.A.

Geological Society of America
Memoir 144
© 1975

Late Cenozoic Basaltic Volcanism and Development of the Rio Grande Depression in the Southern Rocky Mountains

Peter W. Lipman

AND

Harald H. Mehnert

U.S. Geological Survey
Denver Federal Center
Denver, Colorado 80225

ABSTRACT

In the Southern Rocky Mountains, upper Cenozoic basalt flows were erupted widely in areas characterized in middle Tertiary time by predominantly intermediate-composition volcanism. Initiation of basaltic volcanism coincided approximately with the beginning of extensional block faulting that resulted in development of the Rio Grande depression, a major rift structure that separates the stable platforms of the High Plains and the Colorado Plateau. Time relations are especially clear along the San Luis Valley segment of the Rio Grande depression in southern Colorado and northern New Mexico, where 16 basalt flows have been dated by K-Ar methods. Along the west margin of the San Luis Valley, silicic alkalic basalt flows as old as 26 m.y. rest unconformably on a pediment cut on middle Tertiary andesitic and related rocks (35 to 27 m.y. old), and similar basalt 20 to 0.24 m.y. old interfingers with and overlies volcaniclastic alluvial fan deposits (equivalent to Santa Fe Group) that accumulated in the subsiding depression.

Basalt erupted during late Cenozoic block faulting varies in composition with distance from the axis of the northern Rio Grande depression. Tholeiitic rocks are largely confined to the depression, and the basalt types become more alkalic to the west and east. Relatively silicic alkalic basalt, including both undersaturated and saturated types, occurs throughout the region, but very undersaturated alkalic basalt flows were erupted only on the Colorado Plateau and the High Plains. The lateral change from tholeiitic to alkalic basaltic volcanism probably reflects different conditions of magma generation in the mantle that are related to changes in crustal thickness and thermal gradients across the depression.

The compositions and compositional ranges of basalt of the Southern Rocky Mountain region are similar to those of many Pacific islands, but the proportions of basalt types are markedly different. These relations are tentatively interpreted as reflecting origin of both continental and oceanic basalt types under similar P-T conditions, with generation of the Southern Rocky Mountain basalt from lithospheric mantle and the oceanic basalt from relatively shallow asthenosphere.

INTRODUCTION

Late Cenozoic basaltic volcanism occurred widely in the Southern Rocky Mountain region. The stratigraphic, structural, and geochronologic relations of these basalt flows help to unravel the late Cenozoic structural evolution of the region. In addition, petrologic variations among the basalt types, which are interpreted as reflecting different physical conditions at depth during magma generation, permit insight into related tectonic changes deeper within the lower crust and upper mantle. In this paper we summarize new and previously published data that indicate that the large composite Oligocene volcanic field of central and southern Colorado (Steven, 1975) began to be disrupted by block faulting and erosion as early as about 26 m.y. ago. A fundamentally basaltic volcanic suite was emplaced concurrently with the extensional faulting, a volcano-tectonic association that has continued to be intermittently active during most of late Cenozoic time.

REGIONAL VOLCANO-TECTONIC SETTING

The late Cenozoic extensional faulting and related basaltic and bimodal mafic-silicic volcanism were concentrated along the same north-trending axis of the Southern Rocky Mountains that had earlier been a locus of deformation and igneous activity in Late Cretaceous–early Tertiary (Laramide) time and also in the middle Tertiary. By the end of the middle Tertiary volcanic episode, much of the Southern Rocky Mountains appear to have been mantled by a composite volcanic plateau charac-terized by generally subdued topography locally interrupted by primary construc-tional volcanic features.

This volcanic field consisted largely of central volcanoes of intermediate-composi-tion lavas and breccias—andesite, rhyodacite, and quartz latite—surrounded by coalescing aprons of mudflow and other volcaniclastic rocks (Steven, 1975). Rocks of basaltic composition are rare in these middle Tertiary assemblages. Associated with the predominantly intermediate-composition rocks, especially in the San Juan volcanic field, are more silicic ash-flow sheets of quartz latite and low-silica rhyolite erupted from large calderas.

Geophysical evidence and the presence of numerous exposed granitic stocks similar in age to these volcanic rocks point to the underlying presence of shallow batholiths below the volcanic piles (Plouff and Pakiser, 1972; Tweto and Case, 1972). Considerable evidence suggests that these intrusives are comagmatic with the intermediate-composition lavas and with the ash-flow sheets: the intermediate lavas are thought to have sampled the greater part of the batholiths, and the ash-flow sheets sampled their differentiated tops (Lipman and others, 1970, p. 2347; Steven, 1975). This volcano-plutonic association of predominantly intermediate composition-al types has been interpreted to reflect the innermost expression of convergent plate tectonic interactions between the American and eastern Pacific (Farralon) plates (Lipman and others, 1972) through middle Tertiary time, although this interpretation remains controversial (Gilluly, 1971).

In contrast to the predominantly andesitic volcanism of middle Tertiary time, the late Cenozoic volcanic associations of the Southern Rocky Mountain region consist largely of petrologically diverse basaltic lava flows with associated less voluminous silicic rocks. This type of volcanic association, which has been termed "fundamentally basaltic" (Christiansen and Lipman, 1972), includes (1) basalt fields, (2) alkalic fields in which differentiated igneous series commonly can be related to alkali basalt parent magmas, and (3) bimodal association of mafic and silicic rocks—generally basalt and high-silica alkali rhyolite. Similar igneous fields occur in other regions of the world characterized by tectonic extension. This change in volcanic associations and their tectonic settings occurred in much of the western United States during late Cenozoic time and has been interpreted to reflect collision of the East Pacific Rise with a middle Tertiary continental margin trench. It resulted in direct contact of the American and western Pacific plates along a right-lateral transform fault system (McKenzie and Morgan, 1969; Atwater, 1970; Christiansen and Lipman, 1972).

The major late Cenozoic extensional tectonic feature in the Southern Rocky Mountains is the Rio Grande depression (Bryan, 1938; Kelley, 1952, 1956; Chapin, 1971), a major intracontinental rift system 15 to 60 km wide that extends for more than 800 km from central Colorado to Mexico (Figs. 1 and 2). Structurally, the depression consists of a series of en echelon grabens, each bounded on one or both sides by fault-block mountains. On a regional scale, this pattern defines an isolated prong of the Basin and Range province that, in its central part, separates the stable platforms of the Colorado Plateau and the High Plains and extends northward into the heart of the Southern Rocky Mountains.

South of Albuquerque, New Mexico, the Rio Grande depression merges with the Basin and Range province and is a less well-defined structure than to the north. Kelley (1952) restricted the depression in southern New Mexico to the narrow series of structural depressions followed by the modern Rio Grande, thus ignoring the presence of structurally similar parallel basins and horsts to the east and west. Alternatively, the depression has been defined more broadly to include a series of three parallel basins and their intervening horsts that merge indistinguishably to the west with the Basin and Range province (Chapin, 1971). In neither interpretation does the Rio Grande depression constitute a discrete structural feature in its southern part.

In contrast, from Albuquerque north, the Rio Grande depression is a single well-defined structural trough that is the locus of the most intense late Cenozoic block faulting in the Southern Rocky Mountain region. Related late Cenozoic extensional faulting is partly responsible for the present relief of other north-trending mountain ranges in southern and central Colorado, but only one basin parallel to the Rio Grande depression—the Wet Mountain Valley graben, just east of the Sangre de Cristo Mountains (Fig. 2)—was comparably depressed and filled with appreciable thicknesses of upper Tertiary sedimentary rocks (Taylor, 1975; G. R. Scott and R. B. Taylor, 1973, unpub. data).

The broadest and deepest part of the northern Rio Grande depression is the San Luis Valley in southern Colorado (Figs. 1, 2), an asymmetrical graben that displays typical Basin and Range structure. The major bounding fault system is on the eastern side of the graben against the Sangre de Cristo Mountains, and gravity data (Gaca and Karig, 1966) indicate a horst-and-graben structure beneath the sedimentary fill of the valley. Above the level of the valley fill, the western boundary of the San Luis Valley is a dip slope of eastward-tilted volcanic rocks

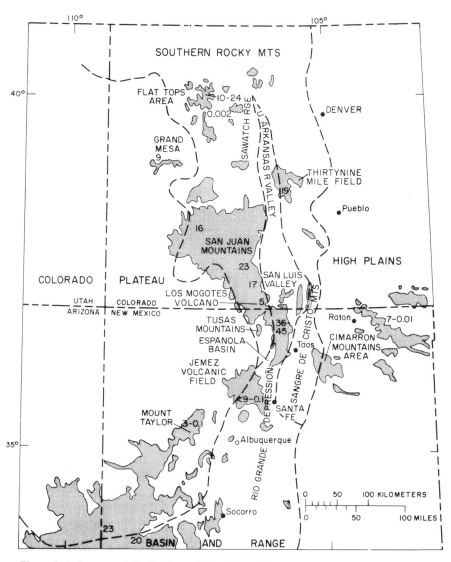

Figure 1. Index map of the Southern Rocky Mountain region, showing general distribution of upper Cenozoic volcanic rocks (light stipple; modifed from Cohee and others, 1961) and major structural provinces. Numbers indicate K-Ar ages (millions of years) of upper Cenozoic basaltic rocks; these are from references cited in text.

of the San Juan Mountains. This dip slope is cut by numerous small north-trending antithetic normal faults, dropped down to the west in opposition to the structural effect of the dip of the rocks. The gravity data indicate that San Luis Valley graben contains as much as 9 km of fill. Because the surface of alluvial fill in the valley is at an elevation of about 2,250 m, the bedrock floor of the graben may in places be as much as 7 or 8 km below sea level, and the total displacement on the eastern bounding fault system adjacent to the upfaulted Sangre de Cristo Mountains (maximum elevation greater than 4,300 m) may be as much as 14 or 15 km. The geometry of the faulting indicates a pattern of crustal extension, characteristic of much of the western United States in late Cenozoic time (Hamilton and Myers, 1966).

North of the San Luis Valley, the Rio Grande depression extends into the upper Arkansas River valley (Van Alstine, 1968), an asymmetrical graben structurally similar to the San Luis Valley except that the major bounding fault system is on the west side of the graben against the Sawatch Range. To the south the San Luis Valley graben is structurally continuous into the Espanola Basin, although a partial topographic separation results from a cluster of late Tertiary volcanoes near the Colorado–New Mexico state line (Lambert, 1966).

BASALT TYPES

A relatively small number of intergradational petrologic types of upper Cenozoic basalt are dominant throughout the Southern Rocky Mountain region (Table 1). Although the proportions of basalt types differ, the compositional range of these basalts in most major and minor elements is similar to that of many oceanic island suites such as those of Hawaii (Hedge and Lipman, 1972). Accordingly, basalt types of the Southern Rocky Mountains are here subdivided into tholeiitic and alkalic suites, as shown on an alkali-silica diagram (Fig. 3A), using the field boundaries that have been empirically determined for Hawaiian basalts (Macdonald and Katsura,

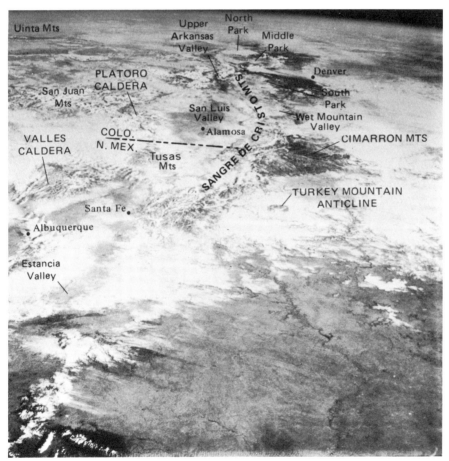

Figure 2. *Apollo 9* photograph looking northwest obliquely across the Rio Grande depression. Photograph courtesy of NASA.

1964). The alkalic suite is further divided into alkali-olivine basalt (<48 percent SiO_2; $\simeq$ ne normative), silicic alkalic basalt (48 to 52 percent SiO_2; $\simeq$ neither ne nor q normative), and basaltic andesite (>52 percent SiO_2; $\simeq$ q normative). Although arbitrary, the Hawaiian field boundary is used because it emphasizes seemingly significant petrographic and geographic distinctions among the Southern

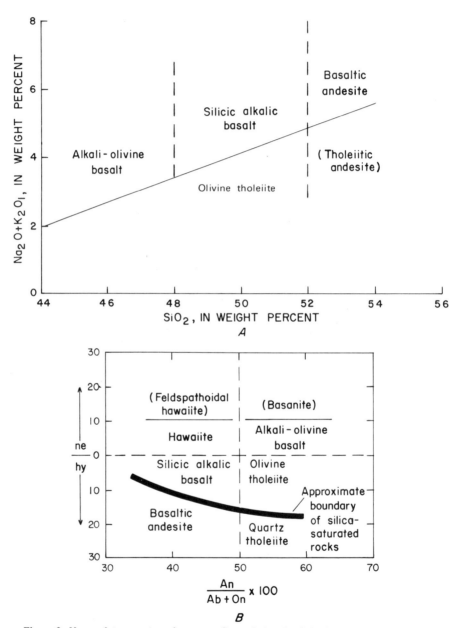

Figure 3. Nomenclature systems for upper Cenozoic basalt of the Southern Rocky Mountain region. A. Chemical classification of basalt based on alkali-silica variations and field boundary between Hawaiian alkalic and tholeiitic basalts (after Macdonald and Katsura, 1964). B. Purely normative classification system (modified from M. G. Best and W. H. Brimhall, 1973, written commun.

Rocky Mountain suites. Alternative boundaries (for example, Kudo and others, 1971) result in slightly different nomenclature for basalts of the region.

The silicic alkalic basalt type is the most widespread and voluminous in the Southern Rocky Mountain region, as well as in the Basin and Range province (Leeman and Rogers, 1970) and along the south and west margins of the Colorado Plateau (Elston and others, 1968, p. 271; Best and others, 1966, 1973, written commun.; Lipman and Moench, 1972). Virtually every petrologist who has dealt with these rocks has emphasized their distinctive high-silica, high-alkali characteristics in comparison with oceanic basalt.

Purely normative classifications, such as that of Yoder and Tilley (1962), have seemed less useful for the Southern Rocky Mountain basalt types because (1) computation problems result from varied ferric/ferrous ratios that significantly affect the apparent degree of silica saturation within related basalt suites (although they result mainly from variable oxidation during eruption and emplacement), and (2) most such normative classifications tend to obscure important variations among the basalt suites. In particular, most representatives of the major alkalic and tholeiitic groups in the Southern Rocky Mountains have neither ne nor q in the norm, and both would be classed as olivine tholeiites by the Yoder-Tilley system, as would most of the important compositionally distinctive abyssal tholeiitic rocks from oceanic ridges (Chayes, 1972; Bass, 1971). Because of its high alkali and silica content, the silicic alkalic basalt of the Southern Rocky Mountains typically has normative plagioclase compositions more sodic than An_{50}, and can be separated fairly well from the tholeiite suite in a normative classification, as suggested by M. Best (1973, written commun.), by splitting the olivine tholeiite group of the Yoder-Tilley scheme into two groups divided by normative plagioclase at An_{50} composition. Especially if ferric/ferrous ratios are adjusted to some uniform value such as 0.4 (Chayes, 1966), a purely normative classification such as that shown in Figure 3B assigns basalt names quite similar to the classification described above (Fig. 3A) based on Macdonald and Katsura's (1964) for Hawaii. However, the artifice of this approach obscures seemingly significant general variations in ferric/ferrous ratios between the major basalt types. These ratios seem highest in the alkalic basalt and lowest in the olivine tholeiite (Table 1).

BASALT CHRONOLOGY IN THE SAN JUAN MOUNTAINS AND DEVELOPMENT OF THE RIO GRANDE DEPRESSION

Development of the Rio Grande structural belt and of related block faulting in Colorado and New Mexico has generally been thought to have begun in late Miocene time and to have culminated in late Pliocene time (Kelley, 1956; Chapin, 1971; Elston and others, 1973). In the San Luis Valley segment of the depression, however, stratigraphic and geochronologic studies—detailed below—demonstrate that the beginning of subsidence in the San Luis Valley segment occurred virtually simultaneously with the initiation of basaltic volcanism in the San Juan Mountains in early Miocene time (about 26 m.y. ago). Tectonic activity in the San Luis Valley has continued into Holocene time, as indicated by low fault scarps that cut alluvial fans of Late Wisconsin age along the western front of the Sangre de Cristo Mountains (Upson, 1939; Scott, 1970).

The time at which late Cenozoic extensional faulting began has proved difficult to establish for many areas of the western United States, and considerable ambiguity remains even for some well-studied areas (for example, compare Elston and others, 1973; Christiansen and Lipman, 1972, p. 260). Evidence for the initiation of basin-range faulting has mostly come from thickness and distribution relations of datable widespread units, especially single ash-flow sheets. Where these sheets predated block faulting, they spread uniformly over areas of subsequent basins and ranges; where they postdated block faulting, they were more or less confined to the basin areas. Time limits on the beginning of extensional faulting are also provided by the oldest sedimentary basin deposits related to the basin-range faulting, but in many basins, the oldest sedimentary fill that accumulated in the deepest part is not necessarily exposed now. Furthermore, fossils or datable volcanic rocks may be sparse or poorly preserved low in the sedimentary section.

An extensive record of evolution of the Rio Grande depression is provided by newly dated basalt flows interlayered with basin-fill sedimentary rocks along the southeast margin of the San Juan Mountains (Fig. 4). A few upper Cenozoic basalt samples from the San Juan volcanic field have been dated previously (Steven and others, 1967; Lipman and others, 1970), but 11 additional K-Ar ages are reported here (Table 2). Most of these basalt flows are relatively silicic alkalic basalt and basaltic andesite, containing sparse phenocrysts of olivine. Together with less voluminous silicic alkalic rhyolite flows and tuffs, they constitute a bimodal mafic-silicic volcanic suite in the San Juan Mountains, typical of the fundamentally basaltic volcanic association in many regions of the western United States. In the San Juan field, the upper Cenozoic basalt and rhyolite have been mapped together as the Hinsdale Formation (Cross and Larsen, 1935; Larsen and Cross, 1956; Steven and others, 1974).

Along the southeast margin of the San Juan field, basalt flows of the Hinsdale Formation and interlayered volcaniclastic sedimentary rocks of the Los Pinos Formation lap unconformably onto an erosion surface cut on Oligocene andesitic lava and breccia (35 to 30 m.y. old) and on overlying ash-flow tuff (30 to 26.5 m.y. old; Fig 5). The ash-flow sheets, including some of the most voluminous units erupted from calderas in the central San Juan Mountains, constituted a little-deformed or eroded volcanic plateau at the end of Oligocene time. They become thinner relatively uniformly in all directions outward from their caldera sources; in particular, all major ash-flow sheets become thinner to the east, toward the area of the San Luis Valley, indicating that this structural feature did not exist as a depositional basin in Oligocene time. In contrast, younger ash-flow sheets that were erupted during subsidence of the Rio Grande depression, such as the Bandelier Tuff erupted from the Valles caldera in the Jemez volcanic field 1.4 to 1.1 m.y. ago, have asymmetrical distribution and thickness patterns, with especially thick accumulations toward the axis of the depression (Smith and Bailey, 1968).

The widespread erosional surface cut on the Oligocene volcanic rocks in the southeastern San Juan Mountains had only subdued local relief, in places with broad valleys as much as 10 to 20 m deep. Older units of the Oligocene volcanic sequence were systematically exposed westward on this early Miocene surface, which indicates tilting to the east during erosion (Fig. 5). We interpret this erosion surface as a pediment formed during initial eastward tilting of the Oligocene San Juan volcanic plateau toward the San Luis Valley in response to downdropping of the valley.

---$\rightarrow$

Figure 4. Distribution of upper Cenozoic basalt in the San Juan Mountains region (modified from Burbank and others, 1935) and collection localities for chemically analyzed and radiometrically dated samples. Dashed lines divide basalt into three regional subgroups that are plotted separately in Figure 8A. Key to sources of chemical analyses: Servilleta Formation: tholeiitic basalts, 1-14—Aoki (1967a, Table 2, nos. 1, 3-10, 13, 15-18); 15, 16—this paper (Table 3, nos. 1, 2); 17-19—Lipman (1969, Table 1, nos. 10-12); alkalic basalts, 20-22—Aoki (1967a,

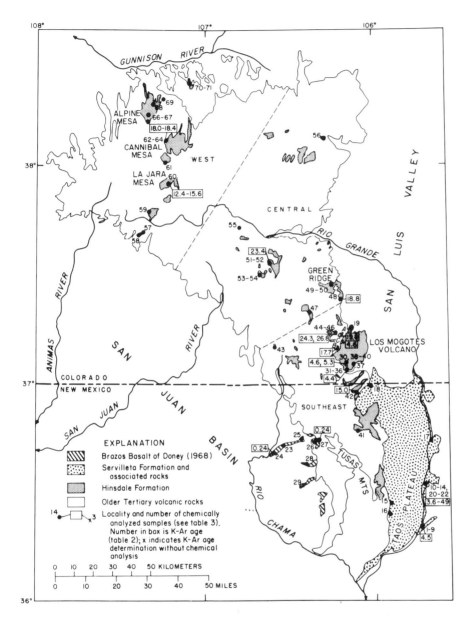

Table 2, nos. 11, 12, 14). Brazos Basalt of Doney, 1968: 23, 24, 27-29—this paper (Table 3, nos. 6-10); 25, 26—Wells (1937, p. 34). Hinsdale Formation: Los Mogotes volcano, 30, 31—Larsen and Cross (1956, Table 25, nos. 3, 5); 32—Lipman (1969, Table 1, no. 2); 33-40—P. W. Lipman (unpub. data); southeast San Juan Mountains, 41—Barker (1958, p. 47, Table 6); 42—this paper (Table 3, no. 3); 43-46—Lipman (1969, Table 1, nos. 3, 7-9); 47—this paper (Table 3, no. 5); central San Juan Mountains, 48—Lipman (1969, Table 1, no. 5); 49, 50—Larsen and Cross (1956, Table 25, nos. 6, 7); 51, 52—Lipman (1969, Table 1, nos. 1, 4); 53, 54—Lipman (1975, Table 11, nos. 13, 14); 55—Lipman (1969, Table 1, no. 6); 56—this paper (Table 3, no. 3); western San Juan Mountains, 57, 58—P. W. Lipman (unpub. data); 59—Larsen and Cross (1956, Table 25, no. 8); 60—Doe and others (1969, App. B); 61—Larsen and Cross (1956, Table 25, no. 9); 62-66—P. W. Lipman (unpub. data); 67, 68—Larsen and Cross (1956, Table 25, nos. 10, 11); 69—Olson and others (1968, Table 1, no. 13); 70, 71—P. W. Lipman (unpub. data).

Resting unconformably on the pediment surface is an eastward-thickening wedge of volcaniclastic sedimentary rocks of the Los Pinos Formation and interlayered basaltic lava flows of the Hinsdale Formation. The Los Pinos Formation consists largely of water-laid tuffaceous sandstone and conglomerate derived from volcanic highlands to the north and west in the San Juan Mountains; thick beds of mudflow conglomerate or breccia are conspicuous, and locally, as in the Tusas Mountains of northern New Mexico, rhyolitic ash-fall tuff and a few weakly welded ash-flow sheets are interlayered within the detrital rocks (Butler, 1946, 1971; Barker, 1958; Bingler, 1968). To the south the volcaniclastic rocks of the Los Pinos Formation are thought to interfinger with predominantly nonvolcanic sedimentary basin-fill sedimentary rocks of the Santa Fe Group (Butler, 1946, 1971). The thickness of the unit of Los Pinos Formation and interlayered basalts varies from near zero along the continental divide in southern Colorado, where the unit occurs only as local valley-filling lenses, to as much as 400 m, where it is exposed in the major east-flowing drainages of the Rio de los Pinos and the Conejos River valley (Steven and others, 1974; Butler, 1971). The thickness of the unit probably also continues to increase eastward beyond present exposures, under covering basalt flows in the San Luis Valley (Fig. 5). Thus, the Los Pinos Formation is interpreted as alluvial-fan deposits that accumulated along the west margin of the deepening San Luis Valley.

Basalt flows of the Hinsdale Formation are interlayered with the Los Pinos alluvial-fan accumulation at several horizons in the southeastern San Juan Mountains (Fig. 4). The oldest recognized basalt flows, exposed in the mouth of Ra Jadero Canyon (Fig. 4, loc. 45) at the base of the volcaniclastic sequence, rest directly on the erosional surface cut on the Masonic Park Tuff, a major ash-flow sheet erupted from the Mount Hope caldera in the central San Juan Mountains about 28.2 m.y. ago (Lipman and others, 1970). Two basalt flows from this stratigraphic zone have yielded K-Ar whole-rock ages of 26.8 ± 0.8 and 24.3 ± 0.8 m.y. (Table 2, nos. 1, 2). The significance of the difference between these two ages is uncertain; stratigraphic relations between the two flows are unknown, owing to discontinuous exposures along the canyon bottom. Two separate flows are clearly represented, as indicated by the contrasting values of K_2O (Table 2). Despite the early age of these flows, they appear scarcely altered except for partial replacement of olivine phenocrysts by iddingsite. These flows, which are typical silicic alkalic basalt of the Hinsdale Formation (SiO_2 of the 26.8-m.y.-old sample is 51.6 percent; complete analysis in Lipman, 1969, Table 2, no. 3), are the oldest dated basalt in the San Juan field and in the Southern Rocky Mountain region.

A K-Ar date of 25.9 m.y. on a local rhyolitic ash-flow tuff, interbedded with conglomerates low in the Los Pinos Formation in northern New Mexico (Bingler,

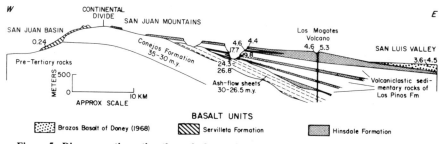

Figure 5. Diagrammatic section through the southern San Juan Mountains. Numbers indicate K-Ar ages (millions of years) of basalt flows and underlying Oligocene volcanic rocks (Table 2; Lipman and others, 1970).

1968, p. 35), also indicates that near the south margin of the San Juan field, Los Pinos detritus began to accumulate within the subsiding Rio Grande depression at a relatively early time.

A slightly younger K-Ar age of 23.4 m.y. was earlier determined from a basalt flow of the Hinsdale Formation in the Beaver Creek area, about 50 km northwest of the Ra Jadero Canyon locality (Fig. 4, loc. 52). This basalt is closely associated with crosscutting silicic alkali rhyolite plug domes of the Hinsdale Formation, one of which has been dated at 22.9 m.y. (Lipman and others, 1970, Table 5), consistent with the K-Ar age of the basalt. The Beaver Creek basalt flow rests unconformably on the Carpenter Ridge Tuff, another major central San Juan ash-flow sheet with an age between 27.8 and 26.7 m.y. (Lipman and others, 1970). Although the isolated Beaver Creek flow cannot be related stratigraphically to alluvial-fan accumulations of the Los Pinos Formation, which wedge out east of Beaver Creek, its early age provides additional evidence that basaltic volcanism in the San Juan field began in early Miocene time.

A basalt flow at the mouth of La Jara Canyon (Fig. 4, loc. A) has yielded a K-Ar age of 19.8 ± 0.6 m.y. (Table 2, no. 4). It rests on a weakly welded thin ash-flow deposit that overlies the Masonic Park Tuff and may be the distal edge of the Carpenter Ridge Tuff. This basalt is distinctly younger than the flows near the base of the Los Pinos Formation in Ra Jadero Canyon only 5 km to the north. This difference in age of basal basalt flows probably reflects the onlapping relations of alluvial-fan deposits of the Los Pinos Formation and interlayered basalt flows onto an eastward-dipping pediment (Fig. 5). Another flow interlayered within the Los Pinos Formation about 20 to 30 m above its base and located on the southeast flank of Green Ridge (Fig. 4, loc. 48) has yielded an only slightly younger K-Ar age of 18.8 ± 0.4 m.y. (Table 2, no. 5). A third flow, similarly interlayered with the Los Pinos Formation in La Jara Canyon, about 10 m above the 19.8-m.y.-old flow, has yielded a K-Ar age of 17.7 m.y. (Table 2, no. 6), in good agreement with the stratigraphic sequence.

Basalt interlayered with the Los Pinos Formation along lower Rio de los Pinos just south of the Colorado–New Mexico state line (Fig. 4, loc. 42) yielded a K-Ar age of 15.0 ± 0.8 m.y. (Table 2, no. 7), in good agreement with ages determined for Hinsdale basalt interlayered at middle levels in the Los Pinos Formation farther north. This age is particularly interesting because the basalt has a distinctive low-potassium tholeiitic composition (Table 3, no. 4) and is the oldest dated tholeiitic basalt in the San Juan area. This flow contains abundant plagioclase phenocrysts, which are uncommon in basalt of the San Juan Mountains and which also serve to distinguish it from the younger aphyric tholeiitic basalt of the Servilleta Formation within the Rio Grande depression.

Previously, workers in northern New Mexico have designated all basalt flows that are interlayered with the volcaniclastic sedimentary rocks of the Los Pinos Formation as the Jarita Basalt Member (Barker, 1958) of the Los Pinos Formation, and the Hinsdale Formation in New Mexico has been restricted to only those basalts overlying the volcaniclastic rocks of the Los Pinos Formation (Barker, 1958; Bingler, 1968; Butler, 1946, 1971). Although locally reasonable, this approach is not feasible for the entire San Juan field, because this distinction between basalt types can only be made for the limited area underlain by the Los Pinos Formation. If followed rigorously, the only basalt of the Hinsdale Formation would be that from Los Mogotes volcano. The few dated basalt samples from the central and western parts of the San Juan field, including rocks from the type area in Hinsdale County, are as old as or older than the Jarita Basalt Member in Rio de los Pinos (Table 2, nos. 12 to 14).

In view of these complications, we suggest that the Los Pinos Formation be restricted to its dominant component of volcaniclastic rocks and that all interlayered basalt flows be assigned to the Hinsdale Formation (Fig. 6). Should subsequent study demonstrate that the basalt flows in northern New Mexico previously assigned to the Jarita Basalt Member have a consistently distinctive tholeiitic composition, then the name Jarita would have utility, either for a separate formation or for a member of the Hinsdale Formation. Farther south in the Tusas Mountains, however, lava flows of the Jarita Basalt Member—including the type area—are petrographically different from those of the Rio de los Pinos locality (Butler, 1946, 1971), and one analyzed sample from the southern area is an alkalic basaltic andesite (Barker, 1958) that is compositionally similar to basaltic andesites of the Hinsdale Formation in Colorado.

The youngest basaltic rocks in the southeastern San Juan Mountains are in a sequence of at least 12 flows erupted from the Los Mogotes shield volcano (Larsen and Cross, 1956, p. 203). These flows are separated from the 17.7-m.y.-old flow in La Jara Canyon (Table 2, no. 6) by an eastward-thickening wedge of Los Pinos Formation as much as 50 m thick. Lowermost flows of the Los Mogotes sequence from La Jara Canyon (Fig. 4, loc. C) and from the Conejos River valley farther south (Fig. 4, loc. 33) yielded whole-rock K-Ar ages of 4.4 ± 0.22 and 4.6 ± 0.26 m.y., respectively (Table 2, nos. 10, 11). These ages are similar to previously obtained K-Ar ages on plagioclase concentrates from a high flow and a dike representing some of the youngest material erupted from the Los Mogotes center (Table 2, nos. 8, 9). The four ages taken together indicate that the Los Mogotes basalt sequence was erupted about 5 m.y. ago within a time span of a few hundred thousand years.

Flows on the eastern flank of the Los Mogotes volcano dip about 5° E. and are onlapped by more gently dipping distinctive olivine tholeiites of the Servilleta Formation (Table 1). The Servilleta basalt flows were seemingly confined to deeper parts of the Rio Grande depression. Numerous samples have yielded dates in the range 3.6 to 4.5 m.y. (Ozima and others, 1967). Sedimentary rocks interlayered with the tholeiitic basalts, which have been included in the Servilleta Formation (Montgomery, 1953; Butler, 1971), are probably contemporaneous and lithologically similar to upper parts of the Santa Fe Group farther south (Galusha and Blick, 1971), where the lower tholeiitic flows wedge out. If these relations are confirmed by further work, we suggest that the Servilleta Formation should probably be restricted to tholeiitic and associated basalts within the Rio Grande depression and that the interlayered sedimentary rocks should be assigned to the Santa Fe Group or its correlatives, just as the older basalts of the Hinsdale Formation are grouped separately from interfingering volcaniclastic sedimentary rocks of the Los Pinos Formation (Fig. 6).

Age relations of the Hinsdale basaltic rocks that interfinger with the Los Pinos Formation demonstrate that these volcaniclastic sedimentary rocks accumulated over the approximate interval of 25 to 5 m.y. ago, spanning most of Miocene and Pliocene time[1]; these relations also confirm previous interpretations (Butler, 1946, 1971) that the Los Pinos Formation is a general facies equivalent of the largely nonvolcanic Santa Fe Group (of Galusha and Blick, 1971), which in its type area is believed, on the basis of a rich vertebrate fauna, to range in age from middle Miocene in its lowest exposed parts to early late Pliocene in its upper

---

[1]Pliocene is here considered as the time span from about 11 m.y. to about 3 m.y. ago (see Fig. 7). Interpretations of the time span are not standardized, and usages by the authors in this volume vary. Some draw Pliocene boundaries at 5 m.y. and 1.8 m.y. ago (Ed.).

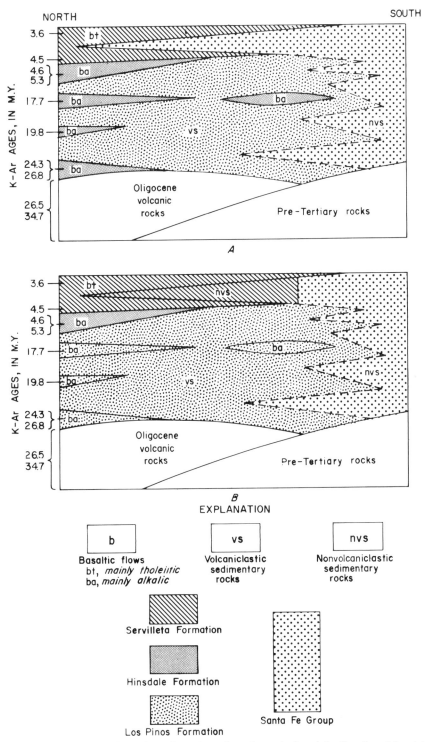

Figure 6. Diagrammatic north-south sections along the east edge of the San Juan Mountains showing alternative stratigraphic nomenclature. A. Nomenclature proposed in this report. B. Nomenclature used by Butler (1971).

part (Galusha and Blick, 1971). Surprisingly, no attempt has been made to calibrate the well-studied vertebrate faunas in the Santa Fe Group by radiometrically dating the numerous interlayered basaltic lavas and rhyolitic ash layers. Such studies will be required for detailed correlation of the basin-fill sequence in the southern San Luis Valley with that in the Espanola Basin.

Most radiometric age determinations on basalt flows of the Hinsdale Formation are from the southeastern San Juan Mountains, where the stratigraphic sequence is most complete and relations between basaltic volcanism and formation of the Rio Grande depression can best be interpreted. Sparse K-Ar data indicate that basaltic volcanism in the western San Juan Mountains occurred concurrently with much of that along the southeast margin of the volcanic field. One flow of Hinsdale basalt from Jarosa Mesa in the west-central San Juan Mountains (Fig. 4, loc. 60) was sampled by Steven and others (1967; Table 2), who reported two somewhat discordant age determinations of 12.4 and 15.6 m.y. on plagioclase concentrates, with a mean of 14.0 m.y. (Table 2, no. 14). Two new determinations on the top and bottom flows of a thick sequence capping Cannibal Mesa (Fig. 4, locs. 62, 63) yielded whole-rock ages of 18.0 ± 0.6 and 18.4 ± 0.6 m.y., respectively (Table 2, nos. 12, 13). This basalt sequence consists of at least 15 separate flows with an aggregate thickness of more than 200 m and is the thickest sequence recognized in the northwestern San Juan Mountains. The basalt flows at Cannibal Mesa are thus approximate time equivalents of intermediate-level flows that intertongue with sedimentary rocks of the Los Pinos Formation at several localities in the southeastern San Juan Mountains. The Cannibal Mesa dates also indicate that, as at the Los Mogotes center, thick local basalt sequences accumulated within geologically brief periods of time in the San Juan field.

Basalt flows younger than the accumulation around the Los Mogotes volcano have not been identified with confidence in the Colorado part of the San Juan volcanic field. Basalt flows near Red Mountain, about 20 km southwest of Green Ridge (Fig. 4, loc. 47), were named the latite of Red Mountain by Larsen and Cross (1956, p. 207), and were mistakenly believed to have flowed down the modern drainage of La Jara Creek in Quaternary time. Recent mapping (Steven and others, 1974) has demonstrated instead that these flows are preserved along the downthrown side of a major fault and that a distinctive sequence of three flows correlates across the fault with basaltic rocks mapped as Hinsdale Formation both by Larsen and Cross (1956, Pl. 1) and by the recent mapping.

A little-eroded small basalt flow in the northwestern San Juan Mountains was tentatively interpreted by Bush and others (1960, p. 463–464) to be of Pleistocene age, but this has not yet been confirmed by radiometric dating.

Pleistocene basalt flows, called the Brazos Basalt by Doney (1968), are present at the south edge of the San Juan field in the Tusas Mountains of northern New Mexico (Fig. 4). These flows are only slightly eroded and have long been considered Pleistocene in age on geomorphic grounds (Larsen and Cross, 1956, p. 207); they have now been dated at two localities, both of which yielded ages of 0.24 m.y. (Table 2, nos. 15, 16). The exact age agreement is considered coincidental, in light of sizable analytical uncertainties, but there seems no doubt about the youthfulness of these flows. In some respects it might be more reasonable to consider them a northern extension of the young Jemez volcanic field (Fig. 1) rather than a southern part of the San Juan field. Flows of the Brazos Basalt were erupted from a north-trending line of vents on strike with young faults extending out from the Jemez field, but they do not differ appreciably from older basalt of the Hinsdale Formation in petrography or chemistry (Table 1).

BASALT CHRONOLOGY ELSEWHERE IN THE SOUTHERN ROCKY MOUNTAIN REGION

Age data on upper Cenozoic basalt elsewhere in the region are relatively sparse, although a few areas have been studied in considerable detail. Available data are summarized here to provide a basis for discussion of regional variations in timing, distribution, and petrology of the basaltic volcanism.

Outside the San Juan volcanic field, the largest areas of upper Cenozoic basalts in the Colorado Rockies are mesa-capping erosional remnants of a formerly more extensive basalt field in the vicinity of the Flat Tops in north-central Colorado (Fig. 1). K-Ar determinations indicate that between 24 and 20 m.y. ago, alkali-olivine and silicic alkalic basalts were erupted in a widespread sequence as much as 230 m thick (Larson, 1968; York and others, 1971; Larson and others, 1975) that interfingers northward with the Browns Park Formation, a basin-fill accumulation that is approximately correlative with the Santa Fe Group. Between 14 and 10 m.y. ago, another widespread unit, predominantly basaltic andesite flows interbedded with fine-grained tuffaceous sedimentary rocks, accumulated to a thickness of as much as 200 m, and at times during the past 7.5 m.y., small amounts of alkali basalt were erupted from isolated centers (Larson and others, 1975).

Basalt as young as Holocene occurs locally in this same central Colorado region. For example, near Dotsero along the Colorado River 20 km southeast of the Flat Tops, a flow and cinder cone of ne-normative alkali-olivine basalt (analysis in Doe and others, 1969) is about 4,000 years old, on the basis of a C^{14} age on included charred wood (Giegengack, 1962). The total age range of these north-central Colorado flows is similar to those in the San Juan field, but the available age data suggest that the times of peak activity may have been significantly different (Fig. 7).

Small erosional remnants of silicic alkalic basalt that may represent outliers of the older basalt flows of the Flat Tops area occur 40 to 60 km (25 to 37 mi) farther northeast in Middle Park, where they overlie lower and middle Miocene basin-fill deposits. These in turn rest on calc-alkalic intermediate-composition volcanic rocks largely of Oligocene age (Izett, 1966, 1968). About 50 km northwest of the Flat Tops in the Elkhead Mountains, approximately equivalent basalt flows include undersaturated alkali-olivine types (Carey, 1955; Ross, 1926). Southwest of the Flat Tops at Grand Mesa (Fig. 1), a K-Ar age of 9.7 m.y. has been obtained on a basalt sample from the middle of a series of flows that aggregate about 250 m in thickness (U.S. Geological Survey, 1966, p. A81); thus, this sequence is at least equivalent in part to middle parts of the basalt section in the Flat Tops area.

In the Thirtynine Mile volcanic field (Fig. 1), intermediate-composition volcanic rocks of Oligocene age are overlain unconformably by gravel and by a sequence of mafic lava flows as much as 350 m thick, the basal flow of which is about 19 m.y. old (Chapin and Epis, 1964; Epis and Chapin, 1968). Although these flows have been designated the "upper andesite" by Epis and Chapin, they consist mainly of silicic alkalic basalt and basaltic andesite and are lithologically similar to flows of related age in the Hinsdale Formation in the San Juan volcanic field (R. C. Epis, 1970, written commun.).

Late Cenozoic fundamentally basaltic volcanic associations in northern New Mexico are lithologically more diverse than those in the Colorado Rockies. A southwest-trending axis of Pliocene and Pleistocene volcanism extends across the Rio Grande depression from the High Plains of northeastern New Mexico, through the Jemez Mountains, to the southeastern margin of the Colorado Plateau (Fig.

1). Basaltic, differentiated alkalic, and bimodal basalt-rhyolite volcanic fields occur along this axis and are associated in a number of places with extensional normal faulting.

The large lava field on the High Plains in northeastern New Mexico (Fig. 1) consists largely of Pliocene to Holocene basaltic lava flows, with which are associated scattered central cones of andesite and rhyodacite (Lee, 1922; Stobbe, 1949; Baldwin and Muehlberger, 1959; Aoki, 1967b; and Stormer, 1972a). The basaltic lava ranges from silicic alkalic basalt that is petrologically similar to basalt of the Hinsdale formation in the San Juan field to distinctive highly alkalic olivine basalt that is low in silica and contains groundmass feldspathoidal minerals. The older basalt flows (Raton and Clayton Basalts of Collins, 1949) cap high mesas and have been extensively eroded, but the younger flows (Capulin Basalts of Collins, 1949) follow present drainages and are only slightly eroded. Several of the older flows have yielded K-Ar ages of 7 to 8 m.y., and other major units were erupted 2 to 3.5 m.y. ago (Stormer, 1972b). One of the youngest basaltic centers of this area, Capulin cone (about 45 km southeast of Raton; Fig. 1), was active between 8000 and 2500 B.C. (Baldwin and Muehlberger, 1959).

Southwest of Raton, in the Cimarron Mountains area (Fig. 1), a large area of little-studied upper Cenozoic mafic lava flows laps across the structural boundary between the High Plains and the Southern Rocky Mountains. These flows, which seem to consist mainly of silicic alkalic basalt and basaltic andesite (Table 3; Aoki and Kudo, 1973), have yielded two late Pliocene K-Ar dates (4.7, 3.3 m.y.; Stormer, 1972b). Unfortunately, no adequate areal geologic context in which to explain the volcanic history of the area has yet been developed.

Basaltic lava flows are also present along the crest of the southern Sangre de Cristo Mountains, just south of the Colorado–New Mexico border (C. L. Pillmore, 1968, written commun.). These flows, which appear to be silicic alkalic types, overlie all older intermediate-composition volcanic rocks in the area but have not

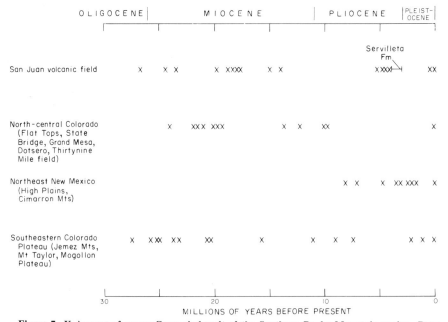

Figure 7. K-Ar ages of upper Cenozoic basalt of the Southern Rocky Mountain region. Data sources discussed in text.

been dated by radiometric methods. They are probably older than the basalt on the High Plains farther east and are approximately correlative with Hinsdale basalt flows of the San Juan Mountains.

Basalt of the Servilleta Formation that flooded the southern San Luis Valley consists mainly of flows of distinctive coarse-grained diktytaxitic olivine tholeiite (Aoki, 1967a; Lipman, 1969). The flows form an impressive volcanic plateau north of Taos, New Mexico, and are as much as 200 m thick in the Rio Grande gorge, where they have been dated as 3.6 to 4.5 m.y. old (Ozima and others, 1967). Tholeiitic flows that appear to be somewhat younger than the dated basalt are present on the gentler slopes that are stepped back from the gorge to the east and west. Interlayered in the Rio Grande gorge with the typical tholeiite are subordinate flows of fine-grained dense silicic alkalic basalt. Cinder-cone vents of similar alkalic flows are exposed along the western margin of the southern San Luis Valley, but it is not clear whether the alkalic basalt flows exposed along the Rio Grande gorge were erupted from vents within the main part of the rift or flowed in from the sides. These flows petrologically resemble nearby silicic alkalic basalt in the southeastern San Juan field. Also exposed along the western margin of the San Luis Valley are several distinctive flows of alkalic basaltic andesite that are characterized by abundant resorbed xenocrysts of quartz and plagioclase. These flows are similar to upper Cenozoic xenocrystic basalt that occurs within the San Juan field (Larsen and others, 1938; Doe and others, 1969; Lipman, 1969, Pls. 1, 2). Some of these xenocrystic flows are definitely older than the olivine tholeiite basalts, but others may be younger than some of the tholeiitic flows.

Also at the southern end of the San Luis Valley, within the main part of the Rio Grande depression, is a cluster of at least 12 little-eroded central volcanoes (Lambert, 1966). Individual cones tend to be monotonously uniform in composition, but nearby cones vary greatly, ranging from dark-gray alkalic olivine andesite to white silicic alkalic rhyolite (P. W. Lipman, unpub. analyses). Several of the cones are onlapped by tholeiitic Servilleta basalt, and one—an alkalic rhyolite plug dome at No Agua, about 60 km northwest of Taos, New Mexico—has yielded a fission-track age on obsidian of 4.8 m.y. (G. A. Izett, 1968, written commun.; Lipman and others, 1970). This indicates that at least some of these San Luis Valley volcanoes are only slightly older than the Servilleta Formation. In association with nearby alkalic basaltic volcanoes of similar age, such as Los Mogotes in the southeastern San Juan Mountains, the volcanoes of the southern San Luis Valley are interpreted as a fractionated igneous series related to an alkalic basaltic parent magma, a typical component of the fundamentally basaltic association (Christiansen and Lipman, 1972).

The Jemez volcanic field on the western margin of the Rio Grande depression in north-central New Mexico (Fig. 1) is an exceptionally complex volcanic assemblage of Pliocene and Quarternary age. The Jemez field has been interpreted (Smith and Bailey, 1968; Smith and others, 1970) as constituting four separate volcanic cycles, the first and last being bimodal basalt-rhyolite sequences and the middle two being continuous differentiation series from basaltic parent magmas. Basaltic volcanism, including both olivine tholeiite and silicic alkalic types, was intermittent in or around margins of the field during each of the four cycles. Rocks of the first cycle were erupted more than 9.1 m.y. ago and are thought to be early Pliocene in age (Late Miocene by some interpretations); climactic ash-flow eruptions of the fourth cycle occurred at 1.4 and 1.1 m.y.; and the youngest volcanic activity in the field probably occurred not more than 100,000 years ago (Bailey and others, 1969).

Petrologically diverse Pliocene and younger basalt and associated more silicic rocks also occur in the vicinity of Mount Taylor (Hunt, 1938; Baker and Ridley, 1970; Lipman and Moench, 1972) near the south margin of the Colorado Plateau, about 50 km west of the Rio Grande depression (Fig. 1). The oldest basalt units in the Mount Taylor area are flows and necks of mainly low-silica ne-normative alkali-olivine basalt (Table 1) that were emplaced prior to formation of the latitic and andesitic Mount Taylor stratovolcano about 2.6 m.y. ago (Bassett and others, 1963). These early flows are probably Pliocene in age, although they could be older (Lipman and Moench, 1972). Lapping onto the flanks of the stratovolcano are flows of intermediate age, which now cap erosional mesas around Mount Taylor and consist of silicic alkalic basalt that is generally similar in petrology to basalt of the San Juan field (Table 1). Later Pleistocene or Holocene basalt of the Mount Taylor area, which occurs near modern drainage levels, includes both alkalic and tholeiitic types. Similarly young alkalic and tholeiitic basalt flows extend southwest from Mount Taylor for several hundred kilometers along the south margin of the Colorado Plateau (Renault, 1970; Kudo and others, 1971; Laughlin and others, 1971, 1972b). Much older silicic alkalic basalt and basaltic andesite also occurs around the south margin of the Colorado Plateau. Samples of these rocks, dated as old as 27.5 m.y., have come from the Mogollon Plateau region of west-central New Mexico (Elston and others, 1973). The basaltic lava and associated more silicic lava and tuff range in age from about 28 to 20 m.y. and overlie predominantly intermediate-composition 28- to 38-m.y.-old rocks (Elston and others, 1973). We interpret them to belong to the fundamentally basaltic late Cenozoic volcanic association.

These data, though fragmentary, suggest that basaltic lava was erupted in parts of the Southern Rocky Mountain region almost continuously through late Cenozoic time, although only brief periods of activity seem to be represented within any small area. In the few areas where age data are adequate, they indicate that thick accumulations of basalt were erupted within short time intervals: at Cannibal Mesa in the western San Juan Mountains about 18 m.y. ago, in the Flat Tops area between about 24 and 20 m.y. ago and again between 14 and 10 m.y. (Larson, 1968; Larson and others, 1975; York and others, 1971), at Los Mogotes volcano in the southeastern San Juans about 5 m.y. ago, within the Rio Grande depression near Taos between 4.5 and 3.6 m.y. ago (Ozima and others, 1967), and on the High Plains in northeastern New Mexico and southern Colorado in the Pleistocene and Holocene Epochs (Stormer, 1972b). Radiometric data document sizable periods of inactivity between episodes of intense basaltic volcanism in the southeastern San Juan Mountains and in the Flat Tops area. Despite this episodicity, however, the available data suggest that there was no regionwide coordination of volcanic events. For instance, the younger period of intense activity in the Flat Tops area, between 14 and 10 m.y. ago, falls within the long period of apparent quiescence of basaltic activity between about 15 and 5 m.y. ago in the southeastern San Juan Mountains (Fig. 7). Moreover, no temporal migration of activity along a geographic trend is discernible from the present data, such as might be expected if the late Cenozoic basaltic activity were related to an asthenosphere "hot spot" source over which the lithosphere moved independently, in the fashion suggested for oceanic island intraplate volcanoes (Wilson, 1963; Morgan, 1971; Jackson and others, 1972). For example, basaltic centers extending in a linear southwest-trending belt along the south side of the Colorado Plateau from the High Plains, through the Cimarron Mountains, Jemez Mountains, and Mount Taylor, to the White Mountains of east-central Arizona (which have been inferred to reflect the trace of a mantle hot spot [Suppe and others, 1973]) seem to have been sporadically active at both northeast and southwest ends from Pliocene (or late Miocene), through Pleistocene, and even into Holocene time. Similarly, there seems to be no ordered time sequence of activity within the Rio Grande depression, despite an apparent longitudinal petrologic variation from alkalic basalt in the southern depression to tholeiitic activity in the northern part (Lipman, 1969, p. 1350; Aoki and Kudo, 1975).

At present it is unclear whether extensional faulting, which seems to have begun about 26 m.y. ago in the San Luis–Arkansas River segments of the northern Rio Grande depression, began later in the south, where the oldest reported fossils in the valley fill are of middle Miocene age (Galusha and Blick, 1971), or whether the lowest parts of the valley fill in the southern areas are hidden or so far unrecognized. The transition from predominantly andesitic to fundamentally basaltic volcanism appears to have occurred in central New Mexico at about the same time as, or even slightly earlier than, in southern Colorado (Christiansen and Lipman, 1972). This suggests a similar volcano-tectonic history for both areas. The oldest dated material interlayered with basin fill in central New Mexico is from 20.6-m.y.-old basalt flows, as reported by Elston and others (1973); this date suggests to those authors that faulting began considerably later than the change in volcanic rock types, but the dated flow may represent only a marginal facies and not necessarily the lowest part of the basin-fill deposits in the area. Alignment of vents of 26 ± 1-m.y.-old rhyolite lava domes along the trends of basin-range faults in the Mogollon Mountains of west-central New Mexico (Ratté and others, 1975) suggests that this tectonic pattern was operative earlier than inferred by Elston and others (1973).

GEOGRAPHIC VARIATIONS IN BASALT COMPOSITION

A few years ago, Lipman (1969) suggested, on the basis of fragmentary data taken mainly from published literature, that upper Cenozoic basalt flows in the Southern Rocky Mountain region vary systematically in petrology with distance from the Rio Grande depression. Basalt and basaltic andesite of alkalic affinities erupted east and west of the depression, whereas tholeiitic basalt accumulated within the depression late in its evolution. The lateral change from tholeiitic to alkalic volcanism was interpreted to reflect different conditions of magma generation in the mantle. More recent studies of basaltic rocks in the region, while generally supporting these interpretations, have shown that the detailed geographic, petrologic, and age relations between rifting and basaltic volcanism are more intricate than originally recognized (Leeman and Rogers, 1970; Renault, 1970; Kudo and others, 1971; Stormer, 1972a, 1972b; Hedge and Lipman, 1972; Lipman and others, 1973; Aoki and Kudo, 1975).

About 200 major-element analyses are now available for upper Cenozoic basaltic rocks of the Southern Rocky Mountain region, including 25 new analyses (Table 3). These analyses define major compositional variations both transversely and longitudinally along the Rio Grande depression.

The transverse variations are especially evident from plots of K_2O or total alkalis versus SiO_2, which show that at any given SiO_2 content, alkali contents of the basaltic rocks are higher on each side of the depression than within it, as observed previously (Lipman, 1969; Lipman and others, 1973). The newly available analyses also indicate that, within geographically restricted areas, total alkalis tend to increase progressively away from the depression. So, basalt of the western San Juan volcanic field is distinctly higher in alkalis than otherwise petrographically similar basalt from eastern parts of the field, and basalt from central areas shows intermediate values (Fig. 8A). A similar pattern seems valid for the upper Cenozoic basalt of northeastern New Mexico, where flows in the Cimarron Mountains area on the east flank of the Sangre de Cristo Mountains are less alkalic than those farther east on the High Plains (Fig. 8B).

These overall transverse variations in alkali contents across the Southern Rocky Mountain region, approximately symmetrical with respect to the Rio Grande depression, seem to persist despite the widely differing ages (from early Miocene to Holocene) of the basalt samples plotted. Thus, in the southern San Juan field, the Pleistocene Brazos Basalt (of Doney, 1968) of the Tusas Mountains is compositionally similar to nearby Miocene basalt of the Hinsdale Formation. However,

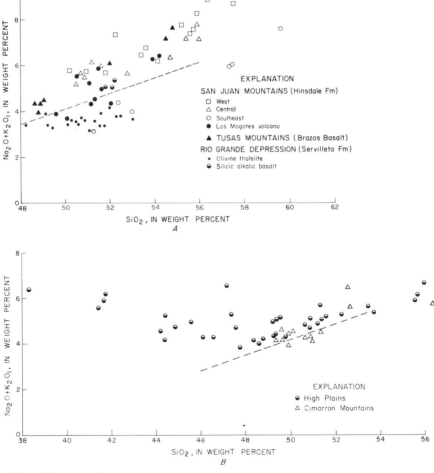

Figure 8. Plots of total alkalis and SiO$_2$ for selected basalt samples of the Southern Rocky Mountain region. Data are from Aoki (1967a, 1967b), Aoki and Kudo (1975), Baldwin and Muehlberger (1959), Larsen and Cross (1956), Lipman (1969), Stormer (1972a), and Table 3. All analyses plotted have been recalculated to 100 percent without H$_2$O or CO$_2$. Dashed line separates fields of Hawaiian tholeiitic and alkalic basalt (Macdonald and Katsura, 1964). A. Basalt of the San Juan Mountains and the Rio Grande depression. B. Basalt east of the Rio Grande depression (Cimarron Mountains and High Plains areas).

as emphasized by Stormer (1972b) and as documented earlier in this paper, distribution both of activity and compositional type of basalt tended to be episodic and asymmetrical to the Rio Grande depression within more restricted time intervals. For example, no analogs of the extremely low-silica alkali-olivine basalt (basanite) of the High Plains have been recognized within the San Juan field to the west of the depression, although similarly alkalic basalt does occur west of the depression in the Mount Taylor area of north-central New Mexico (Lipman and Moench, 1972) and in the Hopi Buttes region of northeastern Arizona (Williams, 1936).

Distributions of the basalt types with respect to the Rio Grande depression are also complicated by the presence within the depression of subordinate amounts of silicic alkalic basalt that petrographically resembles the dominant basalt types to the east and west (Lipman, 1969, p. 1349). Further, the andesitic to rhyolitic central volcanoes of the southern San Luis Valley have major-element compositional trends that follow continuations of trends of the silicic alkalic basalt, rather than those of the tholeiitic basalt. Thus, significant alkalic basaltic and related volcanism, in addition to the tholeiitic activity, emanated from within the depression during its evolution. Probably more significant than the ubiquitous distribution of the alkalic activity is the confinement of olivine tholeiite, low in K_2O, P_2O_5, TiO_2, U, Th, Ba, Rb, and Sr, to the northern and central parts of the depression. This tholeiite rock type is represented by the basalt of the Servilleta Formation within the southern San Luis Valley and the Espanola Basin; farther south in the central part of the depression, similar tholeiitic basalt occurs around the flanks of the Jemez Mountains and in the hills northwest of Albuquerque (Aoki and Kudo, 1973). However, basalt samples from the Cimarron Mountains east of the depression, described as tholeiitic on the basis of a normative classification (Kudo and others, 1971), have notably higher concentrations of the elements listed above than olivine tholeiites of the depression (Fig. 8B).

A similar pattern of transverse compositional variations of basalts in and adjacent to the central and northern Rio Grande depression is also strikingly evident from variations in Th and U contents and Th/K and U/K ratios, which all consistently increase away from the depression (Lipman and others, 1973). The Servilleta olivine tholeiite, which is volumetrically the predominant basalt type in the northern and central parts of the depression, has the lowest K, Th, and U contents of any basalt in the region (Fig. 9). The silicic alkalic basalt, which is the dominant type in the Southern Rocky Mountain region adjacent to the Rio Grande depression, not only has higher K, Th, and U contents than the tholeiitic basalt, but the Th/K and U/K ratios also increase with distance from the Rio Grande depression (Fig. 9). The lowest ratios of these elements occur in the basalt along the southeast margin of the San Juan field on the west margin of the Rio Grande depression and in the sparse similar silicic alkalic basalt interlayered with the tholeiitic basalt farther south within the depression (Fig. 9A). These ratios are higher in the silicic alkalic basalt of the Cimarron and Tusas Mountains, respectively, east and west of the Rio Grande depression; they are even higher in the otherwise petrologically similar basalt of the Mount Taylor area and basalt of the High Plains at even greater distances from the Rio Grande depression. Extremely high Th and U contents, as much as 30 and 8 ppm, respectively, and also high Th/K and U/K ratios, occur only at great distances from the Rio Grande depression: on the High Plains to the east and at Mount Taylor on the southeastern Colorado Plateau (Fig. 9).

South of Albuquerque, New Mexico, where the Rio Grande depression merges with the Basin and Range province, these transverse compositional relations no longer can be recognized. Upper Cenozoic basalt flows both within the depression and on each side seem to be mainly alkalic olivine basalt and basaltic andesite

(Lipman, 1969, p. 1350; Kudo and others, 1971, p. 201; Aoki and Kudo, 1975; Hoffer, 1971), similar to typical upper Cenozoic basalt in the central Basin and Range province to the west (Leeman and Rogers, 1970). Thus, a distinct longitudinal compositional variation of the basalt types within the Rio Grande depression seemingly can be related to changes in the structure of the depression: olivine tholeiite and a transverse compositional gradient of basalt are present only as far south as the depression constitutes a well-defined extensional structure separating the relatively stable regions of the Colorado Plateau and the High Plains.

Petrologically diversified basalt fields that contain tholeiitic types seem closely associated with major structural discontinuities around margins of the region of late Cenozoic extensional (basin-range) faulting in the western United States. Thus, in addition to their occurrence along the Rio Grande depression, basaltic rocks

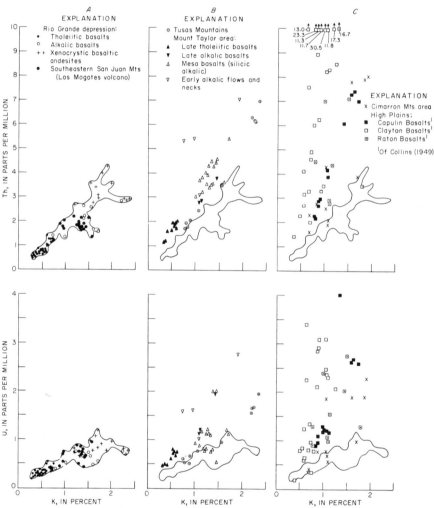

Figure 9. Plots of K, Th, and U contents of basalt of the Southern Rocky Mountain region (from Lipman and others, 1973). A. Basalt of the Rio Grande depression and the southeastern San Juan Mountains. Empirically estimated field boundary of the Rio Grande depression basalt is repeated in B and C. B. Basalt west of the Rio Grande depresssion (Tusas Mountains and Mount Taylor area). C. Basalt east of the Rio Grande depression (Cimarron Mountains and High Plains areas).

of tholeiitic affinities occur around the boundary between the Basin and Range province and the Colorado Plateau of northern New Mexico in the Mount Taylor field (Lipman and Moench, 1972) and in the San Francisco Mountain and Verde River areas of central Arizona (Robinson, 1913; McKee and Anderson, 1971, 1970, written commun.). In the block-faulted Shivwits Plateau area along the west side of the Colorado Plateau, the most abundant basalts (Grand Wash and Hurricane types of Best and Brimhall, 1970) are chemically transitional toward tholeiitic types (Table 1) and have textural features such as subophitic diktytaxitic texture, coarse grain size, and groundmass glass similar to tholeiitic basalts of the Mount Taylor and Rio Grande areas (Best and others, 1966, 1969). Near the western margin of the Basin and Range province, tholeiitic basalt occurs in the Mojave Desert and Owens Valley areas of California (Wise, 1969; Leeman and Rogers, 1970; W. P. Leeman, 1968, written commun.). Upper Cenozoic tholeiitic basalt flows are abundant at the northern edge of the Basin and Range province in the basalt plateaus of northwestern Nevada, northeastern California, and southeastern Oregon (Powers, 1932; Waters, 1962; LeMasurier, 1968; Gunn and Watkins, 1970) and in the Snake River Plain-Yellowstone National Park area (Stone, 1967; Christiansen and Blank, 1969). Tholeiitic basalt flows are seemingly sparse in the interior of the Basin and Range province; of the five analyses of tholeiites plotted in a recent paper on basalt from the Basin and Range province (Leeman and Rogers, 1970), one sample is from the Mount Taylor field, and the others are from marginal regions in northwestern Nevada and eastern California (W. P. Leeman, 1968, written commun.).

Very undersaturated alkalic basalt and basanite also occur mainly around margins of the Basin and Range province, in many of the same basaltic fields as the tholeiites (Mojave Desert—Wise, 1969; western Colorado Plateau—Best and others 1966, 1969; Mount Taylor—Lipman and Moench, 1972), and also in the tectonically stable regions beyond (Hopi Buttes in the middle of the Colorado Plateau—Williams, 1936; High Plains of northeastern New Mexico—Stobbe, 1949; Baldwin and Muehlberger, 1959; Aoki, 1967b). We know of only a single area of Cenozoic basanitic basalt in the interior Basin and Range province: the southern Pancake Range in Nevada (Vitaliano and Harvey, 1965; Trask, 1969; Scott and Trask, 1971). The basanitic basalt samples plotted in the paper by Leeman and Rogers (1970) are from the localities listed above (W. P. Leeman, 1968, written commun.).

In these areas and in most other Cenozoic volcanic areas in the Basin and Range province, however, the dominant basalt type is silicic alkalic basalt similar to the most widespread basalt type of the Southern Rocky Mountain region. Petrologically diverse basaltic suites that probably reflect magma generation over a sizable depth range in the upper mantle seem to have formed most readily in the western United States along major tectonic discontinuities (that is, boundaries of structural provinces). This relation may indicate that the discontinuities extend through the sialic crust and perhaps that they affected the entire lithospheric plate and influenced magma generation at upper mantle depths.

CONDITIONS OF MAGMA GENERATION

The compositional variations among upper Cenozoic basalt flows of the Southern Rocky Mountain region could have resulted from three major types of processes: (1) intermediate- to low-pressure fractional crystallization of the magma prior to eruption, (2) contamination interactions with sialic crust during rise of the magma to the surface, and (3) variable conditions of magma generation and high-pressure fractionation at depth within the upper mantle. By combining mineralogic, elemental, and isotopic compositional data, each of these possibilities can be evaluated under favorable conditions. These alternatives for basalt of the Southern Rocky Mountains and adjacent areas have been discussed at length elsewhere (Lipman, 1969; Doe and others, 1969; Kudo and others, 1971; Stormer, 1972a; Hedge and Lipman, 1972) and will only be summarized briefly here. For most basalt of the Southern Rocky Mountains, low-pressure fractionation and crustal contamination seem to have been relatively unimportant; the major compositional variations are seemingly due to differing degrees and (or) pressures of melting at depth within the upper mantle.

Low-Pressure Fractional Crystallization

An important constraint on any interpretation of basaltic magma genesis in the Southern Rocky Mountain region is that few of the basalt suites appear to have been affected significantly by low-pressure differentiation involving fractional crystallization and separation of phenocrysts. This sort of crystal-liquid fractionation seems to have played a significant part in producing major compositional variations in oceanic-island basaltic shield volcanoes such as Kilauea, Hawaii, where much of the observed variation can be accounted for by olivine fractionation and mixing of different batches of such fractionated lava (Powers, 1955; Murata and Richter, 1966; Wright and Fiske, 1971; Wright, 1971).

With few exceptions, olivine is the only phenocryst phase of the Southern Rocky Mountain basalt flows, and addition or subtraction of olivine cannot account for

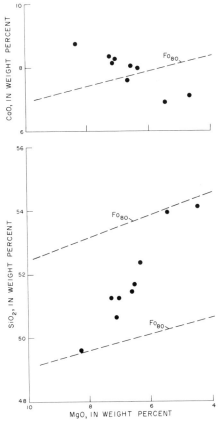

Figure 10. MgO variation diagrams for basalt of Los Mogotes volcano (data from Larsen and Cross, 1956, Table 25, nos. 3, 5; Lipman, 1969, Table 1, no. 2; Table 3). Control lines are for olivine composition of Fo_{80}, the approximate optically determined composition for phenocrysts from Los Mogotes basalt.

observed compositional variations among closely related flows. This relation is especially clear for lava flows of the small 5-m.y.-old Los Mogotes shield volcano in the southeastern San Juan field (Fig. 1). A sequence of at least 12 flows from this center, exposed where the Conejos River has cut across the south flank of the shield, displays a well-defined compositional trend from basaltic andesite, containing about 54 percent SiO_2 at the base, upward to typical silicic alkalic basalt, with about 50 percent SiO_2 at the top. Olivine is the only phenocryst phase throughout, yet the compositional trends are clearly unrelated to olivine control lines, as is readily evident on MgO variation diagrams (Fig. 10). This is in striking contrast to closely related basaltic suites from oceanic-island volcanoes. The necessary interpretation for the Los Mogotes sequence, especially when isotopic and minor-element data that preclude significant crustal contamination are considered, is that high-pressure fractionation is responsible for all major variations in even this limited sequence from a single volcano. Similar arguments are applicable to other compositionally variable basaltic sequences of localized age and geographic distribution, such as the tholeiitic basalt of the Servilleta Formation, which also contains only olivine phenocrysts. Also, an analyzed segregation vein from a Servilleta flow (Aoki, 1967a, Table 1), which clearly represents a residual liquid concentrated during crystallization of the basalt at the surface, defines a composition trend that is discordant to the trends of the overall suite of bulk samples (Aoki, 1967a). In a detailed petrologic study of the basalt of the High Plains area in northeastern New Mexico, Stormer (1972a) similarly concluded that the major compositional variations among the different basalt suites could not have resulted from low-pressure differentiation (or crustal contamination) and must have been due to high-pressure fractionation.

Effects of Crustal Contamination

Compelling evidence from concentration and isotopic data shows that the differences between the three major compositional basalt types of the region—the alkali-olivine basalt, silicic alkalic basalt, and olivine tholeiite—do not result from variable crustal contamination.

First, variations in major-oxide and minor-element concentrations among these main basalt types are such that none could have been generated from another simply by contamination by sialic crust of any plausible composition. For example, addition of sialic material to olivine tholeiite magma, which could feasibly produce concentrations of most major oxides similar to those in the silicic alkalic basalt, cannot account for variations in minor elements such as Sr. Sr content is about twice as high in the silicic alkalic basalt as in the olivine tholeiite (Table 1). Contamination by any likely material would lower Sr concentrations in basalt flows, rather than increase them. It is similarly difficult to derive silicic alkalic basalt by contamination of alkali-olivine basalt, which, though lower in SiO_2, is about the same in alkalis and is much higher in Th and U—elements that would be further increased by crustal contamination. More severe difficulties come in attempting to generate other basalt types by contamination models, for example, olivine tholeiite from alkali-olivine basalt, or vice versa.

The interpretation that crustal contamination was of limited significance in producing the variations among basalt flows of the Southern Rocky Mountains and adjacent regions is supported by Sr isotopic data, which show low and relatively uniform isotopic ratios (Kudo and others, 1971; Leeman, 1970; Hedge and Lipman, 1972). With the exception of one suspect sample, the range in 31 samples from the Southern Rocky Mountain region is only 0.7035 to 0.7050 (Fig. 11)—only about twice the standard deviation of the analytical precision (Hedge and Lipman, 1972).

Furthermore, no correlation is evident between isotopic ratio and geographic position or age or bulk-rock elemental composition. Basalt with low Sr concentrations (or high Rb/Sr ratios) should be especially sensitive to contamination effects, but the range in isotopic ratios of the Sr-poor tholeiitic rocks within the Rio Grande depression (0.7042 to 0.7048) is bracketed by that of Sr-rich alkali-olivine basaltic rocks of the High Plains (0.7035 to 0.7049). Anomalously high Sr isotopic ratios have been observed in a few basalt samples with low Sr concentrations in central New Mexico (Kudo and others, 1971), mainly from a single isotopically heterogeneous flow (Laughlin and others, 1972a). These have reasonably been interpreted as reflecting minor crustal contamination, but the other Southern Rocky Mountain basalt flows that are isotopically homogeneous have been considered essentially unmodified by crustal contamination (Hedge and Lipman, 1972).

In contrast to the Southern Rocky Mountain basalt flows, upper Cenozoic basalt farther south in New Mexico and in adjacent parts of southern Arizona commonly has distinctly lower Sr^{87}/Sr^{86} ratios (Fig. 11), indicating a differing composition of the source region of these magmas (Hedge and Lipman, in prep.).

Special interpretive problems are offered by the petrographically distinctive xenocrystic basaltic andesite types, which have been interpreted in recent years as resulting from one or more processes, including crustal contamination (Doe and others, 1969), high-pressure crystallization (Nicholls and others, 1971), or mixing of mafic and silicic magmas (U.S. Geological Survey, 1971, p. A42). The major-oxide compositions of these rocks are compatible with formation by crustal contamination of the widespread silicic alkalic basalt type. The Sr concentrations of these rocks are too high for the Sr isotopic ratios to be sensitive to the relatively modest degree of contamination required to account for the elemental concentration differences. Nevertheless, Pb isotopic compositions of three xenocrystic basaltic andesite samples from the Southern Rocky Mountains differ from associated

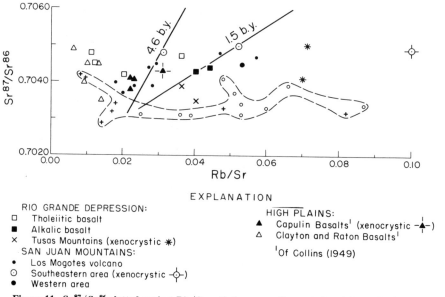

Figure 11. Sr^{87}/Sr^{86} plotted against Rb/Sr ratio for upper Cenozoic basalt from the Southern Rocky Mountain region (from Hedge and Lipman, in prep.). Upper Cenozoic basalt from the Basin and Range province in southern New Mexico and Arizona also plotted (enclosed by dashed line); data from Damon and others (1969, small crosses) and Leeman (1970, small circles).

nonxenocrystic basalt samples and are suggestive of crustal contamination (Doe and others, 1969), although the isotopic variability of seemingly uncontaminated volcanic rocks from the same region is a factor that weakens this argument. Also, the xenocrystic basaltic andesite types commonly have high Rb/Sr ratios that plot separately from Rb-Sr trends defined by groups of nonxenocrystic basalt types of the same area (Hedge and Lipman, 1972). These Rb/Sr relations are more readily interpreted in terms of contamination or magma-mixing models than by the high-pressure crystallization hypothesis, but the origin of these distinctive mafic rocks remains uncertain.

An important further insight into the nature of the source material of the Southern Rocky Mountain basalt is afforded by the Sr and Pb isotopic data just discussed. Figure 11 shows that many of these basaltic rocks do not contain enough Rb relative to Sr to have produced their observed Sr isotopic ratios, even given the entire 4.6 b.y. of the Earth's history. That is, they plot to the left of an isochron for the age of the Earth (Fig. 11). Because a liquid produced by partial melting will have a Rb/Sr ratio greater than that of the residual fraction, the Rb/Sr ratio of the source region for the basalt must have been even lower than the ratios observed in the samples. Similar anomalies have been noted for oceanic ridge basalt (Tatsumoto, 1966) and for a group of Cenozoic basalt types in western Nevada and eastern California (Hedge and Noble, 1971). Such basalt would seem to have been derived from a source that had been depleted in Rb by previous magma generation, as suggested by Gast (1968).

Some understanding of the nature of such a previous fractionation event may be offered by Pb isotope studies of Tertiary volcanic rocks of the San Juan field, which suggest that the source region for these rocks underwent a major geochemical fractionation about 1.7 to 1.8 b.y. ago, the approximate time of major sialic crustal formation in this area (B. R. Doe, 1970, written commun.). A similar geochemical event in the Northern Rocky Mountains, approximately contemporaneous with crustal evolution there about 2.8 b.y. ago, has been inferred from Pb isotopic data for the Absaroka Volcanic Supergroup (Peterman and others, 1970). These relations suggest that major fractionation of the mantle source region of the upper Cenozoic basalt of the Southern Rocky Mountains occurred at the time of development of sialic crust in this area 1.7 to 1.8 b.y. ago and that the composition of the underlying mantle has survived, with only slight compositional modification, to the present.

High-Pressure Fractionation

Crystallization experiments on simplified mafic systems and on actual basaltic materials at high pressures have demonstrated that basaltic magmas can be generated by partial melting of probable mantle constituents and that the composition of the basaltic melt is largely determined by the depth (pressure) of melting and the proportion of melt relative to residuum (Green and Ringwood, 1967; Green, 1968; Ito and Kennedy, 1967; Kushiro, 1968, among others). The nature of the basaltic melt is also strongly influenced by the presence of volatiles within the region of melting (Wyllie, 1971; Kushiro, 1972) and by the degree to which the melt equilibrates with mantle materials as it rises (O'Hara and Yoder, 1967). These experimental results generally suggest that tholeiitic basalt should result when the partial melting is proportionately large and occurs at shallow depth, whereas progressively more alkalic and less saturated melts are to be expected when the melting is of smaller proportion and occurs at greater depth. The role of volatile constituents is thought to be especially significant when the proportion of melt generated is small. These relations have been widely cited to explain the origin of compositionally diverse basalt suites in the western United States (Wise, 1969; Lipman, 1969; Leeman and Rogers, 1970; Baker and Ridley, 1970; Kudo and others, 1971; Stormer, 1972a; Condie and Barsky, 1972) and also to account for basalt variations in oceanic regions (Kuno, 1959, 1968; Jackson and Wright, 1970).

Admittedly, detailed extrapolation of experimentally determined pressure and depth relations for basalt generation can only approximate natural situations because the composition, proportion of melting, and volatile content of the parental mantle material are uncertain. Nevertheless, localization within the northern Rio Grande depression of olivine tholeiite that is thought to represent melting or last equilibration at relatively low pressures (less than 35 km depth, according to Green and Ringwood, 1967) suggests that generation of tholeiitic magmas may be possible within continental regions only where crustal attenuation or other major structural discontinuities permit local upwelling of mantle material and fractionation of basaltic magma at atypically shallow crustal levels. The volumetrically and areally predominant silicic alkalic basaltic rocks are thought to have been generated at greater depths, probably with a lesser degree of partial melting. Accordingly, the alkali-olivine basalt and basanite are interpreted as having been generated at even greater depth, with a still smaller fraction of partial melt and a relatively high volatile content.

Available heat-flow and deep magnetic sounding data indicate that the Southern Rocky Mountain area is one of relatively high heat flow (similar to other regions of crustal extension) between regions of lower heat flow on the Colorado Plateau and the High Plains (Roy and others, 1972; Porath, 1971; Edwards and others, 1973; Decker, 1973). Available seismic data indicate that the Southern Rocky Mountain region is characterized by sialic crust 40 to 55 km thick (Jackson and Pakiser, 1965); this seems too thick to permit generation of tholeiitic magmas in the underlying mantle, but existing data are inadequate to determine whether crustal structure varies locally across the northern and central Rio Grande depression. A longitudinal seismic profile down the axis of the depression would probably be required.

The Sr and Pb isotopic data for basalts of the Southern Rocky Mountain region indicate that any mantle upwelling along the Rio Grande depression must have been limited, because the source region for generation of all the basalt types seems to have been within that part of the lithospheric upper mantle of the American plate that has been little modified geochemically since formation of the sialic crust in this region about 1.7 b.y. ago. This lithospheric source region for the basalt magmas contrasts with inferred zones of magma generation in oceanic regions, where the lithosphere is much thinner and magma generation probably occurs at its base or within the underlying asthenosphere (Fig. 12).

Considering their contrasting environments of magma generation, the similarities in compositional range between the basalt in the Southern Rocky Mountains and basalt in ocean basins are remarkable. The most significant contrast—the volumetric prevalence of silicic alkalic basalt in the Southern Rocky Mountains, in contrast with the dominant tholeiitic basalt of ocean floors and the lower parts of oceanic-island volcanoes—probably reflects this difference in environment of magma generation. Diapiric upwelling, bringing hot deeper mantle material to a level where melting could begin by intersection of the goetherm with the peridotite solidus, presumably was slower within lower parts of continental lithosphere than within the less viscous oceanic asthenosphere and retarded both the volume of magma generated and the proportion of melting that could occur at any level within an upwelling system.

The upper Cenozoic basalt flows of southern New Mexico and Arizona that are relatively nonradiogenic in Sr (Fig. 11) may represent an intermediate site of magma generation in relation to the lithosphere-asthenosphere boundary. These flows, which occur in an early formed part of the Basin and Range province (Christiansen and Lipman, 1972), were erupted in regions previously characterized by middle Tertiary andesitic volcanic activity seemingly derived from mantle source regions but characterized by notably higher Sr^{87}/Sr^{86} ratios (Damon and others, 1969). This apparent change in Sr isotopic composition of a mantle source region could have resulted from gradual late Cenozoic upwelling of asthenospheric mantle with isotopic characteristics similar to source regions of oceanic basaltic magmas to sufficiently shallow levels in the southern Basin and Range province that it could provide a source for the younger basaltic magmas of this region. Similar asthenospheric upwelling does not seem to have been sufficient to influence magma generation farther north in more recently activated parts of the Basin and Range province or in the restricted zone of extensional faulting along the Rio Grande depression in the Southern Rocky Mountains. In both regions the upper Cenozoic volcanic rocks are relatively radiogenic in Sr, as are earlier Tertiary and Mesozoic igneous rocks of the same areas (Kistler and Peterman, 1972; Hedge and Lipman, 1972).

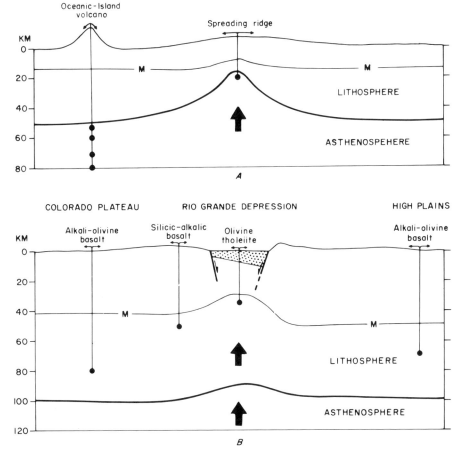

Figure 12. Diagrammatic cross sections contrasting environments of basaltic magma generation along a spreading ocean ridge (A) and in the Southern Rocky Mountain region (B). Dots indicate inferred sites of melting; large arrows indicate varying amounts of upwelling. Vertical exaggeration, ~2×.

ACKNOWLEDGMENTS

This summary of present knowledge of late Cenozoic basaltic volcanism in the Southern Rocky Mountain region leans heavily on results of varied geochemical studies undertaken jointly with U.S. Geological Survey colleagues, including Carl Bunker, Bruce Doe, and Carl Hedge. Data recently acquired by others have been generously made available in preprint form, especially by K. Aoli, M. G. Best, and A. M. Kudo. We also benefited from extended discussions on the Cenozoic geologic history of the Southern Rocky Mountains with T. A. Steven, J. C. Ratté, A. P. Butler, R. C. Epis, and G. R. Scott, among many others.

REFERENCES CITED

Aoki, Ken-ichiro, 1967a, Petrography and petrochemistry of latest Pliocene olivine-tholeiites of Taos area, northern New Mexico, U.S.A.: Contr. Mineralogy and Petrology, v. 14, no. 3, p. 191-203.

——1967b, Alkaline and calc-alkaline basalts from Capulin Mountain, northeastern New Mexico, U.S.A.: Japanese Assoc. Mineralogists, Petrologists and Econ. Geologists Jour., v. 58, no. 4, p. 143-151.

Aoki, K., and Kudo, A. M., 1975, Major element variations of late Cenozoic basalts of New Mexico: New Mexico Univ. Pubs. Geology, no. 8 (in press).

Atwater, Tanya, 1970, Implications of plate tectonics for the Cenozoic tectonic evolution of western North America: Geol. Soc. America Bull., v. 81, no. 12, p. 3513-3536.

Bailey, R. A., Smith, R. L., and Ross, C. S., 1969, Stratigraphic nomenclature of volcanic rocks in the Jemez Mountains, New Mexico: U.S. Geol. Survey Bull. 1274-P, 19 p.

Baker, I., and Ridley, W. I., 1970, Field evidence and K, Rb, Sr data bearing on the origin of the Mt. Taylor volcanic field, New Mexico, U.S.A.: Earth and Planetary Sci. Letters, v. 10, no. 1, p. 106-114.

Baldwin, Brewster, and Muehlberger, W. R., 1959, Geologic studies of Union County, New Mexico: New Mexico Bur. Mines and Mineral Resources Bull. 63, 171 p.

Barker, Fred, 1958, Precambrian and Tertiary geology of Las Tablas quadrangle, New Mexico: New Mexico Bur. Mines and Mineral Resources Bull. 45, 104 p.

Bass, M. N., 1971, Occurrence of transitional abyssal basalt: Lithos, v. 5, no. 1, p. 57-67.

Bassett, W. A., Kerr, P. F., Schaeffer, O. A., and Stoenner, R. W., 1963, Potassium-argon ages of volcanic rocks near Grants, New Mexico: Geol. Soc. America Bull., v. 74, no. 2, p. 221-226.

Best, M. G., and Brimhall, W. H., 1970, Late Cenozoic basalt types in the western Grand Canyon region, in Hamblin, W. K., and Best, M. G., eds., The western Grand Canyon district: Utah Geol. Soc., Guidebook to Geology of Utah, no. 23, p. 57-74.

Best, M. G., Hamblin, W. K., and Brimhall, W. H., 1966, Preliminary petrology and chemistry of late Cenozoic basalts in the western Grand Canyon region: Brigham Young Univ. Research Studies Geology Ser., v. 13, p. 109-123.

Best, M. G., Brimhall, W. H., and Hamblin, W. K., 1969, Late Cenozoic basalts on the western margin of the Colorado Plateaus, Utah and Arizona: Brigham Young Univ. Research Studies Geology Ser., Rept. 69-1, 39 p.

Bingler, E. C., 1968, Geology and mineral resources of Rio Arriba County, New Mexico: New Mexico Bur. Mines and Mineral Resources Bull. 91, 158 p.

Bryan, Kirk, 1938, Geology and ground-water conditions of the Rio Grande depression in Colorado and New Mexico, in Regional planning, Pt. 6, Rio Grande joint investigation in the upper Rio Grande basin: Washington, Natl. Resources Comm., v. 1, pt. 2, sec. 1, p. 197-225.

Burbank, W. S., Lovering, T. S., Goddard, E. N., and Eckel, E. B., 1935, Geologic map of Colorado: U.S. Geol. Survey, scale 1:500,000.

Bush, A. L., Marsh, O. T., and Taylor, R. B., 1960, Areal geology of the Little Cone quadrangle, Colorado: U.S. Geol. Survey Bull. 1082-G, p. 423-492.

Butler, A. P., Jr., 1946, Tertiary and Quaternary geology of the Tusas-Tres Piedras area,

New Mexico [Ph.D. dissert.]: Cambridge, Mass., Harvard Univ., 188 p.

——1971, Tertiary volcanic stratigraphy of the eastern Tusas Mountains, southwest of the San Luis Valley, Colorado-New Mexico, in New Mexico Geol. Soc. Guidebook, 22d Field Conf., San Luis Basin, Colorado: p. 289-300.

Carey, B. D., Jr., 1955, The Elkhead Mountains volcanic field, northwestern Colorado, in Intermountain Assoc. Petroleum Geologists Guidebook, 6th Ann. Field Conf. 1955: p. 44-46.

Chapin, C. E., 1971, The Rio Grande rift, Pt. I—Modifications and additions, in New Mexico Geol. Soc. Guidebook, 22d Field Conf., San Luis Basin, Colorado: p. 191-202.

Chapin, C. E., and Epis, R. C., 1964, Some stratigraphic and structural features of the Thirtynine Mile volcanic field, central Colorado: Mtn. Geologist, v. 1, no. 3, p. 145-160.

Chayes, Felix, 1966, Alkaline and subalkaline basalts: Am. Jour. Sci., v. 264, no. 2, p. 128-145.

——1972, Silica saturation in Cenozoic basalt: Royal Soc. London Philos. Trans., ser. A, v. 271, no. 1213, p. 285-296.

Christiansen, R. L., and Blank, H. R., Jr., 1969, Volcanic evolution of the Yellowstone Rhyolite Plateau and the eastern Snake River Plain, in Symposium on volcanoes and their roots: Oxford, England, Internat. Assoc. Volcanic Chemistry Earth's Interior, p. 220-221.

Christiansen, R. L., and Lipman, P. W., 1972, Cenozoic volcanism and plate tectonic evolution of the western United States—[Pt.] II, late Cenozoic: Royal Soc. London Philos. Trans., ser. A. v. 271, no. 1213, p. 249-284.

Cohee, G. V., chm., and others, 1961, Tectonic map of the United States, exclusive of Alaska and Hawaii: U.S. Geol. Survey and Am. Assoc. Petroleum Geologists, scale 1:2,500,000 [1962].

Collins, R. F., 1949, Volcanic rocks of northeast New Mexico: Geol. Soc. America Bull., v. 60, no. 6, p. 1017-1040.

Condie, K. C., and Barsky, C. K., 1972, Origin of Quaternary basalts from the Black Rock Desert region, Utah: Geol. Soc. America Bull., v. 83, no. 2, p. 333-352.

Cross, C. W., and Larsen, E. S., 1935, A brief review of the geology of the San Juan region of southwestern Colorado: U.S. Geol. Survey Bull. 843, 138 p.

Damon, P. E., and collaborators, 1969, Correlation and chronology of ore deposits and volcanic rocks: U.S. Atomic Energy Comm. Progress Rept. COO-689-120, 137 p.

Decker, E. R., 1973, Geothermal studies in the Southern Rocky Mountain region, 1971-73: Geol. Soc. America, Abs. with Programs (Rocky Mountain Sec.), v. 5, no. 6, p. 475-476.

Doe, B. R., Lipman, P. W., Hedge, C. E., and Kurasawa, Hajime, 1969, Primitive and contaminated basalts from the Southern Rocky Mountains, USA: Contr. Mineralogy and Petrology, v. 21, no. 2, p. 142-156.

Doney, H. H., 1968, Geology of the Cebolla quadrangle, Rio Arriba County, New Mexico: New Mexico Bur. Mines and Mineral Resources Bull. 92, 114 p.

Edwards, C. L., Reiter, M. A., and Weidman, C., 1973, Geothermal studies in New Mexico and southern Colorado [abs.]: EOS (Am. Geophys. Union Trans.), v. 54, p. 463.

Elston, W. E., Coney, P. J., and Rhodes, R. C., 1968, A progress report on the Mogollon Plateau volcanic province, southwestern New Mexico, in Cenozoic volcanism in the Southern Rocky Mountains: Colorado School Mines Quart., v. 63, no. 3, p. 261-287.

Elston, W. E., Damon, P. E., Coney, P. J., Rhodes, R. C., Smith, E. I., and Bikerman, M., 1973, Tertiary volcanic rocks, Mogollon-Datil province, New Mexico, and surrounding region: K-Ar dates, patterns of eruption, and periods of mineralization: Geol. Soc. America Bull., v. 84, no. 7, p. 2259-2274.

Epis, R. C., and Chapin, C. E., 1968, Geologic history of the Thirtynine Mile volcanic field, central Colorado, in Cenozoic volcanism in the Southern Rocky Mountains: Colorado School Mines Quart., v. 63, no. 3, p. 51-86.

Gaca, J. R., and Karig, D. E., 1966, Gravity survey in the San Luis Valley area, Colorado: U.S. Geol. Survey Open-File Rept.

Galusha, Ted, and Blick, J. C., 1971, Stratigraphy of the Santa Fe Group, New Mexico: Am. Mus. Nat. History Bull., v. 144, no. 1, 127 p.

Gast, P. W., 1968, Trace element fractionation and the origin of theoleiitic and alkaline magma types: Geochim. et Cosmochim. Acta, v. 32, no. 10, p. 1057-1086.

Giegengack, R. F., Jr., 1962, Recent volcanism near Dotsero, Colorado [M.S. thesis]: Boulder, Colo., Univ. Colorado, 43 p.

Gilluly, James, 1971, Plate tectonics and magmatic evolution: Geol. Soc. America Bull., v. 82, no. 9, p. 2383-2396.

Green, D. H., 1968, Origin of basaltic magmas, in Hess, H. H., and Poldervaart, A., eds., Basalts—The Poldervaart treatise on rocks of basaltic composition, Vol. 2: New York and London, Interscience Pubs., Inc., p. 835-862.

Green, D. H., and Ringwood, A. E., 1967, The genesis of basaltic magmas: Contr. Mineralogy and Petrology, v. 15, no. 2, p. 103-190.

Gunn, B. M., and Watkins, N. D., 1970, Geochemistry of the Steens Mountain Basalts, Oregon: Geol. Soc. America Bull., v. 81, no. 5, p. 1497-1516.

Hamilton, Warren, and Myers, W. B., 1966, Cenozoic tectonics of the western United States: Rev. Geophysics, v. 4, no. 4, p. 509-550.

Hedge, C. E., and Lipman, P. W., 1972, Upper Cenozoic basalts of the Southern Rocky Mountain region: P. 2, Strontium isotopes and the geochemistry of Sr, Rb, and Ba: Geol. Soc. America, Abs. with Programs (Cordilleran Sec.), v. 4, no. 3, p. 169.

Hedge, C. E., and Noble, D. C., 1971, Upper Cenozoic basalts with high Sr^{87}/Sr^{86} and Sr/Rb ratios, southern Great Basin, western United States: Geol. Soc. America Bull., v. 82, no. 12, p. 3503-3510.

Hoffer, J. M., 1971, Mineralogy and petrology of the Santo Tomas-Black Mountain basalt field, Potrillo volcanics, south-central New Mexico: Geol. Soc. America Bull., v. 82, no. 3, p. 603-611.

Hunt, C. B., 1938, Igneous geology and structure of the Mount Taylor volcanic field, New Mexico: U.S. Geol. Survey Prof. Paper 189-B, p. 51-80.

Ito, Keisuke, and Kennedy, G. C., 1967, Melting and phase relations in a natural peridotite to 40 kilobars: Am. Jour. Sci., v. 265, no. 6, p. 519-538.

Izett, G. A., 1966, Tertiary extrusive volcanic rocks in Middle Park, Grand County, Colorado, in Geological Survey research 1966: U.S. Geol. Survey Prof. Paper 550-B, p. B42-B46.

——1968, Geology of the Hot Sulphur Springs quadrangle, Grand County, Colorado: U.S. Geol. Survey Prof. Paper 586, 79 p.

Jackson, E. D., and Wright, T. L., 1970, Xenoliths in the Honolulu volcanic series, Hawaii: Jour. Petrology, v. 11, p. 405-430.

Jackson, E. D., Silver, E. A., and Dalrymple, G. B., 1972, Hawaiian-Emperor Chain and its relation to Cenozoic circumpacific tectonics: Geol. Soc. America Bull., v. 83, no. 3, p. 601-618.

Jackson, W. H., and Pakiser, L. C., 1965, Seismic study of crustal structure in the Southern Rocky Mountains, in Geological Survey research 1965: U.S. Geol. Survey Prof. Paper 525-D, p. D85-D92.

Kelley, V. C., 1952, Tectonics of the Rio Grande depression of central New Mexico, in New Mexico Geol. Soc. Guidebook, 3d Field Conf., Rio Grande Country, central New Mexico, 1952: p. 93-105.

——1956, The Rio Grande depression from Taos to Santa Fe, in New Mexico Geol. Soc. Guidebook, 7th Field Conf., 1956: p. 109-114.

Kistler, R. W., and Peterman, Z. E., 1972, Variations in Sr, Rb, K, Na, and initial Sr^{87}/Sr^{86} in Mesozoic granitic rocks in California: Geol. Soc. America, Abs. with Programs (Cordilleran Sec.), v. 4, no. 7, p. 562-563.

Kudo, A. M., Aoki, K. I., and Brookins, D. G., 1971, The origin of Pliocene-Holocene basalts of New Mexico in the light of strontium-isotopic and major-element abundances: Earth and Planetary Sci. Letters, v. 13, no. 1, p. 200-204.

Kuno, H., 1959, Origin of Cenozoic petrographic provinces of Japan and surrounding area: Bull. Volcanol., v. 20, p. 37-76.

——1968, Differentiation of basalt magmas, in Hess, H. H., and Poldervaart, A., eds., Basalt—The Poldervaart treatise on rocks of basaltic composition, Vol. 2: New York and London, Interscience Pubs., Inc., p. 623-688.

Kushiro, Ikuo, 1968, Compositions of magmas formed by partial zone melting of the Earth's upper mantle: Jour. Geophys. Research, v. 73, no. 2, p. 619-634.

——1972, Effect of water on the composition of magmas formed at high pressures: Jour. Petrology, v. 13, no. 2, p. 311-334.

Lambert, Wayne, 1966, Notes on the Late Cenozoic geology of the Taos-Questa area, New Mexico, in New Mexico Geol. Soc. Guidebook, 17th Field Conf., Taos-Raton-Spanish Peaks country, New Mexico and Colorado, 1966: p. 43-50.

Larsen, E. S., Jr., and Cross, Whitman, 1956, Geology and petrology of the San Juan region, southwestern Colorado: U.S. Geol. Survey Prof. Paper 258, 303 p.

Larsen, E. S., Jr., Irving, J., Gonyer, F. A., and Larsen, E. S., III, 1938, Petrologic results of a study of the minerals from the Tertiary volcanic rocks of the San Juan region, Colorado, Pts. 8-10: Am Mineralogist, v. 23, p. 417-429.

Larson, E. E., 1968, Miocene and Pliocene rocks in the Flat Tops Primitive Area, Colorado, in Guidebook for the high altitude and mountain basin deposits of Miocene age in Wyoming and Colorado Field Conf.: Boulder, Colo., Univ. Colorado Museum.

Larson, E. E., Ozima, Minoru, and Bradley, W. C., 1975, Late Cenozoic basic volcanism in northwestern Colorado and its implications concerning tectonism and the origin of the Colorado River system in Curtis, Bruce, ed., Cenozoic history of the Southern Rocky Mountains: Geol. Soc. America Mem. 144, p. 155-178.

Laughlin, A. W., Brookins, D. G., Kudo, A. M., and Causey, J. C., 1971, Chemical and strontium isotopic investigations of ultramafic inclusions and basalt, Bandera crater, New Mexico: Geochim. et Cosmochim. Acta, v. 35, no. 1, p. 107-113.

Laughlin, A. W., Brookins, D. G., and Carden, J. R., 1972a, Variations in the initial strontium ratios of a single basalt flow: Earth and Planetary Sci. Letters, v. 14, no. 1, p. 79-82.

Laughlin, A. W., Brookins, D. G., and Causey, J. D., 1972b, Late Cenozoic basalts from the Bandera lava field, Valencia County, New Mexico: Geol. Soc. America Bull., v. 83, no. 5, p. 1543-1551.

Lee, W. T., 1922, Description of the Raton, Brilliant, and Koehler quadrangles, with a section on igneous rocks, by J. B. Mertie, Jr.: U.S. Geol. Survey Geol. Atlas 214.

Leeman, W. P., 1970, The isotopic composition of strontium in late-Cenozoic basalts from the Basin-Range province, western United States: Geochim, et Cosmochim. Acta, v. 34, no. 8, p. 857-872.

Leeman, W. P., and Rogers, J.J.W., 1970, Late Cenozoic alkali-olivine basalts of the Basin-Range province, USA: Contr. Mineralogy and Petrology, v. 25, p. 1-24.

LeMasurier, W. E., 1968, Crystallization behavior of basalt magma, Santa Rosa Range, Nevada: Geol. Soc. America Bull., v. 79, no. 8, p. 949-972.

Lipman, P. W., 1969, Alkalic and tholeiitic basaltic volcanism related to the Rio Grande depression, southern Colorado and northern New Mexico: Geol. Soc. America Bull., v. 80, no. 7, p. 1343-1353.

——1975, Evolution of the Platoro caldera complex and related volcanic rocks, southeastern San Juan Mountains, Colorado: U.S. Geol. Survey Prof. Paper 852 (in press).

Lipman, P. W., and Moench, R. H., 1972, Basalts of the Mount Taylor volcanic field, New Mexico: Geol. Soc. America Bull., v. 83, no. 5, p. 1335-1344.

Lipman, P. W., Steven, T. A., and Mehnert, H. H., 1970, Volcanic history of the San Juan Mountains, Colorado, as indicated by potassium-argon dating: Geol. Soc. America Bull., v. 81, no. 8, p. 2329-2352.

Lipman, P. W., Prostka, H. J., and Christiansen, R. L., 1972, Cenozoic volcanism and plate tectonic evolution of the western United States—[Pt.] I, early and middle Cenozoic: Royal Soc. London Philos. Trans., ser. A, v. 271, no. 1213, p. 217-248.

Lipman, P. W., Bunker, C. M., and Bush, C. A., 1973, Potassium, thorium, and uranium contents of upper Cenozoic basalts of the Southern Rocky Mountain region, and their relation to the Rio Grande depression: U.S. Geol. Survey Jour. Research, v. 1, no. 4, p. 387-401.

Macdonald, G. A., and Katsura, T., 1964, Chemical composition of Hawaiian lavas: Jour. Petrology, v. 5, pt. 1, p. 82-133.

McKee, E. H., and Anderson, C. A., 1971, Age and chemistry of Tertiary volcanic rocks in north-central Arizona and relation of the rocks to the Colorado Plateaus: Geol. Soc. America Bull., v. 82, no. 10, p. 2767-2782.

McKenzie, D. P., and Morgan, W. J., 1969, Evolution of triple junctions: Nature, v. 224, no. 5215, p. 125-133.

Montgomery, A., 1953, Pre-Cambrian geology of the Picuris Range, north-central New Mexico: New Mexico Bur. Mines and Mineral Resources Bull. 30, 89 p.

Morgan, W. J., 1971, Convection plumes in the lower mantle: Nature, v. 230, no. 5288, p. 42-43.

Murata, K. J., and Richter, D. H., 1966, The settling of olivine in Kilauean magma as shown by lavas of the 1959 eruption: Am. Jour. Sci., v. 264, no. 3, p. 194-203.

Nicholls, J., Carmichael, I.S.E., and Stormer, J. C., Jr., 1971, Silica activity and P total in igneous rocks: Contr. Mineralogy and Petrology, v. 33, no. 1, p. 1-20.

O'Hara, M. J., and Yoder, H. S., Jr., 1967, Formation and fractionation of basic magmas at high pressures: Scottish Jour. Geology, v. 3, pt. 1, p. 67-117.

Olson, J. C., Hedlund, D. C., and Hansen, W. R., 1968, Tertiary volcanic rocks in the Powderhorn-Black Canyon region, Gunnison and Montrose Counties, Colorado: U.S. Geol. Survey Bull. 1251-C, p. C1-C29.

Ozima, M., Kono, M., Kaneoka, I., Kinoshita, H., Kobayashi, Kayuo, Nagata, Takesi, Larson, E. E., and Strangway, D. W., 1967, Paleomagnetism and potassium-argon ages of some volcanic rocks from the Rio Grande gorge, New Mexico: Jour. Geophys. Research, v. 72, no. 10, p. 2615-2622.

Peck, L. C., 1964, Systematic analysis of silicates: U.S. Geol. Survey Bull. 1170, 89 p.

Peterman, Z. E., Doe, B. R., and Prostka, H. J., 1970, Lead and strontium isotopes in rocks of the Absaroka volcanic field, Wyoming: Contr. Mineralogy and Petrology, v. 27, no. 2, p. 121-130.

Plouff, Donald, and Pakiser, L. C., 1972, Gravity study of the San Juan Mountains, Colorado, in Geological Survey research 1972: U.S. Geol. Survey Prof. Paper 800-B, p. B183-B190.

Porath, H., 1971, Magnetic variation anomalies and seismic low-velocity zone in the western United States: Jour. Geophys. Research, v. 76, no. 11, p. 2643-2648.

Powers, H. A., 1932, The lavas of the Modoc Lava Bed quadrangle, California: Am. Mineralogist, v. 17, no. 7, p. 253-294.

——1955, Composition and origin of basaltic magma of the Hawaiian Islands: Geochim. et Cosmochim. Acta, v. 7, nos. 1-2, p. 77-107.

Ratté, J. C., Gaskill, D. L., Eaton, G. P., Peterson, D. W., Stotelmyer, R. B., and Meeves, H. C., 1975, Mineral resources of the Gila Primitive Area and Gila Wilderness, Catron and Grant Counties, New Mexico: U.S. Geol. Survey Bull. (in press).

Renault, Jacques, 1970, Major-element variations in the Potrillo, Carrizozo, and McCartys basalt fields, New Mexico: New Mexico Bur. Mines and Mineral Resources Circ. 113, 22 p.

Robinson, H. H., 1913, The San Franciscan volcanic field, Arizona: U.S. Geol. Survey Prof. Paper 76, 213 p.

Ross, C. S., 1926, A Colorado lamprophyre of the verite type: Am. Jour. Sci., 5th ser., v. 12, p. 217-229.

Roy, R. F., Blackwell, D. D., and Decker, E. R., 1972, Continental heat flow, in Robertson, E. C., ed., The nature of the solid earth: McGraw-Hill Book Co., p. 506-543.

Scott, D. H., and Trask, N. J., 1971, Geology of the Lunar Crater volcanic field, Nye County, Nevada: U.S. Geol. Survey Prof. Paper 599-I, 22 p.

Scott, G. R., 1970, Quaternary faulting and potential earthquakes in east-central Colorado, in Geological Survey research 1970: U.S. Geol. Survey Prof. Paper 700-C, p. C11-C18.

Shapiro, Leonard, 1967, Rapid analysis of rocks and minerals by a single-solution method, in Geological Survey research 1967: U.S. Geol. Survey Prof. Paper 575-B, p. B187-B191.

Shapiro, Leonard, and Brannock, W. W., 1962, Rapid analysis of silicate, carbonate, and phosphate rocks: U.S. Geol. Survey Bull. 1144-A, 56 p.

Smith, R. L., and Bailey, R. A., 1968, Stratigraphy, structure, and volcanic evolution of the Jemez Mountains, New Mexico [abs.], in Cenozoic volcanism in the Southern Rocky Mountains: Colorado School Mines Quart., v. 63, no. 3, p. 259-260.

Smith, R. L., Bailey, R. A., and Ross, C. S., 1970, Geologic map of the Jemez Mountains, New Mexico: U.S. Geol. Survey Misc. Geol. Inv. Map I-571.

Steven, T. A., 1975, Middle Tertiary volcanic field in the Southern Rocky Mountains, in Curtis, Bruce, ed., Cenozoic history of the Southern Rocky Mountains: Geol. Soc. America Mem. 144, p. 75-94.

Steven, T. A., Mehnert, H. H., and Obradovich, J. D., 1967, Age of volcanic activity in the San Juan Mountains, Colorado, in Geological Survey research 1967: U.S. geol. Survey Prof. Paper 575-D, p. D47-D55.

Steven, T. A., Lipman, P. W., Hail, W. J., Jr., Barker, Fred, and Luedke, R. G., 1974, Geologic map of the Durango quadrangle, southwestern Colorado: U.S. Geol. Survey Misc. Geol. Inv. Map I-764.

Stobbe, H. R., 1949, Petrology of volcanic rocks of northeastern New Mexico: Geol. Soc. America Bull., v. 60, no. 6, p. 1041-1095.

Stone, G. T., 1967, Petrology of upper Cenozoic basalts of the western Snake River Plain, Idaho [Ph.D. thesis]: Boulder, Colo., Univ. Colorado, 392 p.

Stormer, J. C., Jr., 1972a, Mineralogy and petrology of the Raton-Clayton volcanic field, northeastern New Mexico: Geol. Soc. America Bull., v. 83, no. 11, p. 3299-3322.

——1972b, Ages and nature of volcanic activity on the southern High Plains, New Mexico and Colorado: Geol. Soc. America Bull., v. 83, no. 8, p. 2443-2448.

Suppe, John, Powell, C., and Berry, R., 1973, Regional topography, seismicity, volcanism, and the present-day tectonics of the western United States, in Kovach, R. L., and Nur, A., eds., Proceedings of the conference on tectonic problems of the San Andreas fault system: Stanford Univ. Pubs. Geol. Sci., v. 13, 494 p.

Tatsumoto, M., 1966, Genetic relations of oceanic basalts as indicated by lead isotopes: Science, v. 153, no. 3740, p. 1094-1101.

Taylor, R. B., 1975, Neogene tectonism in south-central Colorado, in Curtis, Bruce, ed., Cenozoic history of the Southern Rocky Mountains: Geol. Soc. America Mem. 144, p. 211-226.

Trask, N. J., 1969, Ultramafic xenoliths in basalt, Nye County, Nevada, in Geological Survey research 1969: U.S. Geol. Survey Prof. Paper 650-D, p. D43-D48 [1970].

Tweto, Ogden, and Case, J. E., 1972, Gravity and magnetic features as related to geology in the Leadville 30-minute quadrangle, Colorado: U.S. Geol. Survey Prof. Paper 726-C, p. C1-C31 [1973].

Upson, J. E., 1939, Physiographic subdivisions of the San Luis Valley, southern Colorado: Jour. Geology, v. 47, no. 7, p. 721-736.

U.S. Geological Survey, 1966, Geological Survey research 1966: U.S. Geol. Survey Prof. Paper 550-A, 385 p. [1967].

——1971, Geological Survey research 1971: U.S. Geol. Survey Prof. Paper 750-A, 418 p. [1972].

Van Alstine, R. E., 1968, Tertiary trough between the Arkansas and San Luis Valleys, Colorado, in Geological Survey research 1968: U.S. Geol. Survey Prof. Paper 600-C, p. C158-C160.

Vitaliano, C. J., and Harvey, R. D., 1965, Alkali basalt from Nye County, Nevada: Am. Mineralogist, v. 50, nos. 1-2, p. 73-84.

Waters, A. C., 1962, Basalt magma types and their tectonic associations—Pacific Northwest of the United States, in The crust of the Pacific Basin: Am. Geophys. Union Geophys. Mon. 6, p. 158-170.

Wells, R. C., 1937, Analyses of rocks and minerals from the laboratory of the U.S. Geological Survey, 1914-1936: U.S. Geol. Survey Bull. 878, 134 p.

Williams, Howel, 1936, Pliocene volcanoes of the Navajo-Hopi country: Geol. Soc. America Bull., v. 47, no. 1, p. 111-172.

Wilson, J. T., 1963, A possible origin of the Hawaiian Islands: Canadian Jour. Physics, v. 41, no. 6, p. 863-870.

Wise, W. S., 1969, Origin of basaltic magmas in the Mojave Desert area, California: Contr. Mineralogy and Petrology, v. 23, no. 1, p. 53-64.

Wright, T. L., 1971, Chemistry of Kilauea and Mauna Loa lava in space and time: U.S.

Geol. Survey Prof. Paper 735, 40 p. [1972].

Wright, T. L., and Fiske, R. S., 1971, Origin of the differentiated and hybrid lavas of Kilauea Volcano, Hawaii: Jour. Petrology, v. 12, no. 1, p. 1-65.

Wyllie, P. J., 1971, Role of water in magma generation and initiation of diapiric uprise in the mantle: Jour. Geophys. Research, v. 76, no. 5, p. 1328-1338.

Yoder, H. S., Jr., and Tilley, C. E., 1962, Origin of basalt magmas—An experimental study of natural and synthetic rock systems: Jour. Petrology, v. 3, pt. 3, p. 342-532.

York, D., Strangway, D. W., and Larson, E. E., 1971, Preliminary study of a Tertiary magnetic transition in Colorado: Earth and Planetary Sci. Letters, v. 11, no. 4, p. 333-338.

MANUSCRIPT RECEIVED BY THE SOCIETY MAY 8, 1974

Geological Society of America
Memoir 144
© 1975

Late Cenozoic Basic Volcanism in Northwestern Colorado and Its Implications Concerning Tectonism and the Origin of the Colorado River System

Edwin E. Larson

*Department of Geological Sciences and
Cooperative Institute for Research in Environmental Sciences
University of Colorado
Boulder, Colorado 80302*

Minoru Ozima

*Geophysical Institute
University of Tokyo
Tokyo, Japan*

William C. Bradley

*Department of Geological Sciences
University of Colorado
Boulder, Colorado 80302*

ABSTRACT

Upper Cenozoic terrestrial basin-fill sedimentary and basic volcanic rocks are common in the 20,700 km² Basalt area, which includes parts of the Gore, Sawatch, and southern Park Ranges, Elk Mountains, Grand Mesa, and White River Plateau. Principally on the basis of whole-rock K-Ar ages from basalt flows, the rocks can be placed in four groups. Group 1 rocks attain a thickness of 210 m and range in age from 24 to 20 m.y. (early Miocene). They consist primarily of flows of alkali-olivine basalt or of basalt flows interlayered with cross-bedded sandstone of the Browns Park Formation. Group 2 rocks, 14 to 9 m.y. old (late Miocene and perhaps early Pliocene), attain a maximum thickness of 180 m and are composed largely of basalt, basaltic andesite, and fine-grained tuffaceous fluvial, lacustrine, and eolian sedimentary rocks containing a rich vertebrate fauna indicative of a semiarid climate and a steppe vegetation. Toward the end stages of the accumulation

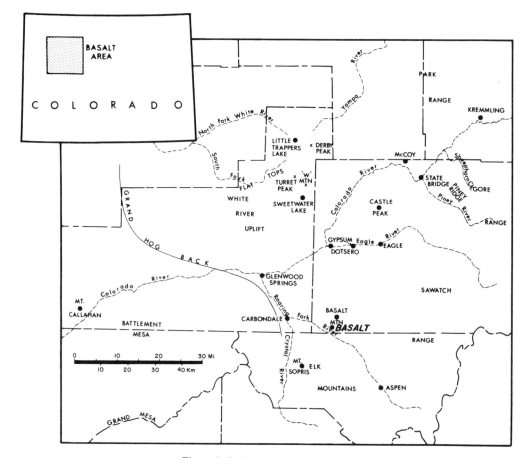

Figure 1. Index map of Basalt area.

of group 2 rocks, about 10 m.y. ago, the region was subjected to major tectonism, including uplift, reactivation of some Laramide (Late Cretaceous-early Tertiary) structural features, and creation of new warps, folds, and faults. The initiation of the Colorado River system apparently occurred at that time in response to increased precipitation stemming from an increase in elevation and relief. Downcutting began at this time and has continued intermittently to the present. By about 8 m.y. ago (late Miocene or early Pliocene, depending on boundary interpretation), the Roaring Fork River had downcut 600 m and had formed a broad flood plain, upon which thin alkali-olivine basalt flow units (group 3 rocks) were erupted. Downcutting appears to have been slow between about 8 and 1.5 m.y. ago; since then, an additional 300 m of valley deepening has occurred. Sporadic volcanism since 1.5 m.y. has accounted for several small cinder cones and flows of nepheline-normative alkali basalt (group 4 rocks); the last eruption occurred about 4,000 yr ago.

INTRODUCTION

Volcanic rocks have been extruded sporadically in northwestern Colorado during the past 25 m.y. These and associated sedimentary rocks provide data on the structural and physiographic development of this region during the latter half of Cenozoic time. The area under discussion covers approximately 20,700 km² and includes the White River Plateau, parts of the Park, Gore, and Sawatch Ranges, the Elk Mountains, and Grand Mesa (see Fig. 1 for pertinent geographic names). For simplification, this area is hereafter termed the Basalt area, in reference to the centrally located town of Basalt. The name is particularly appropriate because most of the conclusions in this paper have resulted from investigation of basic lava flows.

A generalized geologic map of the Basalt area is shown in Figure 2. Because the scale is small, many geologic elements have been eliminated or minimized: faults and fold axes have been omitted, and prevolcanic rocks are simply designated as Precambrian, Paleozoic, Mesozoic, and Tertiary. The map is a product of several months of field studies (extending over a five-year period), combined with data from other sources.[1] Field mapping was particularly concentrated in the Flat Tops Primitive Area, the State Bridge area, and the area southeast of Glenwood Springs. Also included were extensive studies of paleomagnetism, which proved to be especially useful in the correlation of volcanic units. In addition, selected lava flows were dated by whole-rock K-Ar methods, chemically analyzed, and petrographically studied. Sample descriptions and age data are given in Table 1, and percentages of the major metallic oxides are given in Table 2.

THE ROCKS

Basic volcanic rocks, in places interlayered with sedimentary rocks, are found in many parts of the area (see Fig. 2) and range in age from about 24 m.y. to 4,000 yr. On the basis of age and stratigraphic position, the rocks can be grouped into four units: group 1 (24 to 20 m.y. B.P., early Miocene); group 2 (14 to 9 m.y. B.P., late Miocene and perhaps early Pliocene); group 3 (about 8 m.y. B.P., late Miocene or early Pliocene, depending on boundary interpretation); and group 4 (since 1.5 m.y. B.P., Pleistocene). As additional radiometric dates become available, regrouping may be necessary.

Group 1 Rocks: 24 to 20 M.Y. Old

The principal occurrences of rocks of this age in the Basalt area are in the Flat Tops Primitive Area on the east flank of the White River Plateau, near State Bridge on the west flank of the Gore Range, and north of Gypsum (Fig. 2).

In the Flat Tops Primitive Area, the maximum thickness of rocks of this group is about 270 m; they consist of alkali-olivine basalt flows, eolian sandstone, and basal conglomerate. In the southern and western parts of the Flat Tops Primitive Area, the section is composed predominantly of flow-on-flow basalt, but toward the north and northeast basal conglomerate appears, and the flows become increasingly interlayered with cross-bedded eolian sandstone that Kucera (1962) considered to be part of the Browns Park Formation. For example, at Turret

---

[1] Bass and Northrop, 1963; Beckett, 1955; Benson and Bass, 1955; Brennan, 1969; Bryant, 1971, 1972; Donnell, 1961; Donnell and Yeend, 1968a, 1968b, 1968c, 1968d, 1968e; Donner, 1949; Freeman, 1972a, 1972b; Gaskill and Godwin, 1966; Gaskill and others, 1967; Gates, 1950; Giegengack, 1962; Godwin, 1968; Hanshaw, 1958; Holt, 1961; Hubert, 1954; Izett, 1968; Kucera, 1962; Mackay, 1953; Mallory and others, 1966; McElroy, 1953; Mull, 1960; Murray, 1962, 1966; Mutschler, 1968, 1969; Poole, 1954; Schmidt, 1961; Sharps, 1962; Stauffer, 1953; Steinbach, 1956; Taggart, 1962; Trask, 1956; Tweto and others, 1970; Wanek, 1953; Welder, 1954; Yeend, 1969.

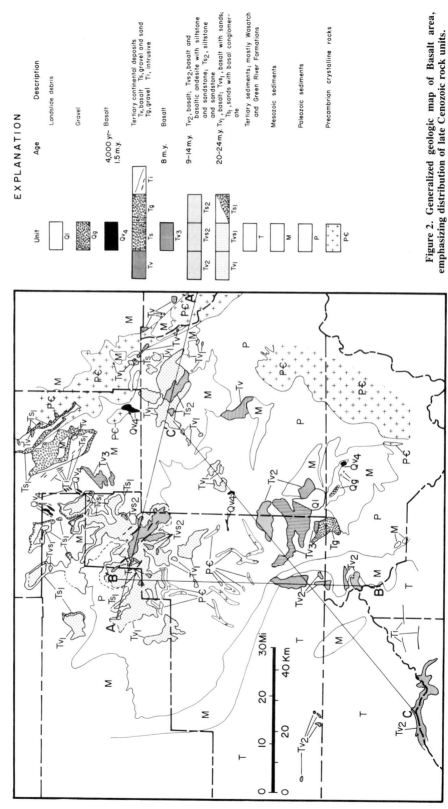

Figure 2. Generalized geologic map of Basalt area, emphasizing distribution of late Cenozoic rock units.

EXPLANATION

Unit	Age	Description
Ql		Landslide debris
Qg		Gravel
Qv4	4,000 yr– 1.5 m.y.	Basalt
Tv Ts Tg Ti		Tertiary continental deposits Tv, basalt Ts, gravel and sand Tg, gravel Ti, intrusive
Tv3	8 m.y.	Basalt
Tv2 Tvs2 Ts2	9–14 m.y.	Tv2, basalt; Tvs2, basalt and basaltic andesite with siltstone and sandstone; Ts2, siltstone and sandstone
Tv1 Tvs1 Ts1	20–24 m.y.	Tv1, basalt; Tvs1, basalt with sands; Ts1, sands with basal conglomerate
T		Tertiary sediments; mostly Wasatch and Green River Formations
M		Mesozoic sediments
P		Paleozoic sediments
P€		Precambrian crystalline rocks

Peak (see Fig. 1) the section contains only basalt flows, whereas 16 km to the north the section consists of basal conglomerate and equal amounts of sandstone and basalt, and still farther north the sandstone becomes dominant. These relations are shown schematically in Figure 3. The sequence of flows thins to the southwest across the crest of the White River Plateau, a Laramide structural feature exposing resistant Precambrian crystalline rocks and lower to middle Paleozoic strata. The onlapping relation has been verified by paleomagnetic studies and is shown in Figures 3 and 6.

Individual basalt flows most commonly are 3 to 15 m thick, although where local ponding occurred, single flows may be as much as 60 m thick. Feeder dikes and vent systems indicate that flows were fed by magma rising along northwest-trending fractures, parallel to the Laramide structural grain. Near vents, interflow material is abundant and consists of reddened cinders, bombs, and scoria. K-Ar ages have been determined for eight flows in one stratigraphic sequence at Turret Peak and Little Trappers Lake. Ages range from about 23 m.y. at the base to 20 m.y. at the top of the section (Table 1). At least five reversals of the Earth's magnetic field are recorded in this sequence.

In the area around State Bridge, group 1 rocks reach a maximum thickness of about 330 m and are composed almost entirely of alkali-olivine basaltic lava flows. Whole-rock K-Ar ages have been determined for 12 flows exposed at Yarmony Mountain (just north of State Bridge) by York and others (1971). Flows near the base are about 24 m.y. old; the upper flows show some variation but average about 21 m.y. in age. Paleomagnetic studies by W. J. Brennan (1971, personal commun.) indicate that the basalt sections on Piney Ridge and Castle Peak (see Fig. 1) are essentially identical to those exposed on Yarmony Mountain. The monotony of the lava section at Yarmony Mountain is interrupted only by the occurrence of a prominent 12-m-thick sedimentary unit that lies between the fifth and sixth flows. The lower one-fourth of the clastic unit is a cobble conglomerate composed largely of Precambrian rocks; the upper three-fourths is a coarse, unsorted unit (possibly a debris flow) that consists mostly of basic to acidic volcanic fragments, which G. A. Izett (1973, personal commun.) finds strongly reminiscent of the Rabbit Ears Volcanics in Middle Park (Izett, 1968).

Poorly sorted volcanic and Precambrian debris, which appears to be closely related to this unit, can be found about 14 km northeast of State Bridge, along both sides of Sheephorn Creek (Fig. 1). The poorly stratified material, totaling about 120 m in maximum exposure, rests on an irregularly beveled surface cut into pre-Tertiary rocks, including Precambrian crystalline rocks. The coarse debris is capped by two to four flows of alkali-olivine basalt (see Table 2). Another flow of similar composition is intercalated in the upper part of the sedimentary section. A sample taken from the capping section yields a whole-rock K-Ar age of about 23 m.y. (see Table 1). On the bases of the age, similarity of interbedded coarse clastic debris, and similarity in normal paleomagnetic polarity, the lava-capped section along Sheephorn Creek is correlated with the lower part of the section exposed at Yarmony Mountain. A small outlier, composed of only two to three lava flows, rests directly on the Precambrian crystalline rocks just northwest of Sheephorn Creek and across the Colorado River. Because these flows are alkalic in composition (see Table 2) and have a normal paleomagnetic polarity, we have tentatively correlated this outlier with the flow section at Sheephorn Creek.

On the north side of and 370 m above the Eagle River near Gypsum, an alkali-olivine basalt unit consisting of at least two flows is exposed (see Fig. 2 and Table 2). Whole-rock K-Ar analysis of a sample from the upper flow yields an age of about 22 m.y. (see Table 1). The flow unit discontinuously caps much

TABLE 1. RADIOMETRIC DETERMINATIONS

Description	Lat (N.)	Long (W.)	Collected by	Analyzed by	Location	Magnetic polarity	K (%)	Ar^{40} (radiogenic) (moles/g)	Ar^{40} air/total (%)	Age (m.y.)	Generally expected maximum error in determination
Fine-grained nepheline-normative basalt, fresh	39°56′55″	106°43′18″	E. Larson	M. Ozima, Univ. Tokyo	McCoy flow, middle, along dirt road to mesa cap	N	2.04; 2.02	2.17×10^{-12}; 2.42×10^{-12}	98	0.64	±0.2
Do.	39°18′16″	106°54′53″	E. Larson & F. Mutschler	do.	10 mi north of Aspen on north side of road along Roaring Fork River, lowest exposed flow	R	2.16	7.58×10^{-12}	91	1.98	±0.3
Do.	39°18′18″	106°54′53″	do.	do.	Same as above, but 5th flow up from base	R	2.53	5.69×10^{-12}	84	1.25	±0.2
Fine-grained basalt, fresh	39°24′35″	107°5′52″	do.	do.	Roadcut on road north of El Jebel, first fresh flow above weathered basalt	N	1.5	2.10×10^{-11}	56	7.86	±0.4
Do.	39°25′21″	107°8′58″	do.	do.	Roadcut on road north of Catharine in flow resting on Minturn Formation	N	0.949	1.47×10^{-11}	51	8.68	±0.4
Do.	39°26′29″	107°2′36″	do.	do.	Roadway on jeep trail winding up north side of Basalt Mountain	N	1.87	2.93×10^{-11}	49	8.78	±0.4
Do.	39°32′00″	107°15′20″	E. Larson	do.	Roadway at crest of Lookout Mountain, east of Glenwood Springs	N	1.08	1.95×10^{-11}	74	10.1	±0.5
Fine-grained basalt, relatively fresh	39°31′52″	107°3′01″	do.	do.	Poorly exposed outcrop south of Cottonwood Pass road just west of summit	N	1.80	3.58×10^{-11}	93	11.1	±1.0
Fine-grained, dense basaltic andesite, fresh	39°57′52″	107°7′52″	do.	do.	Cliff face of southwest side of flat-topped Derby Peak	R	2.17; 2.11	3.932×10^{-11}; 3.217×10^{-11}		10.3	±0.5
Do.	39°52′43″	107°10′32″	do.	do.	On northwest side of "W" Mountain, top flow	N	1.78; 1.79	2.906×10^{-11}; 3.089×10^{-11}		9.6	±0.5
Do.	39°52′43″	107°10′32″	do.	do.	Same as above, but 2nd flow down	N	1.35; 1.38	2.987×10^{-11}	26		
Do.	39°52′43″	107°10′32″	do.	do.	Same as above, but 3rd flow down, equal to cap flow on Turret Peak to west	N	2.58	6.06×10^{-11}	45	13.17	±0.5
Fine-grained, dense basalt, fresh	39°56′40″	107°15′00″	do.	do.	Top flow on Trappers Peak	N	2.69; 2.29	6.21×10^{-11}; 5.42×10^{-11}	60	13.4	±0.5
Fine-grained, diktytaxitic basalt flow, fresh	39°50′48″	107°11′13″	do.	do.	South side Turret Peak, 27th flow up from exposed base	N	1.350; 1.375	4.77×10^{-11}	74; 75	20.63	±0.8
Do.	39°50′48″	107°11′13″	do.	do.	South side Turret Peak, 20th flow up from exposed base	R	1.612; 1.615	5.16×10^{-11}; 5.60×10^{-11}	57; 53	20.05	±0.8
Do.	39°50′48″	107°11′13″	do.	do.	South side Turret Peak, 14th flow up from exposed base	N	2.74; 2.70	10.06×10^{-11}; 9.735×10^{-11}	56; 53	19.9	±0.8
Fine-grained, dense basalt flow, fresh	39°50′48″	107°11′13″	do.	do.	South side Turret Peak, 10th flow up from exposed base	R	0.832; 0.828	2.755×10^{-11}	61	18.5	±0.8
Do.	39°50′48″	107°11′13″	do.	do.	South side Turret Peak, 6th flow up from exposed base	N	0.859; 0.823	3.44×10^{-11}; 3.14×10^{-11}	65; 66	22.2	±1.0
Fined-grained, dense basalt flow, relatively fresh	39°50′48″	107°11′13″	do.	do.	South side Turret Peak, 4th flow up from exposed base	N	1.35	5.024×10^{-11}	47	20.7	±1.0
Do.	39°50′48″	107°11′13″	do.	do.	South side Turret Peak, 3rd flow up from exposed base	N	1.18; 1.24	4.711×10^{-11}	42	21.4	±1.0
Do.	39°59′37″	107°12′4″	do.	do.	Lowest exposed flow just north of Little Trappers Lake	R	0.857; 0.845	3.412×10^{-11}	63	23.0	±1.0
Fine-grained, dense basalt, fresh	39°57′00″	106°32′00″	do.	G. Curtis, Univ. Calif.	North side of Sheephorn Creek, near town of Radium, top flow	N	2.37; 2.311		79	23.2	±1.0
Do.	39°55′45″	107°00′00″	do.	do.	In valley side, ~5.5 km northwest of Gypsum	N	0.9674; 0.9648		54	22.1	±1.0

of the upland terrane north of the Eagle River for several miles (Fig. 4). Near the river, where exposure is best, the flows can be seen to overlie 1.5 to 3 m of fluvial gravel that contains clasts of Precambrian rocks. Bedrock beneath the gravel and flows is Eagle Valley Evaporite and Maroon Formation; the former is responsible for the Eagle River diapiric anticline (Benson and Bass, 1955; Mallory, 1971). Adjacent to the Eagle River, the basalt flows now dip 10° to 25° northward, away from the valley (as shown in Fig. 4), indicating that salt intrusion and anticlinal development have certainly occurred since extrusion of the flows 22 m.y. ago. Because the Minturn Formation, particularly that part containing sizable amounts of evaporites, is so prone to sliding, slumping, and collapse, the basalt-flow and gravel unit has been so greatly broken and structurally disturbed that it is difficult to tell just what its relation is to the basalt flows and underlying gravel exposed at Castle Rock. From their similar age and the similar occurrence of underlying gravels, it appears most likely that both flows were part of the same pulse of extrusives that were poured onto a gently rolling stream-cut surface. It appears most probable that the undated basalt-flow unit capping Bellyache Mountain southeast of Eagle is part of this same pulse of eruptive activity.

Thus far, our studies indicate that the thick flow sequence in the Flat Tops Primitive Area, in the area around State Bridge, and in the volcanic outliers near Gypsum and Radium all are essentially coeval and that they are remnants of a volcanic sequence that at one time was much more widespread than at present (see Fig. 6). It is easy to visualize a volcanic plain 20 m.y. ago that extended from the Gore Range to the White River Plateau, and probably southward to the Sawatch Range and Elk Mountains.

Group 1 rocks accumulated in a broad, shallow basin that lay along the west side of the Park Range. For convenience, we refer to this basin as the Steamboat basin, in reference to Steamboat Springs, which occupies a central position (located just north of the Basalt area). Volcanism was restricted to the southern part of the area, but elsewhere the basin received sediment of the Browns Park Formation (Kucera, 1962, 1968; Sharps, 1962; Buffler, 1967; Segerstrom and Young, 1972). In many places along the east side of the basin, sedimentation began with a basal conglomerate and continued with eolian sandstone, which is interbedded with the volcanic rocks (Fig. 3). Along the south side of the basin, sedimentation, if it occurred, was restricted to the basal conglomerate; infilling of the basin was principally the result of volcanism. The basal conglomerate is particularly helpful in outlining the basin. Figure 5 shows an isopach map of the southern extent of this unit. It has a maximum thickness of 110 m close to the Park Range and thins toward the southwest. We believe the thinning results from depositional pinchout against the edge of the basin, because the volcanic rocks similarly thin toward the southwest (Fig. 3); the zero-isopach line in Figure 5 represents the inferred southern margin of the Steamboat basin at that time. Clasts in the unit are chiefly of Precambrian rock types, and their grain size decreases markedly away from the Park Range. All of this suggests that the basal conglomerate was a fanglomerate deposited along the western flank of the Park Range by streams draining the range. Axial drainage within the Steamboat basin was probably northward (suggested by Fig. 5), and fine-grained alluvium may have been an important source for the eolian sands (G. A. Izett, 1973, personal commun.).

Group 2 Rocks: 14 to 9 M.Y. Old

Group 2 rocks are most voluminous in the Flat Tops Primitive Area, where they attain a thickness of 200 m and consist of basalt and basaltic andesite flows

TABLE 2. CHEMICAL ANALYSES

Flow	SiO$_2$	Al$_2$O$_3$	FeO	Fe$_2$O$_3$	TiO$_2$	MnO	CaO	MgO	Na$_2$O	K$_2$O	P$_2$O$_5$	CO$_2$	F	Cl	S	H$_2$O$^-$	H$_2$O$^+$	Total H$_2$O	Age group (m.y.)
Willow Peak flow	48.8	16.4	3.6	8.0	1.5	0.09	7.8	7.0	3.2	2.7	0.46	<0.1	0.06	0.01				0.2	<1.5
Dotsero flow	48.2	15.9	6.4	5.1	1.4	0.13	8.1	7.6	3.2	2.7	0.37	<0.1	0.10	0.03	<0.05			0.2	<1.5
Dotsero bomb	49.9	15.1	6.6	5.5	1.5	0.08	8.0	7.6	3.4	2.8	0.18	<0.1	0.07	<0.01	<0.05			0.1	<1.5
Rock Creek flow	48.5	15.5	5.8	5.1	1.8	0.11	8.5	6.5	3.9	2.5	0.62	<0.1	0.11	0.04	<0.05			0.3	<1.5
Aspen flow	49.0	15.7	6.7	3.8	1.5	0.11	7.4	7.0	3.8	2.7	0.62	<0.1	0.10	0.04	<0.05			0.2	<1.5
Roaring Fork terrace	51.5	15.1	7.8	3.7	1.4	0.07	7.3	7.3	3.1	1.1	0.11	<0.1	0.08	<0.01	<0.05			0.1	~8.
Roaring Fork terrace	52.4	15.5	7.0	3.3	1.4	0.07	7.4	6.5	3.1	1.7	0.16	<0.1	0.08	<0.01				0.4	~8.
Top Trappers Peak	57.6	16.2	1.9	5.7	1.2	0.05	5.5	3.3	4.6	2.8	0.25	<0.1	0.09	<0.01				1.1	14-9
Trappers Peak, #2 down	50.0	14.2	6.0	6.2	1.8	0.10	8.3	7.8	3.6	0.9	0.17	<0.1	0.09	<0.01				0.7	14-9
Trappers Peak, #3 down	57.4	14.4	3.9	4.4	1.0	0.05	6.3	5.3	3.8	2.5	0.10	<0.1	0.09	<0.01				1.4	14-9
Trappers Peak, #4 down	55.0	14.5	4.2	4.9	1.3	0.08	6.2	5.3	4.3	2.3	0.17	<0.1	0.09	<0.01				0.6	14-9
Trappers Peak, base	53.5	14.7	5.5	3.9	1.5	0.05	6.7	5.6	4.0	1.7	0.22	<0.1	0.09	<0.01				0.5	14-9
"W" Mountain, top flow	53.2	15.8	4.7	4.3	1.6	0.14	7.0	5.2	4.1	2.3	0.52	<0.05				0.26	0.94		14-9
Derby Peak, top flow	51.9	15.0	5.3	3.8	1.4	0.15	7.6	7.4	3.7	2.3	0.63	<0.05				0.22	0.55		14-9
Derby Peak, middle flow	54.6	14.5	5.8	3.3	1.4	0.13	7.3	5.9	3.2	2.1	0.44	0.12				0.28	1.0		14-9
Derby Peak, basal flow	54.1	14.2	4.8	3.7	1.4	0.13	7.0	6.4	3.7	2.6	0.56	0.05				0.49	0.71		14-9
Turret Peak, top flow	52.3	15.3	2.9	6.9	1.5	0.08	7.4	5.6	4.3	2.4	0.30	0.51	0.09	<0.01				0.9	14-9
Turret Peak, #2 down	52.0	15.5	2.5	8.0	1.6	0.08	8.3	5.5	3.6	1.9	0.44	0.22	0.07	<0.01				1.1	14-9
Mount Orno, near top flow	48.7	14.2	2.5	8.8	1.8	0.19	8.4	5.8	3.5	2.2	1.2	<0.05				1.1	1.3		24-20
Turret Peak,																			
#27 up from base	45.8	13.9	1.8	10.2	1.8	0.18	9.0	7.4	3.1	1.6	1.0	0.18				1.4	1.7		24-20
#26 up from base	49.5	14.9	4.4	7.5	1.8	0.09	8.7	6.3	3.4	1.6	0.5	0.38	0.08	<0.01				1.8	24-20
#20 up from base	48.5	15.0	2.6	7.1	1.6	0.13	10.4	4.8	3.0	1.9	0.75	1.6				1.0	1.4		24-20
#19 up from base	51.0	15.1	2.6	8.2	1.8	0.08	8.7	5.6	3.4	1.9	0.50	0.22	0.13	<0.01				2.2	24-20
#14 up from base	48.4	13.6	2.5	8.4	2.0	0.09	9.1	6.3	2.5	3.0	0.62		0.2	<0.01				3.2	24-20
#12 up from base	48.7	13.8	5.1	6.3	2.0	0.09	8.7	9.1	3.1	1.4	0.62	0.35	0.16	<0.01				3.0	24-20
#10 up from base	50.1	15.5	5.9	6.1	1.4	0.1	8.9	5.9	3.0	0.96	0.35	<0.05	0.07	<0.01		0.92	1.6		24-20
#7 up from base	50.6	16.6	2.1	11.6	1.7	0.05	8.1	2.8	3.5	1.1	0.25	<0.1	0.08	<0.01				3.0	24-20
#6 up from base	51.3	16.3	2.4	8.0	1.4	0.13	9.5	3.8	3.2	0.95	0.48					1.3	1.3		24-20
#4 up from base	48.2	15.8	1.7	12.0	1.6	0.25	9.1	3.4	3.1	1.2	0.56	<0.05				1.3	1.3		24-20
#3 up from base	49.0	16.2	1.6	11.6	1.4	0.11	9.5	3.7	3.1	1.2	0.40	<0.05	0.07	<0.01		1.4	1.3		24-20
Little Trappers Lake, base flow	47.5	14.3	4.5	6.3	1.5	0.16	10.3	5.0	3.0	1.0	0.40	2.8				1.3	1.8		24-20
Yarmony Mountain flows																			
#10 up from base	45.7	14.2	3.4	8.6	2.5	0.09	8.2	7.9	3.2	2.0	0.69	0.2	0.11	0.01				2.8	24-20
#7 up from base	49.6	15.3	5.7	5.2	1.4	0.05	9.1	6.2	3.1	0.7	0.17	1.1	0.06	<0.01				2.6	24-20
#5 up from base	53.8	14.5	3.9	5.6	1.2	0.05	6.8	5.8	3.7	1.9	0.26	0.4	0.08	0.01				1.6	24-20
#3 up from base	46.9	14.4	2.6	7.8	1.4	0.12	9.8	5.3	3.0	1.0	0.34	2.3	0.08	<0.01				4.7	24-20
basal flow	49.6	15.4	3.9	6.7	1.6	0.06	7.9	6.4	3.2	1.4	0.34	0.2	0.08	<0.01				2.6	24-20
Gypsum flow	47.9	14.9	6.0	7.0	2.4	0.07	8.1	5.6	3.3	1.9	0.69	<0.1	1.0	<0.01				1.4	24-20
Radium flow	48.5	13.8	5.6	5.0	2.1	0.10	9.0	8.0	3.1	2.6	1.0	<0.1		<0.02				1.5	24-20
Sheephorn Creek, top flow	47.7	13.2	5.4	6.7	2.2	0.11	9.2	7.3	3.2	2.8	1.4	<0.1		<0.02				2.0	24-20
Sheephorn Creek, #2 down	49.1	14.0	3.9	6.7	2.0	0.08	7.9	7.0	3.1	3.1	0.70	0.2	0.22	<0.02				1.6	24-20

(see Table 2) interbedded with siltstone and some sandstone (Fig. 3). The flows occur singly (separated by sedimentary rocks) more often than in flow-on-flow sequences, are commonly 15 to 23 m thick, and in many places display a platy structure. Vent systems are well exposed; certain of the topographic highs, such as Shingle, Marvine, and Sheep Peaks, are the remnants of vent superstructures. The orientations of numerous dikes and the alignment of vents indicate that magma rose along northwest-trending fractures, just as with the group 1 volcanic flows. Four flows, from "W" Mountain and Derby Peak (see Fig. 1), yielded K-Ar dates that range from about 13 to 10 m.y. B.P. (Table 1). The topmost flow on Trappers Peak was dated by whole-rock K-Ar methods at 13.5 m.y. B.P. (Table 1).

Interlayered with the flows in the Flat Tops Primitive Area are buff-colored, fine-grained, rather massively bedded sedimentary rocks that strongly resemble the nearby Troublesome and North Park Formations (see Izett, 1968). The deposits include fluvial sedimentary rock, loess, and air-fall ash and contain a Barstovian (late Miocene) vertebrate fauna that includes species of horse, camel, rhinoceros, antelope, turtle, and rodent (Robinson, 1968). Vertebrate fossils have been recovered from Trappers Peak, "W" Mountain, Little Marvine Peak, and Derby Peak; the latter, at an elevation of 3,660 m, is the highest known vertebrate locality in Colorado (P. Robinson, 1971, personal commun.). The fossils are extremely important because they indicate a semiarid climate and a steppe flora that provided nourishment for browsing and grazing animals (Robinson, 1968; 1973, personal commun.); this is in dramatic contrast to the climate, flora, and fauna that exist today at these localities.

In the area around State Bridge (Figs. 1, 2), Brennan (1969) has described about 150 m of North Park-type sediment that rests conformably on group 1 basalt flows. Lava flows are entirely lacking in this section, but its Barstovian vertebrate fauna is similar to that in the Flat Tops Primitive Area (Brennan, 1969; P. Robinson, 1970, personal commun.), suggesting that the sedimentary rocks in these two areas are contemporaneous.

These rocks are the southernmost group 2 deposits of the Steamboat basin. Similar rocks farther north have a thickness of one hundred to several hundred meters and have been included in the upper part of the Browns Park Formation (Kucera, 1962, 1968; Sharps, 1962; Buffler, 1967; Segerstrom and Young, 1972). Only in the southern part of the basin are the sedimentary rocks interlayered with lava flows, just as is true of the group 1 rocks.

Group 2 rocks southwest of the Steamboat basin consist of basic lava flows of restricted areal extent. Although many outliers are shown on Figure 2, only four of these have thus far been dated (the rest represent our best guess at correlation).

Just east of Glenwood Springs, at an elevation of 2,900 m, is a flat-topped ridge

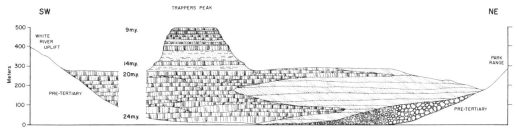

Figure 3. Restored diagrammatic cross section from White River Plateau and southern part of Flat Tops Primitive Area on southwest to Yampa Valley and southern Park Range on northeast. Shows general character and relations of rocks that filled in Steamboat basin from about 24 to 9 m.y. ago.

capped by two to three basalt flows, which overlie beveled sandstones of the Maroon Formation (see Fig. 6). The fact that the flow unit is so thin, yet is widespread, indicates that it was highly fluid and erupted onto a relatively flat surface. A K-Ar age for the uppermost flow exposed just west of Cottonwood Pass is 11.1 ± 1.0 m.y. (Table 1). A whole-rock K-Ar age for the uppermost flow (in what appears to be a three-flow unit), on Lookout Mountain southeast of Glenwood Springs, is 10.1 ± .5 m.y. (Table 1). Disconformably overlying the flows is stream gravel that includes cobbles of Tertiary basaltic volcanic and Precambrian crystalline rocks. The latter could have come from the Park, Gore, or Sawatch Ranges, a minimum transport distance of about 48 km. These are the oldest deposits we have found that clearly indicate an established Colorado River system.

Basalt Mountain (Figs. 1, 2) is an old shield volcano 430 m high that erupted 20 or more flows in a short period of time, 8.78 ± 0.4 m.y. ago (Table 1). Erosion and landslides have exposed the section in a cliff at the south end of the mountain. All flows are composed of light- to dark-gray basalt, are 3 to 9 m thick, dip radially away from the peak at angles of 3° to 5°, and have essentially the same paleomagnetic pole direction (north-northwest and down). The flows rest on a surface of low relief eroded into Mesozoic rocks, now at a level of approximately 2,900 m. Flows from Basalt Mountain must have been more extensive at one time. A broad terrace near the Roaring Fork River, just northwest of Basalt Mountain (Tv_3 on Fig. 2), is capped by lava that is slightly younger than that of Basalt Mountain and has a different paleomagnetic direction. This lowland area evidently did not exist at the time Basalt Mountain was active.

The fourth date comes from Grand Mesa (Figs. 1, 2), where flow-on-flow basalts have accumulated to a thickness of about 240 m (Marvin and others, 1966). A sample collected from near the middle of the section has given a whole-rock K-Ar date of 9.7 ± 0.5 m.y. B.P. (Marvin and others, 1966). Paleomagnetic directions from six flows exposed at Lands End (the western extremity of Grand Mesa) are essentially coincident, indicating that these flows were erupted in quick succession.

Battlement Mesa, to the north and at about the same elevation, is capped by 900 m of flow-on-flow basalts that resemble those on Grand Mesa (J. R. Donnell, 1973, personal commun.). An additional basalt outlier (perhaps two flows) occurs across the Colorado River at Mount Callahan (Donnell, 1961; 1973, personal commun.). The volcanic flows in all three areas are presumed to be roughly the same age, but corroborative dating is needed. Even if they are contemporary, it is not known whether they occurred in separated locations or are remnants of a former widespread volcanic plain. Basalt dikes east of Grand Mesa (Fig. 2), presumed to have fed the nearby flows, and widespread basalt-rich alluvial deposits that flank Grand and Battlement Mesas (Yeend, 1969; Donnell and Yeend,

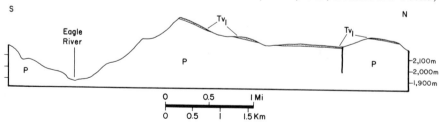

Figure 4. Structural cross section showing surface cut by ancestral Eagle River, then capped by thin group 1 lava. Cap has subsequently been uparched near Eagle River valley by salt intrusion. Symbols and patterns as in Figure 2.

1968a, 1968b, 1968c, 1968d, 1968e, 1968f; Hail, 1972a, 1972b), indicate that the basalt outliers were at one time substantially larger than at present.

An additional point of interest is that all three volcanic remnants locally overlie gravels that include pebbles of Precambrian rock types (Nygren, 1935; Hunt, 1969; J. R. Donnell, 1973, personal commun.; O. Tweto, 1973, personal commun.). Donnell (1973, personal commun.) has said he believes these gravels were deposited by the Colorado River system, but Tweto (1973, personal commun.) wondered whether they might represent only local reworking of older conglomeratic units, such as those in the Wasatch Formation. This is a problem worthy of additional study.

Group 3 Rocks: About 8 M.Y. Old

Rocks of this group are recognized definitely only south-southeast of Glenwood Springs and recognized possibly south of the Yampa River (Fig. 2). Although available dates are only slightly younger than the youngest group 2 dates, these rocks are treated separately because a significant amount of erosion occurred during the group 2 to group 3 time interval.

Group 3 rocks southeast of Glenwood Springs consist of one to two basalt flows that occur on both sides of and 300 m above the Roaring Fork River (Fig. 2). The flows have an alkali-olivine composition and have been dated at about 8 m.y. B.P. (Tables 1 and 2). Their thin, widespread, relatively horizontal nature indicates extrusion onto a broad valley bottom of the Roaring Fork and its tributaries; this valley developed after the rivers had eroded 600 m below the nearby gravel-

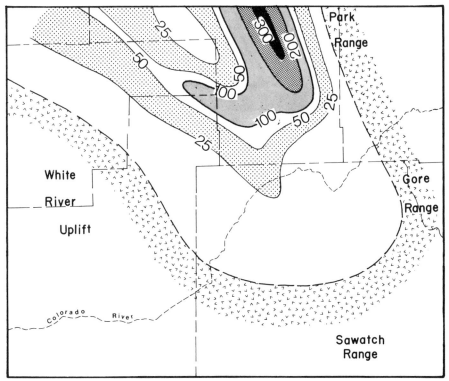

Figure 5. Generalized isopach map of conglomerate at base of Browns Park Formation. Zero isopach is estimated terminus of conglomerate and provides general outline of Steamboat basin in this part of area. Numbers are in feet.

capped 10.1- to 11.1-m.y.-old basalt flows (see Figs. 6, 7). The 8-m.y.-old flows have been broken into myriad blocks, which have been differentially jostled to produce an irregular topography with closed depressions as much as 45 m deep and 600 m across. Bedrock here is chiefly the Eagle Valley Evaporite (see Bass and Northrop, 1963; Mallory, 1971), and we attribute the chaotic dinnerware-fracture pattern to flowage, expansion, and solution of the evaporite. The fracturing makes it impossible to tell whether the flows dip gently away from the river, as might be expected if this area is a continuation of the Cattle Creek salt anticline described by Mallory (1966, 1971).

Alluvial gravel of unknown age overlies the basalt flows (Fig. 2). In particular, the deposit situated between the Roaring Fork and Crystal Rivers is anomalously thick (more than 460 m) and appears to be composed chiefly of debris flows from Mount Sopris and of alluvium. It is entirely possible that late Cenozoic movement on the Cattle Creek salt anticline could have influenced the deposition of the gravel.

Group 4 Rocks: 1.5 M.Y. Old or Less

Volcanic rocks of this group were erupted during the final phases of valley development. Pending corroboration by additional dates, there appears to have been a volcanic hiatus between approximately 8 and 1.5 m.y. ago.

The earliest known group 4 volcanism occurred about 1.5 m.y. ago in the Roaring Fork River valley 16 km north-northwest of Aspen (Fig. 2 and Table 1). Magma rose along the Crystal River fault and broke out on the side of the valley, producing a cinder cone and a number of small flows that total about 150 m in maximum thickness. Exposure of the base is poor, but V. Freeman (1969, personal commun.) reported finding underlying gravel with rock types appropriate for the Roaring Fork River. Similar gravel occurs across the river at about the same elevation (Fig. 2), with a thickness that increases downvalley from 6 to 90 m. All of this indicates eruption about 1.5 m.y. ago onto a Roaring Fork flood plain, with subsequent downcutting amounting to 300 m (see Fig. 8).

Along Rock Creek, just northeast of McCoy (Figs 1, 2), volcanism produced two cinder cones and a basalt flow, which spread out on an old Rock Creek flood plain. The flow has been dated by whole-rock K-Ar analysis as 0.64 ± 0.2 m.y. old. The freshness of topographic form of one of the cinder cones and a normal magnetic polarity corroborate this age.

Even younger eruptions are recorded near the junction of the Colorado and Eagle Rivers (Figs. 1, 2) by volcanic flows that overlie essentially modern topography. West of the junction, high-level volcanism of uncertain age produced a cinder cone (known as Willow Peak) and a basalt flow that descended the upper part of a small valley, tributary to the Colorado. East of the junction, a volcanic vent 370 m above the Eagle River produced cinders and a basalt flow that descended to the valley bottom and evidently forced the Eagle River against its south valley margin (Giegengack, 1962; Bass and Northrop, 1963). The flow is extremely fresh, and wood collected from the base of the cinders yields a C^{14} date of 4,150 ± 300 yr B.P. (Giegengack, 1962). This appears to be the youngest volcanism in Colorado.

STRUCTURAL IMPLICATIONS

The Laramide orogeny, occurring in the period from about 70 to 45 m.y. ago (Late Cretaceous to middle Eocene; Tweto, 1975), produced many large-scale tectonic features in the western two-thirds of Colorado. Principle elements in the Basalt area include the elongated Sawatch Range and Park-Gore Range uplifts, the domical White River Plateau and associated Grand Hogback monocline, and numerous generally north- to northwest-trending folds and normal and reverse faults (see Kucera, 1962; Bass and Northrop, 1963; Brennan, 1969; Tweto and others, 1970; Tweto, 1975). Between the uplifts were structural sags. Intrusive activity was restricted to the Colorado mineral belt, a northeast-trending zone that transects the southern part of the Basalt area (Elk Mountains and Sawatch Range, Fig. 1).

Intrusive activity and associated local uplift continued in the Elk Mountains and Sawatch Range during Oligocene time (see Mutschler, 1968), but what happened at this time in the rest of the Basalt area is unknown because no rock record remains. In other parts of Colorado, Laramide relief had been extensively reduced by or during Oligocene time (see Epis and Chapin, 1975; Scott, 1975; Taylor, 1975).

The Basalt area was tectonically quiet during Miocene time. Sedimentary rocks and lava flows that accumulated in the Steamboat basin are generally conformable (although disconformities may be present). By 9 or 10 m.y. ago, relief was low, sedimentary rocks were fine grained, the climate was semiarid, and the Steamboat basin was connected with other nearby basins (Robinson, 1968, 1970, 1972).

Basalt area volcanic rocks allow rather precise dating of the tectonism that followed. We presume that canyon cutting was initiated by uplift, and the youngest precanyon flows are the ones either southeast of Glenwood Springs (10.1 ± 0.5 m.y. old) or in Basalt Mountain (8.8 ± 0.5 m.y. old). Similarly, the youngest deformed lava flow in the Flat Tops area is 9.6 ± 0.5 m.y. old. A minimum age is provided by the 8-m.y.-old flows that were erupted after the Roaring Fork River had cut a valley 600 m deep. Because of the uncertainties involved, we conclude that tectonism occurred approximately 10 m.y. ago in this area.

In part, tectonism reactivated Laramide structural features. This is shown by tilting and displacement of group 1 and group 2 rocks, which indicate relative uplift of the Park, Gore, and Sawatch Ranges, the Elk Mountains, and the White River Plateau. Rocks that now occur at an elevation of 3,800 m in the Flat Tops area are at least 600 m higher than similar rocks just to the east (Fig. 6). Tilted lava flows define a gentle downwarp between the White River Plateau on the north and the Sawatch Range-Elk Mountains on the south, indicating renewed relative uplift of these Laramide positive elements. Flows north of the Colorado River dip 3° to 4° south, and those south of the river dip north (shown in part on Fig. 6); it appears that the river follows this downwarp. Finally some of the small folds and faults in the northern Flat Tops area seem to represent reactivated Laramide structures (Kucera, 1962).

Late Cenozoic relative uplift was not uniform, however. Some structural elements that were strongly positive in Laramide time lagged behind in the later movements. Kucera (1962) noted that the basal conglomerate of the Browns Park Formation occurs at an elevation of 3,500 m in the Flat Tops Primitive Area, 460 m above its source, the southern Park Range. Kucera (1962) also reported other late Cenozoic structural features that have a sense of relative movement opposite to what they had in Laramide time. Buffler (1967) and Hansen (1965) reached similar conclusions for features in the northern Park Range and Uinta Mountains, respectively.

Some late Cenozoic structural features have no Laramide ancestry. The State Bridge syncline, with more than 600 m of structural relief, has no counterpart in older rocks (Fig. 6). Also, certain of the northwest-trending folds and faults at the north end of the White River Plateau are strictly of late Cenozoic age. Group 1 rocks near Gypsum are at an elevation about 760 m lower than are flows of similar age on Castle Peak, 9.6 km to the northeast (Fig. 6), and they appear to be downdropped along northwest-trending faults.

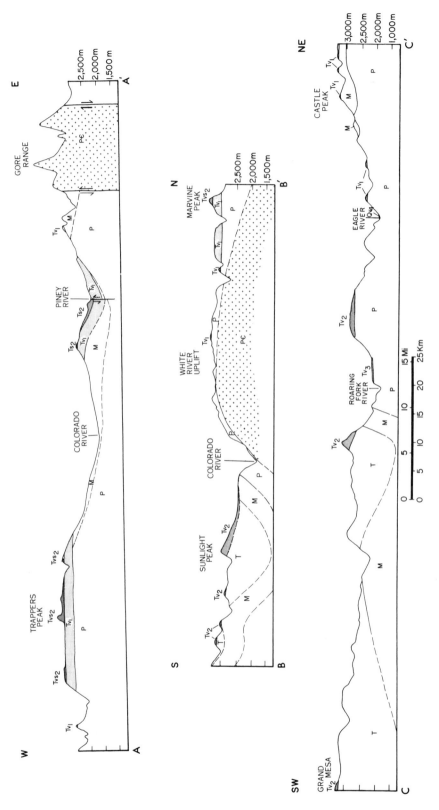

Figure 6. Structural cross sections of areas shown in Figure 2. Symbols and patterns as in Figure 2. Top: A–A', from western part of Flat Tops Primitive Area on west to Gore Range on east. Note sag of groups 1 and 2 rocks in State Bridge syncline, at Piney River. Middle: B–B', from Elk Mountains on south to northwestern part of Flat Tops Primitive Area on north. Groups 1 and 2 rocks of Flat Tops area lap out southward. Note northward dip of group 2 volcanic rocks south of Colorado River into structural low at position of Colorado River. Bottom: C–C', from Grand Mesa on southwest to Castle Peak on northeast. Gentle structural low near center of cross section is revealed in elevations of groups 1 and 2 rocks. Note position of group 3 rocks on old terrace incised by Roaring Fork River.

In addition to late Cenozoic differential movements, we presume that some unknown amount of regional uplift also occurred. Our reasons for thinking this are (1) a Miocene fauna belonging to a semiarid steppe environment occurs at 3,660 m elevation, and (2) there is evidence for a sudden increase in the size and vigor of rivers in the area (discussed in the next section).

Plutonism accompanied late Cenozoic tectonism. In the Elk Mountains, the Treasure Mountain stock was emplaced 12.4 ± 0.6 m.y. ago, causing 1,500 m of doming (Mutschler, 1968). Toward the north end of the Steamboat basin, intrusive rocks at Hahns Peak and nearby localities yield ages in the general range of 9 to 12 m.y. (see Segerstrom and Young, 1972, p. 41).

Although the main orogenic pulse seems to have occurred approximately 10 m.y. ago, there is evidence that suggests activity has continued during the past 1 m.y. or so. Canyon development (discussed in the next section) during the past 10 m.y. has occurred at a variable rate: it was rapid between 10 and 8 m.y. ago, negligible between 8 and 1.5 m.y. ago, and rapid again in the past 1.5 m.y. *If tectonism has been influential in canyon development, one could accept this as evidence of Quaternary activity.* Support comes from the late Quaternary faulting reported by Kucera (1962) and Tweto and others (1970).

Late Cenozoic tectonism was not unique to the Basalt area; it occurred in other parts of Colorado too (see Epis and Chapin, 1975; Izett, 1975; Scott, 1975; Taylor, 1975; Segerstrom and Young, 1972). Other parts of the western United States also experienced orogenic activity at approximately this time: the Sierra Nevada (Axelrod, 1957), the Basin and Range province (Eardley, 1963; Larson, 1965), the Colorado Plateau (McKee and McKee, 1972), and the Southern Rocky Mountains (Cook, 1960; Izett, 1975; Scott, 1975; Taylor, 1975).

GEOMORPHIC IMPLICATIONS

The age of the rivers and why they flow where they do are problems that have intrigued observers of the upper Colorado River and its tributaries ever since John Powell made his epic voyages more than a hundred years ago. Powell himself started the literature (1875), and its growth right up to the present time (see Hunt's 1969 synthesis) indicates that the problems have not yet been completely solved. A scarcity of Tertiary deposits that can be identified as belonging to through going rivers and the difficulty in dating the deposits that do exist are two reasons why debate continues. Data presented in this and other papers in this volume yield an improved understanding of the development of the Colorado River between Middle Park and Glenwood Springs and encourage resurrection of an old hypothesis relating to climate.

In Middle Park, the oldest deposits that indicate a large through going river occur near the top of the Troublesome Formation and are dated as 11.2 ± 1.8 m.y. (Izett, 1975). West of the Gore and Park Ranges, the oldest identified Colorado River gravels lie just southeast of Glenwood Springs on top of basalt flows dated as 10.1 ± 0.5 m.y. old. Still farther downstream, gravels locally underlie the lava flows on Grand Mesa, Battlement Mesa, and Mount Callahan (Nygren, 1935; J. R. Donnell, 1973, personal commun.; Hunt, 1969, p. 74; O. Tweto, 1973, personal commun.), and a Grand Mesa flow has been dated as 9.7 ± 0.5 m.y. old (Marvin and others, 1966). However, it is not known whether the gravels of the Grand Mesa area are related to a through going river or are of local significance; therefore, they are not considered further at this time.

The foregoing does allow us to say that the Colorado River was flowing from Middle Park to Glenwood Springs, in about its present geographic position, approximately 10 m.y. ago. In this we concur with Hunt (1969). From 10 m.y.

on, the existence of the Colorado River system is easy to document; 600 m of downcutting had already occurred along the Roaring Fork River when flows were erupted about 8 m.y. ago, and late Cenozoic gravels can be found within many valleys.

By contrast, unequivocal evidence for a Colorado River older than 10 m.y. has yet to be produced. It is a curious fact that older Colorado River sedimentary rocks have not been recognized anywhere, even though Miocene deposits were accumulating in basins that now lie athwart or near the present river. For example, the Colorado River crosses the basin site of the Troublesome Formation in Middle Park, yet the only deposits that can be identified as belonging to the Colorado River are found near the top of the stratigraphic unit (Izett, 1968, 1975). Even if the Colorado had been transporting only sand and mud at that time, it should have left a recognizable record in the Troublesome Formation. Similarly, the Miocene sedimentary deposits that accumulated in the Steamboat basin, now dissected by the Colorado River near State Bridge, have no characteristics that indicate the existence of a large river at that location and time (Brennan, 1969). Nor do the sedimentary rocks farther to the northwest give any sign that a Miocene Colorado River might have flowed into the Steamboat basin (Kucera, 1962, 1968). Hunt (1969, p. 70-71) reported finding prevolcanic gravel on north flank of the White River Plateau which he interpreted as belonging to an Oligocene Colorado River; however, detailed mapping of this region by one of us (Larson) failed to locate any gravels other than the basal conglomerate of the Browns Park Formation, which is the initial phase of the Steamboat basin fill, as described by Kucera (1962, 1968), Sharps (1962), Buffler (1967), and Segerstrom and Young (1972), rather than a deposit of the Colorado River.

A postulation of a southwesterly course for a Miocene Colorado River, a course similar to that of today, is refuted by volcanic evidence. Group 1 lava flows near State Bridge rest on either bedrock or a few feet of the basal conglomerate, and their similarity to lava sequences of the same age farther west and southwest lead us to conclude that eruption took place onto a widespread surface of low relief rather than into a valley.

The question remains: Where was the Colorado River in Miocene time? Might it have left Middle Park at a different location? The present structural and topographic barriers were already in existence in Miocene time, making it unlikely that an escape route could have existed that was superior to the low saddle in the Gore and Park Ranges (Izett, 1973, personal commun.). Might the record be unobservable because of either burial or erosion? One can only answer that the best mapping to date has failed to turn up evidence of a Miocene river following the course of the Colorado. An alternative suggestion worthy of consideration is that there were no Miocene rivers the size of the Colorado River.

The physical and biological characteristics of upper Miocene sedimentary rocks in Middle Park, North Park, and the Steamboat basin indicate that relief was low and that a semiarid climate maintained a steppe flora and fauna (Kucera, 1962, 1968; Buffler, 1967; Robinson, 1968, 1970, 1972, and 1973, personal commun.; Izett, 1968, 1975; Brennan, 1969). The deposits include loess, eolian sand, fluvial and lacustrine sedimentary rocks, and air-fall ash and lava flows. Precipitation was insufficient to sustain either large rivers or permanent lakes. The fluvial sedimentary deposits themselves suggest that streams were small and ephemeral, being sometimes flashy, but more often sluggish, when deposition occurred. Scanty precipitation was evidently related to low relief, which reached a minimum when basins had been so filled that they were depositionally interconnected, and fauna and flora could readily migrate from basin to basin (Robinson, 1968, 1970, 1972).

Relief was low at this time in other parts of Colorado, too (see other papers in this volume). All of this is consistent with the view of Hunt (1956, 1969).

This period of regional quiescence was ended by tectonism approximately 10 m.y. ago, both here and elsewhere in Colorado (see other papers in this volume). Major uplift of the Grand Canyon region may also have occurred at this time (McKee and McKee, 1972). Tectonism increased both elevation and relief, which must have had a significant effect on climate. Precipitation became sufficient to sustain large, perennial rivers. These rivers, flowing on locally improved gradients, have been responsible for the cycle of canyon development that continues at the present time. This idea that today's rivers are the product of increased precipitation resulting from late Tertiary uplift was proposed by Blackwelder in 1934 (p. 561); it is an idea that merits renewed interest. Initiation of the upper Colorado River about 10 m.y. ago would support Lucchitta's (1972) proposed timetable for establishment of the lower Colorado River.

The Colorado River is a consequent stream between Middle Park and Glenwood Springs. Its route in Middle Park is assumed to be that of the ephemeral water courses of late Miocene time, perhaps modified somewhat by tectonism. Downstream from Middle Park, the river follows downwarps in the Miocene units. From State Bridge to McCoy, the Colorado follows the State Bridge syncline (Figs. 2, 6), as was noted by Hunt (1969). Downstream from McCoy, the river follows a downwarp in group 1 and group 2 lava flows (shown in part in Fig. 6). We believe the late Miocene water courses, west of the Gore and Park Ranges, turned northwestward into the Steamboat basin and that tectonism blocked that route and forced the Colorado to spill into the aforementioned structural sags. In addition to being a consequent stream, the Colorado River has been superimposed (onto pre-Miocene rocks and structures) and is probably antecedent as well (to part of the late Cenozoic tectonism). Can one imagine a more fitting memorial to Powell than a consequent-superimposed-antecedent Colorado River?

Taking 10 m.y. as the approximate time when canyon development began, the Colorado River has since downcut 980 m just east of Glenwood Springs and more than 330 m near State Bridge; and the Eagle River near Gypsum has incised more than 370 m during that time. Rates of downcutting, however, have been far from uniform. The lower Roaring Fork River valley was already 600 m deep and quite open when group 3 lava flows were erupted into it about 8 m.y. ago (Figs. 6, 7), giving for that period an average rate of downcutting of about 0.3 m/1,000 yr. By contrast, the interval from about 8 to 1.5 m.y. ago is one in which little additional downcutting seems to have occurred. In the Roaring Fork valley, group 3 and group 4 lava flows and associated sedimentary deposits both define a valley bottom approximately 300 m above today's river (Figs. 7, 8). It is not known

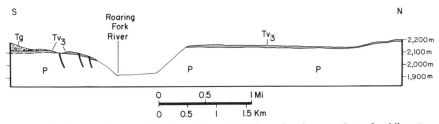

Figure 7. Structural cross section from just north of Mount Sopris on south to a few kilometers north of Basalt Mountain on north. Group 3 flow, now dissected by Roaring Fork River, is about 8 m.y. old. It flowed onto old valley bottom of Roaring Fork River. Canyon of present Roaring Fork was cut during past 1.5 m.y. Symbols and patterns as in Figure 2.

if the Colorado River had a similar history, but it is hard to imagine the two rivers following greatly different timetables. The average rate of downcutting in the Roaring Fork drainage during the past 1.5 m.y. is about 0.21 m/1,000 yr. Similarly, Rock Creek has downcut about 150 m during the past 0.6 m.y. for an average rate of 0.24 m/1,000 yr. The cause of renewed downcutting is unknown, but the evidence for recent deformation (Kucera, 1962; Tweto and others, 1970) makes renewed tectonism a likely possibility.

VOLCANIC VOLUMES AND PETROLOGY

The basaltic eruptive flows in the study area are petrologically interesting because they provide a record of intermittent magma generation and magma variation within a relatively restricted area during the past 24 m.y. It is evident that the discharge of eruptive material has diminished with time. Group 1 volcanic flows are commonly several hundreds of feet thick and are areally extensive; the flows probably were continuous over an area of about 4,660 km^2. The original volume of group 2 volcanic rocks is uncertain because it is not known to what extent the present remnants were formerly connected. Basalts less than 10 m.y. old (groups 3 and 4) are extremely restricted; in volume they total less than 1 percent of all the material erupted during the past 24 m.y.

Chemical analyses from 40 volcanic rocks (Table 2) show that, in general, group 1 flows in the Flat Tops Primitive Area contain from 47 to 51 percent SiO_2 and 1 to 2 percent K_2O. In view of the abundance of alkali oxides, these flows have been termed alkali olivine basalt. Figure 9 shows variations of Na_2O plus K_2O, and CaO, with respect to silica. It appears that in the section exposed at Turret Peak in the Flat Tops Primitive Area, the alkali content increases abruptly from about 4 to 5 wt percent during the 24- to 20-m.y. B.P. time span, whereas the silica content remains constant (Table 2). The increase in alkali content is due primarily to an increase of K_2O. CaO, on the other hand, remains constant throughout the section at about 9 percent. Petrographic study indicates that xenocrysts of quartz containing pyroxene reaction rims are not uncommon in these flows, and it might be thought that some of the increse in K_2O resulted from contamination. However, inasmuch as the SiO_2 content does not show any similar increase for those rocks that are richer in alkalis, contamination appears to be minimal. Lipman and Mehnert (1975) offer evidence that the compositional variation primarily reflects differences in depth of magma generation. Eggler (1973, personal commun.), on the other hand, believes that the magma character is controlled principally by variation in volatile content in the zone of melting and percentage of the peridotitic mantle that is melted.

Chemical analyses obtained from five flows on Yarmony Mountain near State Bridge (Table 2) indicate that this flow section is also composed predominantly of alkali olivine basalt and basaltic andesite. The three group 1 flows from the area around Sheephorn Creek are basaltic but are unusual for their high K_2O content (2.6 to 3.1 percent) and relatively high contents of MgO and P_2O_5.

Group 2 flows in the Flat Tops Primitive Area have SiO_2 values that generally increase upward in the section from about 52 percent near the base (basaltic andesite) to nearly 57 percent at the top (andesite). As shown in Figure 9, there is a regular increase in alkali content and a decrease in CaO content with increase in silica— probably indicative of a differentiation trend.

Only two samples of group 3 basalts were analyzed, and both have analyses close to that of a typical alkali olivine basalt (Table 2). Magma of this age again seems to have been piped directly from the mantle.

The known group 4 flows all have about 48 to 49 percent SiO_2 and 2.5 to 2.8 percent K_2O (see Table 2). As shown in Figure 10, total alkalis have a narrow range of 5.9 to 6.5 wt percent and CaO ranges between 7.4 and 8.5 wt percent. Because of the large percentage of alkalis, all group 4 basalts are nepheline normative. Reasons why the latest phase of volcanism is uniformly so rich in alkalis are as yet unclear.

SUMMARY OF GEOLOGIC HISTORY [2]

1. Laramide tectonism occurred between 70 and 45 m.y. ago and produced a number of large, elongate to domical positive elements (Park, Gore, and Sawatch Ranges, and White River Plateau), as well as numerous north- and northwest-trending folds and normal and reverse faults. Granitic intrusions produced local positive elements in the Sawatch Range and Elk Mountains.

2. Between 34 and 29 m.y. ago, during the Oligocene Epoch, stocks and other intrusive bodies were again emplaced in the Sawatch Range and Elk Mountains.

3. Between 24 and 10 m.y. ago, during the Miocene Epoch (and early Pliocene, by some interpretations), the Steamboat basin was filled in with sediments and, at its southern end, with basic lava flows. Early Miocene extrusions (24 to 20 m.y. ago) produced a flow-on-flow sequence of alkali olivine basalts. Later extrusions of late Miocene and possibly early Pliocene age (14 to 9 or 10 m.y. ago) produced single flows of basalt and basaltic andesite, suggesting that differentiation had occurred between the two eruptive phases.

Sedimentary rocks show a fining-upward sequence, with a basal conglomerate, a middle sandstone, and an upper siltstone. No record of a large river system such as the Colorado has been found. Miocene streams evidently were small, ephemeral, and chiefly of local derivation. Fossils indicate that the region had a semiarid climate and a steppe vegetation. By approximately 10 m.y. ago (late Miocene or early Pliocene, depending on time boundary interpretation), relief was low and the Steamboat basin was interconnected with other nearby basins.

4. Intrusion of the Treasure Mountain granitic stock 12.4 m.y. ago (late Miocene) caused 1,500 m of structural doming in the Elk Mountains.

5. Approximately 10 m.y. ago, major tectonism reactivated some Laramide structural features, created some new warps, folds, and faults, and elevated the region. Differential uplift amounted to as much as 760 m. The Gore and Sawatch Ranges, Elk Mountains, and White River Plateau were differentially raised; the Park Range, if it was raised, was raised less than some of the other elements.

The increase in elevation and relief caused an increase in precipitation, which in turn initiated the upper Colorado River system. The Colorado followed a consequent course through the Middle Park depositional basins and a consequent course to Glenwood Springs along downwarps in the Miocene lava flows (the latter reach was adopted because of tectonic disruption of the more natural drainage into the Steamboat basin).

6. By approximately 8 m.y. ago (variously interpreted as late Miocene or early Pliocene), the Roaring Fork River had downcut 600 m and had carved out a broad valley bottom, now covered by thin, 8-m.y.-old alkali olivine basalt flows.

7. From about 8 to 1.5 m.y. ago, when an eruption of nepheline-normative alkali basalt occurred, the Roaring Fork River maintained its valley bottom of approximately the same level; little additional downcutting occurred.

---

[2]This summary incorporates ideas from many people but omits references in order to preserve readability.

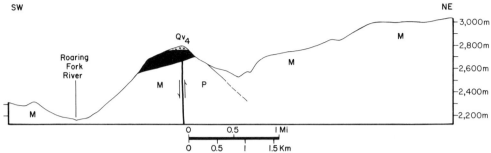

Figure 8. Structural cross section across Roaring Fork River about 16 km (10 mi) north of Aspen. Group 4 cinder cone and flow unit, which erupted into ancestral Roaring Fork valley, is dated at 1.5 m.y. Incision since eruption has amounted to about 305 m (1,000 ft) and has produced narrow V-shaped valley. Symbols and patterns as in Figure 2.

Figure 9. Silica versus alkali and calcium oxide plots for volcanic rocks of groups 1 through 3. Y, samples from Yarmony Mountain; SH, samples from Sheephorn Creek. Circles and hexagons, 24 to 20 m.y. old (group 1); triangles, 14 to 9 m.y. old (group 2); squares, 7 to 8 m.y. old (group 3). Top: silica versus alkali weight percentages. Group 1 rocks show abrupt increase in alkali content, but SiO_2 remains relatively constant. Group 2 rocks, in Flat Tops Primitive Area, increase along trend indicative of differentiation with time, seemingly from more alkalic parent basalt; analyzed group 3 rocks are indicative of return to primitive mantle-derived magma. Bottom: silica versus calcium oxide weight percentages. Most group 1 rocks show relatively small variation in SiO_2, whereas CaO varies from about 7 to 10.5 percent. Group 2 rocks, however, follow what appears to be a differentiation trend up into andesite. Group 3 flows have values nearly coincident with group 1.

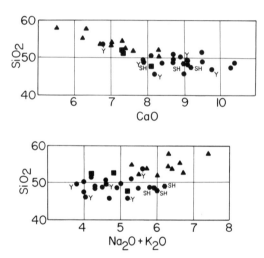

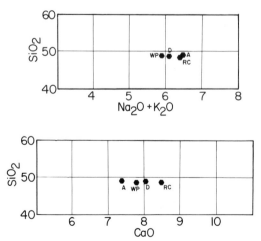

Figure 10. Silica versus alkali and calcium oxide plots for group 4 rocks. D, Dotsero flow; A, Aspen flow; RC, Rock Creek flow; WP, Willow Peak flow. Top: silica versus alkali weight percentages; note that all flows have nearly 6 percent or more total alkalis. Bottom: silica versus calcium oxide weight percentages.

8. During the past 1.5 m.y. (Pleistocene and Holocene), the Roaring Fork River has cut an inner valley 300 m deep; during the past 0.64 m.y., Rock Creek has incised a narrow canyon about 150 m deep. Possibly the downcutting was related to renewed tectonism, inasmuch as local areas give evidence of late Quaternary faulting. Volcanism during this time has produced minor amounts of cinders and flows of nepheline-normative alkali basalts; the most recent eruption took place about 4,000 yr ago near the junction of the Colorado and Eagle Rivers.

REFERENCES CITED

Axelrod, D. I., 1957, Late Tertiary floras and the Sierra Nevada uplift: Geol. Soc. America Bull., v. 68, p. 19-45.

Bass, N. W., and Northrop, S. A., 1963, Geology of Glenwood Springs quadrangle and vicinity, northwestern Colorado: U.S. Geol. Survey Bull. 1142-J, 74 p.

Beckett, R. L., 1955, Geology of the Red Canyon area, Eagle County, Colorado [M.S. thesis]: Boulder, Univ. Colorado, 36 p.

Benson, J. C., and Bass, N. W., 1955, Eagle River anticline, Eagle County, Colorado: Am. Assoc. Petroleum Geologists Bull., v. 39, p. 103-106.

Blackwelder, Eliot, 1934, Origin of the Colorado River: Geol. Soc. America Bull., v. 45, p. 551-566.

Brennan, W. J., 1969, Structural and surficial geology of the west flank of the Gore Range, Colorado [Ph.D. thesis]: Boulder, Univ. Colorado, 109 p.

Bryant, Bruce, 1971, Geologic map of the Aspen quadrangle, Pitkin County, Colorado: U.S. Geol. Survey Geol. Quad. Map GQ-933, scale, 1:24,000.

——1972, Geologic map of the Highland Peak quadrangle, Pitkin County, Colorado: U.S. Geol. Survey Geol. Quad. Map GQ-932, scale 1:24,000.

Buffler, R. T., 1967, The Browns Park Formation and its relationship to the late Tertiary geologic history of the Elkhead region, northwestern Colorado-south-central Wyoming [Ph.D. thesis]: Berkeley, Univ. California, Berkeley, 148 p.

Cook, H. J., 1960, New concepts of Late Tertiary major crustal deformations in the Rocky Mountain region of North America: Internat. Geol. Cong., 21st Copenhagen 1960, Rept., pt. 12, p. 198-212.

Donnell, J. R., 1961, Tertiary geology and oil-shale resources of the Piceance Creek basin between Colorado and White Rivers, northwestern Colorado: U.S. Geol. Survey Bull. 1082-L, p. 835-891.

Donnell, J. R., and Yeend, W. E., 1968a, Geologic map of the South Mamm Peak quadrangle, Garfield and Mesa Counties, Colorado: U.S. Geol. Survey Open-File Map, scale 1:24,000.

——1968b, Geologic map of the Rulison quadrangle, Garfield County, Colorado: U.S. Geol. Survey Open-File Map, scale 1:24,000.

——1968c, Geologic map of North Mamm Peak quadrangle, Garfield County, Colorado: U.S. Geol. Survey Open-File Map, scale 1:24,000.

——1968d, Geologic map of Hawxhurst Creek quadrangle, Garfield and Mesa Counties, Colorado: U.S. Geol. Survey Open-File Map, scale 1:24,000.

——1968e, Geologic map of the Grand Valley quadrangle, Garfield County, Colorado: U.S. Geol. Survey Open-File Map, scale 1:24,000.

——1968f, Geologic map of Housetop Mountain quadrangle, Garfield and Mesa Counties, Colorado: U.S. Geol. Survey Open-File Map, scale 1:24,000.

Donner, H. F., 1949, Geology of the McCoy area, Eagle and Routt Counties, Colorado: Geol. Soc. America Bull., v. 60, p. 1215-1248.

Eardley, A. J., 1963, Structural geology of North America (2d ed.): New York, Harper and Row, 743 p.

Epis, R. C., and Chapin, C. E., 1975, Geomorphic and tectonic implications of the post-Laramide late Eocene erosion surface in the Southern Rocky Mountains, *in* Curtis, Bruce, ed., Cenozoic history of the Southern Rocky Mountains: Geol. Soc. America Mem. 144, p. 45–74.

Freeman, V. L., 1972a, Geologic map of the Woody Creek quadrangle, Pitkin and Eagle Counties, Colorado: U.S. Geol. Survey Geol. Quad. Map GQ-967, scale 1:24,000.

——1972b, Geologic map of the Ruedi quadrangle, Pitkin and Eagle Counties, Colorado: U.S. Geol. Survey Geol. Quad. Map GQ-1004, scale 1:24,000.

Gaskill, D. L., and Godwin, L. H., 1966, Geologic map of the Marble quadrangle, Gunnison and Pitkin Counties, Colorado: U.S. Geol. Survey Geol. Quad. Map GQ-512, scale 1:24,000.

Gaskill, D. L., Godwin, L. H., and Mutschler, F. E., 1967, Geologic map of the Oh-be-joyful quadrangle, Gunnison County, Colorado: U.S. Geol. Survey Geol. Quad. Map GQ-578, scale 1:24,000.

Gates, O., 1950, The stratigraphy and structure of the Radium area, Colorado [M.S. thesis]: Boulder, Univ. Colorado, 69 p.

Giegengack, R. F., Jr., 1962, Recent volcanism near Dotsero, Colorado [M.S. thesis]: Boulder, Univ. Colorado, 44 p.

Godwin, L. H., 1968, Geologic map of the Chair Mountain quadrangle, Gunnison and Pitkin Counties, Colorado: U.S. Geol. Survey Geol. Quad. Map GQ-704, scale 1:24,000.

Hail, W. J., Jr., 1972a, Reconnaissance geologic map of the Hotchkiss area, Delta and Montrose Counties, Colorado: U.S. Geol. Survey Misc. Geol. Inv. Map I-698, scale 1:48,000.

——1972b, Reconnaissance geologic map of the Cedaredge area, Delta County, Colorado: U.S. Geol. Survey Misc. Geol. Inv. Map I-697, scale 1:48,000.

Hansen, W. R., 1965, Geology of the Flaming Gorge area, Utah-Colorado-Wyoming: U.S. Geol. Survey Prof. Paper 490, 196 p.

Hanshaw, B. B., 1958, Structural geology of the west side of the Gore Range, Eagle County, Colorado [M.S. thesis]: Boulder, Univ. Colorado, 51 p.

Holt, H. E., 1961, Geology of the lower Blue River area, Summit and Grand Counties, Colorado [Ph.D. thesis]: Boulder, Univ. Colorado, 107 p.

Hubert, J. F., 1954, Structure and stratigraphy of an area east of Brush Creek, Eagle County, Colorado [M.S. thesis]: Boulder, Univ. Colorado, 104 p.

Hunt, C. B., 1956, Cenozoic geology of the Colorado Plateau: U.S. Geol. Survey Prof. Paper 279, 99 p.

——1969, Geologic history of the Colorado River, *in* The Colorado River region and John Wesley Powell: U.S. Geol. Survey Prof. Paper 669-C, p. 59–130.

Izett, G. A., 1968, Geology of the Hot Sulphur Springs quadrangle, Grand County, Colorado: U.S. Geol. Survey Prof. Paper 586, 79 p.

——1975, Late Cenozoic sedimentation and deformation in northern Colorado and adjoining areas, *in* Curtis, Bruce, ed., Cenozoic history of the Southern Rocky Mountains: Geol. Soc. America Mem. 144, p. 179–210.

Kucera, R. E., 1962, Geology of the Yampa district, northwest Colorado [Ph.D. thesis]: Boulder, Univ. Colorado, 675 p.

——1968, Geomorphic relationship of Miocene deposits in the Yampa district, northwest Colorado, *in* Univ. Colorado Museum guidebook for field conference on the high altitude and mountain basin deposits of miocene age in Wyoming and Colorado: 16 p.

Larson, E. E., 1965, The structure, stratigraphy, and paleomagnetics of the Plush area, SE Lake County, Oregon [Ph.D. thesis]: Boulder, Univ. Colorado, 166 p.

Lipman, P. W., and Mehnert, H. W., 1975, Late Cenozoic basaltic volcanism and development of the Rio Grande depression in the Southern Rocky Mountains, *in* Curtis, Bruce, ed., Cenozoic history of the Southern Rocky Mountains: Geol. Soc. America Mem. 144, p. 119–154.

Lucchitta, Ivo, 1972, Early history of the Colorado River in the Basin and Range province: Geol. Soc. America Bull., v. 83, p. 1933–1948.

Mackay, I. H., 1953, Geology of the Thomasville-Woods Lake area, Eagle and Pitkin Counties, Colorado: Colorado School Mines Quart., v. 48, no. 4, 76 p.

Mallory, W. W., 1966, Cattle Creek anticline, a salt diapir near Glenwood Springs, Colorado, *in* Geological Survey research 1966: U.S. Geol. Survey Prof. Paper 550-B, p. 12-15.

——1971, The Eagle Valley Evaporite, northwest Colorado—A regional synthesis: U.S. Geol. Survey Bull. 1311-E, 37 p.

Mallory, W. W., Post, E. V., Ruane, P. J., and Lehmbeck, W. L., 1966, Mineral resources of the Flat Tops Primitive Area, Colorado: U.S. Geol. Survey Bull. 1230-C, 30 p.

Marvin, R. F., Mehnert, H. H., and Mountjoy, W. M., 1966, Age of basalt cap on Grand Mesa, *in* Geological Survey research 1966; U.S. Geol. Survey Prof. Paper 550-A, p. A81.

McElroy, J. R., 1953, Geology of the Derby Creek area, Eagle, Routt, and Garfield Counties, Colorado [M.S. thesis]: Boulder, Univ. Colorado, 78 p.

McKee, E. D., and McKee, E. H., 1972, Pliocene uplift of the Grand Canyon region—Time of drainage adjustment: Geol. Soc. America Bull., v. 83, p. 1923-1932.

Mull, C. G., 1960, Geology of the Grand Hogback monocline near Rifle, Colorado [M.S. thesis]: Boulder, Univ. Colorado, 189 p.

Murray, F. N., 1962, The geology of the Grand Hogback monocline near Meeker, Colorado [M.S. thesis]: Boulder, Univ. Colorado, 139 p.

——1966, Stratigraphy and structural geology of the Grand Hogback monocline, Colorado [Ph.D. thesis]: Boulder, Univ. Colorado, 219 p.

Mutschler, F. E., 1968, Geology of the Treasure Mountain dome, Gunnison County, Colorado [Ph.D. thesis]: Boulder, Univ. Colorado, 240 p.

——1969, Geologic map of the Snowmass Mountain quadrangle, Gunnison and Pitkin Counties, Colorado: U.S. Geol. Survey Geol. Quad. Map GQ-853, scale 1:24,000.

Nygren, W. E., 1935, An outline of the general geology and physiography of the Grand Valley district [M.S. thesis]: Boulder, Univ. Colorado, 109 p.

Poole, F. G., 1954, Geology of the southern Grand Hogback area, Garfield and Pitkin Counties, Colorado [M.S. thesis]: Boulder, Univ. Colorado, 128 p.

Powell, J. W., 1875, Exploration of the Colorado River of the west and its tributaries: Washington, D.C., U.S. Govt. Printing Office, 291 p.

Robinson, P., 1968, Comments on the smaller mammals of Miocene age from Middle Park, Colorado, *in* Univ. Colorado Museum Guidebook for field conference on the high altitude and mountain basin deposits of Miocene age in Wyoming and Colorado: 7 p.

——1970, The Tertiary deposits of the Rocky Mountains—A summary and discussion of unsolved problems: Contr. Geology, v. 9, p. 86-96.

——1972, Tertiary history, *in* Mallory, W. W., ed., Geologic atlas of the Rocky Mountain region: Denver, Rocky Mtn. Assoc. Geologists, p. 233-242.

Schmidt, P. B., 1961, The geology of the State Bridge area, Colorado [M.S. thesis]: Boulder, Univ. Colorado, 67 p.

Scott, G. R., 1975, Cenozoic surfaces and deposits in the eastern part of the Southern Rocky Mountains and their recognition, *in* Curtis, Bruce, ed., Cenozoic history of the Southern Rocky Mountains: Geol. Soc. America Mem. 144, p. 227-248.

Segerstrom, Kenneth, and Young, E. J., 1972, General geology of the Hahns Peak and Farwell Mountain quadrangles, Routt County, Colorado: U.S. Geol. Survey Bull. 1349, 63 p.

Sharps, Seymour L., 1962, Geology of Pagoda quadrangle, northwestern Colorado [Ph.D. thesis]: Boulder, Univ. Colorado, 282 p.

Stauffer, John E., 1953, Geology of an area west of Wolcott, Eagle County, Colorado [M.S. thesis]: Boulder, Univ. Colorado, 56 p.

Steinbach, R. C., 1956, Geology of the Azure area, Grand County, Colorado [M.S. thesis]: Boulder, Univ. Colorado, 90 p.

Taggart, J. N., 1962, Geology of the Mount Powell quadrangle, Colorado [Ph.D. thesis]: Cambridge, Mass., Harvard Univ., 239 p.

Taylor, R. B., 1975, Neogene tectonism in south-central Colorado, *in* Curtis, Bruce, ed., Cenozoic history of the Southern Rocky Mountains: Geol. Soc. America Mem. 144, p. 211-226.

Trask, N. J., Jr., 1956, Geology of the Buford area, Rio Blanco County, Colorado [M.S.

thesis]: Boulder, Univ. Colorado, 90 p.

Tweto, Ogden, 1975, Laramide (Late Cretaceous-early Tertiary) orogeny in the Southern Rocky Mountains, in Curtis, Bruce, ed., Cenozoic history of the Southern Rocky Mountains: Geol. Soc. America Mem. 144, p. 1-44.

Tweto, Ogden, Bryant, Bruce, and Williams, F. E., 1970, Mineral resources of the Gore Range-Eagles Nest Primitive Area and vicinity, Summit and Eagle Counties, Colorado: U.S. Geol. Survey Bull. 1319-C, 127 p.

Wanek, Leo J., 1953, Geology of an area east of Wolcott, Eagle County, Colorado [M.S. thesis]: Boulder, Univ. Colorado, 62 p.

Welder, George E., 1954, Geology of the Basalt area, Eagle and Pitkin Counties, Colorado [M.S. thesis]: Boulder, Univ. Colorado, 72 p.

Yeend, W. E., 1969, Quaternary geology of the Grand and Battlement Mesas area, Colorado: U.S. Geol. Survey Prof. Paper 617, 50 p.

York, D., Strangway, D. W., and Larson, E. E., 1971, Preliminary study of a Tertiary magnetic transition in Colorado: Earth and Planetary Sci. Letters, v. 11, p. 333-338.

MANUSCRIPT RECEIVED BY THE SOCIETY MAY 8, 1974

Geological Society of America
Memoir 144
© 1975

Late Cenozoic Sedimentation and Deformation in Northern Colorado and Adjoining Areas

GLEN A. IZETT

U.S. Geological Survey
Federal Center
Denver, Colorado 80225

ABSTRACT

Miocene sedimentary rocks in northern Colorado record evidence of late Cenozoic deformation, including folding, uplift, and normal faulting. Faults with late Cenozoic movements are localized along zones of Laramide faulting, and many have movements in an opposite direction from their Laramide movements. The Miocene formations in northern Colorado and their stratigraphic equivalents in Wyoming and Nebraska include the Browns Park, North Park, and Troublesome of northwest Colorado and the Arikaree and Ogallala of northeast Colorado. These formations, which formerly were much more extensive, are mainly nonorogenic eolian and fluvial siltstone and sandstone as much as 900 m thick. In the White River Plateau, Grand Mesa, State Bridge, and Middle Park areas, the sediments are interlayered with, or intruded by, basalts that are remnants of a much more extensive volcanic field than is preserved today.

Deformation accompanied and followed deposition of Miocene sediments and basalts, as shown by (1) deposition of Miocene rocks in a paleovalley cut prior to 25 m.y. along the axis of the Uinta arch and normal faulting later than 9 m.y. ago in the eastern Uinta Mountains, (2) major uplift later than 10 m.y. ago of Miocene rocks of the White River Plateau and folding of Miocene basalt in the State Bridge area, (3) faulting of Miocene rocks on the west flank of the Park Range, (4) faulting of Miocene rocks indicating renewed deformation along the trace of the Williams Range thrust (Laramide ancestry) in Middle Park, (5) faulting of Miocene rocks along the Blue River, suggesting uplift of the Gore Range, and (6) sharp folding of Miocene rocks in the North Park syncline and faulting in Saratoga valley.

INTRODUCTION

Past interpretations of the depositional and deformational geologic history of latest Mesozoic and Cenozoic time in the Southern Rocky Mountains and adjoining areas have stressed events during Laramide time (Late Cretaceous through late Eocene). This was a time of major orogenesis (Tweto, 1975) and of igneous activity with associated metallization whose economic importance drew the attention of the early geologic workers in the region. Because it influenced depositional and tectonic patterns of middle and late Cenozoic time, Laramide orogenesis was given further special attention. Consequently the late Cenozoic geologic history of the Southern Rocky Mountains has been neglected, although there have been excellent geologic descriptions of the late Tertiary deformation in the eastern Uinta Mountains (Powell, 1876; Sears, 1924a; Bradley, 1936) and the Rio Grande depression (Bryan, 1938).

In the past two decades it has become increasingly evident that many areas of the Southern Rocky Mountains and adjoining areas have been very much affected by late Cenozoic tectonism. The time of the tectonism has been dated through the study of fossil mammals included in deformed rock and through the use of radiometric age determinations on volcanic rocks interlayered with or capping upper Cenozoic rocks. The structural style of late Cenozoic tectonism in the Southern Rocky Mountains is similar to that of the widespread, large-scale tensional or normal faulting and associated volcanism that is evident elsewhere in the Western Cordillera, especially in the Great Basin.

This paper brings together some new evidence and briefly summarizes some of the significant geologic work from the past that documents uplift, folding, collapse, and normal faulting of upper Cenozoic rocks at several places in northern Colorado.

MIOCENE ROCKS

Because of their fairly wide distribution, Miocene rocks are particularly well suited for the documentation of late Cenozoic deformation in northern Colorado. For the most part, the Miocene sedimentary rocks are fine-grained nonorogenic deposits, although locally conglomerate and coarse-grained sandstone make up parts of the sequence. In the region discussed here, the Miocene sedimentary rocks include the Browns Park, North Park, Troublesome, Arikaree, and Ogallala Formations. Areas underlain by these rocks are shown in Figures 1 and 2, and a correlation chart showing the estimated radiometric ages of the formations is given in Figure 3.

In northwestern Colorado, two major lithologic types are represented in these beds: (1) gray to brown fine-grained sandstone and clayey sandstone of the Browns Park Formation of early to late Miocene age, and (2) orange-gray sandstone and clayey-silty sandstone of the North Park and Troublesome Formations chiefly of late Miocene age. Locally sandstone similar to that in the lower Miocene Arikaree Formation of southwestern Nebraska can be found in northwest Colorado, and in parts of the area, other rock types make up important stratigraphic units in the Miocene beds. Conglomerate and conglomeratic sandstone containing clasts as large as several meters in diameter of locally derived older rocks occur in some areas at the base of the Miocene formations and at higher stratigraphic levels. Impure limestone and varicolored mudstone are other minor rock types.

In northeastern Colorado, Miocene rocks can be divided into two parts. The lower part is well-sorted, very fine grained brown to gray sandstone and siltstone characterized by (1) conspicuous amounts of bluish-gray magnetite grains, (2) a heavy mineral assemblage composed chiefly of volcanically derived minerals (Denson, 1969), and (3) a large content (10 to 20 percent) of volcanically derived feldspar. These lower Miocene rocks are very similar lithologically to rocks of the Arikaree Group of Nebraska. The upper part of the Miocene sequence is poorly sorted, commonly gray to orange-gray, very fine to coarse grained sandstone and conglomerate that contains a mineral assemblage composed of a large percentage of heavy minerals interpreted by Denson (1969) to have been derived from plutonic and metamorphic rocks of Precambrian age. These sandstones and conglomerates are typical of the Ogallala Formation elsewhere in the Great Plains.

An important feature of the Miocene sedimentary rocks is the widespread occurrence of lenticular beds of rhyolitic vitric ash at many stratigraphic levels. These volcanic ashes, which reflect closely spaced episodes of volcanism far to the west, chiefly in the Great Basin, offer great potential for correlation and radiometric dating of the Miocene rocks in widely separated regions. Not only are there many discrete ash beds in Miocene rocks, but the Miocene sandstone also contains large amounts of glass shards.

Miocene volcanic rocks consisting of mafic lava flows and dikes and silicic to intermediate flows and intrusive masses are interlayered with or cut across Miocene sedimentary rocks in some areas of northwest Colorado (Larson and others, 1975). These Miocene volcanic rocks, together with some of late Oligocene age, are shown in Figure 1. Where these rocks have been radiometrically dated, they provide valuable information concerning the timing of late Cenozoic deformational events. The volcanic rocks of the Southern Rocky Mountains (Lipman and Mehnert, 1975; Larson and others, 1975; Steven, 1975) are remnants of a widespread Cenozoic volcanic field that covered large areas of northern Colorado and included the volcanic rocks of the Elkhead Mountains near Steamboat Springs, the Flat Top basalt field in the White River Plateau, the Rabbit Ears Volcanics in the Rabbit Ears Range, and volcanic rocks of the Specimen Mountain–Cameron Pass area.

The upper Cenozoic terrestrial sedimentary rocks of Colorado and surrounding areas previously were assigned either a Miocene or Pliocene age, on the basis of comparison of their included fossil mammals with those in Miocene and Pliocene stratotypes in Europe. New K-Ar age determinations on Miocene and Pliocene rocks in Europe and North America, coupled with the uncertainty of correlations between late Cenozoic mammalian faunas of Europe and North America, cast doubt on some of the previous assignments. The Miocene Series, for the purposes of this paper, arbitrarily includes all the rocks deposited or emplaced from 29 to 5 m.y. ago (Arikareean through Hemphillian land mammal ages). The basis for using 29 m.y. to mark the beginning of Miocene time is recent K-Ar age dating of ash beds near the base of the Gering Sandstone of the Arikaree Group of southwest Nebraska (Obradovich and others, 1973). The Gering has been considered by some paleontologists to be earliest Miocene in age, based on comparison of its included fossil land mammals with those in the Miocene of Europe (Wood and others, 1941, Pl. 1). Berggren (1969) suggested that the Oligocene-Miocene time boundary is 22.5 m.y. B.P., based on a correlation of radiometrically dated volcanic rocks interlayered with planktonic foraminifer-bearing marine rocks in California with similar fossils and rocks in the type lower Miocene of the Aquitaine basin in France. If the base of the Miocene Series is about 22.5 m.y. old, then the lower part of the Arikaree Formation would be Oligocene in age (see also discussion of Arikaree stratigraphy below).

Using 5 m.y. as the upper age limit of the Miocene Epoch is a consequence of seemingly reliable K-Ar age determinations (8 to 6 m.y.) reported by Van Couvering

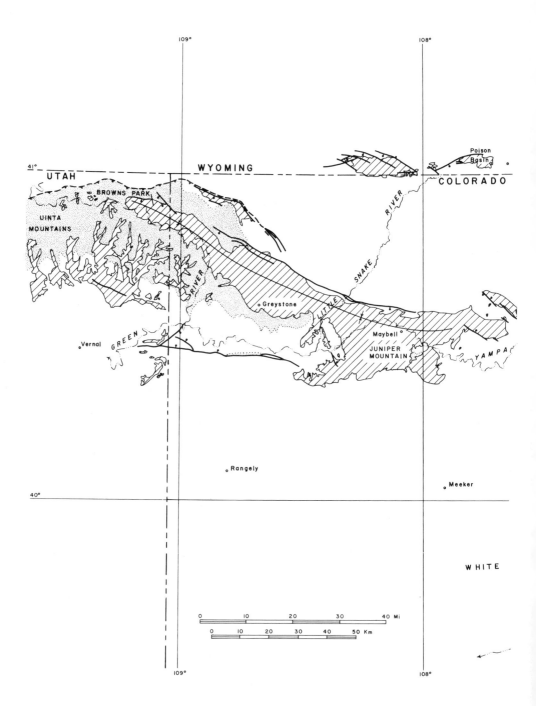

Figure 1. Geologic sketch map of northwestern Colorado and adjoining areas, showing distribution of Precambrian rocks (light stipple), pre-Miocene sedimentary rocks (no pattern), Oligocene and Miocene volcanic rocks (black), and Miocene sedimentary rocks (ruled). In this and succeeding figures, solid lines indicate faults with late Cenozoic movement, and dashed lines indicate Laramide faults. Circled numbers indicate localities mentioned in text. Sources for mapping: west of Craig—Bergin (1959), Buffler (1967), Burbank and others (1935), Dyni (1968), Hansen (1965), Hansen and others (1960), Izett (unpub. map), McKay (1974), McKay and Bergin (1974), Roehler (unpub. map), Sears (1962), Stokes and Madsden (1961); east of Craig

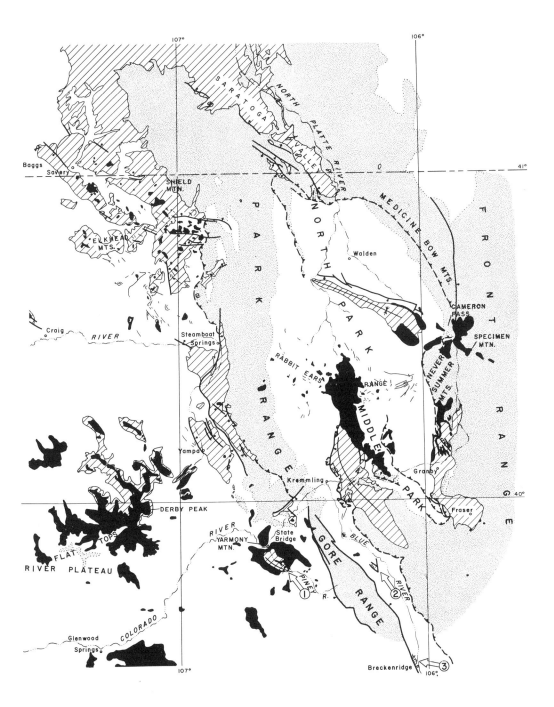

and west of Park Range—Blackstone (1953), Brennan (1969), Buffler (1967), Burbank and others (1935), Donner (1949), Kucera (1962), Mallory and others (1966), Segerstrom and Young (1973), Snyder (unpub. map); Middle and North Parks and adjoining areas—Burbank and others (1935), Corbett (1965), Gorton (1953), Hail (1965, 1968), Houston (1968), Izett (1968, 1973), Izett and Barclay (1972), Kinney (1970a, 1970b, 1970c), Kinney and Hail (1970a, 1970b), Kinney and others (1970), Love and others (1955), McCallum (unpub. map), Montagne (1955), Snyder (unpub. map), Stark and others (1949), Taylor (1975a), Theobald (1965), Tweto and others (1970), Tweto (unpub. map).

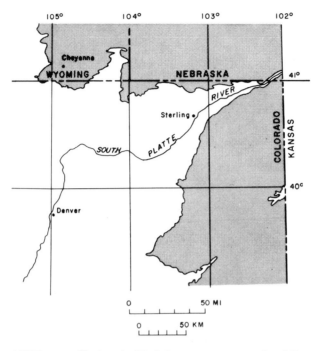

Figure 2. Sketch map of northeastern Colorado, showing distribution of Miocene sedimentary rocks (shaded).

(1972) on tuffs interbedded in uppermost marine Miocene rocks (Messinian) in Europe and the K-Ar age (7.6 m.y.) of granodiorite fragments in the base of lowermost Pliocene marine rocks on the island of Elba (Tongiorgi and Tongiorgi, 1964; Funnell, 1964). Other papers that bear on the Miocene and Pliocene boundary (at about 5 m.y. B.P.) include those by Gill and McDougall (1973) and Berggren (1972, p. 206). As a result of using the 5-m.y. date and the end of the Hemphillian land mammal age to mark the end of Miocene time, some beds, such as the upper parts of the Ogallala, Browns Park, and North Park Formations, which for many years have been considered to be of Pliocene or possible Pliocene age, are here arbitrarily assigned to the Miocene. Pliocene rocks (5 to 2 m.y. old) are seemingly rare in northern Colorado, and Pliocene time (approximately Blancan land mammal age) was seemingly a time of uplift and erosion.

A twofold stratigraphic division of Miocene time (see Sato and Denson, 1967, p. C42; Izett, 1968, p. 46) is used in this report; the top of the lower division, in terms of stratigraphic units of western Nebraska, would be at the top of the Marsland Formation of Lugn (1939), and the base of the upper division would be at the base of the Sheep Creek Formation of Lugn (1939). Lewis (1968, p. 75-76) has recently proposed a twofold faunal division of the Miocene Epoch.

West of the Park Range

Miocene sedimentary rocks west of the Park Range have been assigned to either the Browns Park or North Park Formations. The Browns Park Formation, named by Powell (1876, p. 44), comprises about 550 m of upper Tertiary sedimentary rocks that overlie all the older rocks with marked angular unconformity. Prior to latest Cenozoic erosion, the formation must have covered large areas, buried bedrock highs, and extended from basins to high on bordering mountain flanks. The formation now is limited to large outcrop areas west of Craig, Colorado, outcrops on the north and south flanks of the Uinta Mountains (Fig. 4), and outcrop areas on the west flank of the Park Range, along the upper Yampa River, near the crest of the White River Plateau, and in the Poison Basin–Baggs area (Fig. 1). The thickest deposits fill downfaulted blocks and paleovalleys of major Oligocene rivers, such as the valley that was cut along the crestline of the eastern Uinta arch and was filled in Miocene time by the Browns Park Formation (Fig. 4).

In the large outcrop areas west of Craig, Colorado, the Browns Park Formation mainly consists of eolian and fluviatile sandstone. At many places, such as along the Little Snake River at the north margin of the Browns Park outcrop, Arikaree-like sandstone beds occur in the lower part of the formation. East of Craig and west of the Park Range, three types of the Miocene rocks are found: (1) Arikaree-like sandstone crops out near the Three Forks Ranch east of Savery, Wyoming, and along the Yampa River near Yampa, Colorado; (2) loose, cross-bedded white eolian sandstone occurs in the Elkhead Mountains, in the Poison Basin area, and in the Savery, Wyoming, area (Buffler, 1967); (3) Miocene sedimentary rocks composed of orange-gray siltstone and sandstone lie on the west side of the Park Range and have been variously assigned to the North Park and Browns Park Formations. The name North Park Formation, rather than Browns Park Formation, should be applied to these orange-gray siltstones and sandstones, for they are lithologically and faunally similar to the North Park Formation at its type locality. These western

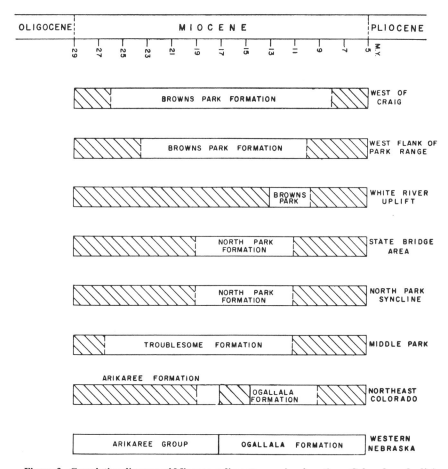

Figure 3. Correlation diagram of Miocene sedimentary rocks of northern Colorado and adjoining areas, showing estimated times of deposition.

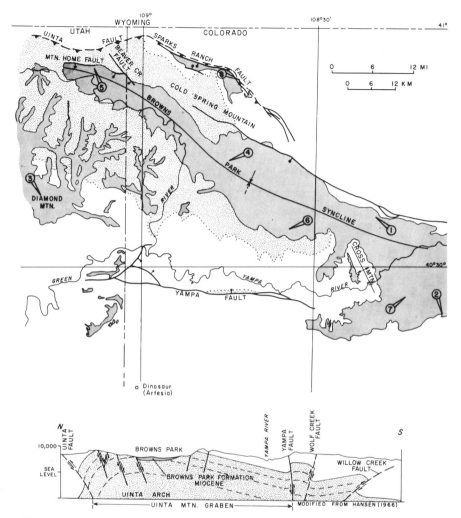

Figure 4. Geologic sketch map and diagrammatic north-south cross section showing distribution of Precambrian (stippled) and Miocene (shaded) rocks of eastern Uinta Mountains and adjoining areas. Dashed lines, solid lines, and circled numbers as in Figure 1.

exposures of the North Park Formation occur along the Piney River near State Bridge, Colorado (Brennan, 1969), and near Radium, Colorado, about 13 km northeast, at Shield Mountain north of Steamboat Springs (Buffler, 1967), and just south of Steamboat Springs along U.S. Highway 40 (Fig. 1).

The age assignments of the Browns Park and North Park Formations, which are critical to dating late Cenozoic structural events west of the Park Range, are mainly based on evidence from fossil mammals (described later) and radiometric age determinations on interlayered volcanic ash beds and associated volcanic rocks (summarized in Table 1). The evidence indicates that the Browns Park Formation ranges from about 26 m.y. to 9 m.y. (early into late Miocene) in age and that the North Park Formation west of the Park Range is of late Miocene age by the definition used here.

Fossil mammals, including the mastodon *"Serridentinus" fricki*, a rhinoceros

TABLE 1. RADIOMETRIC AGE DETERMINATIONS SIGNIFICANT TO AGE OF
BROWNS PARK AND NORTH PARK FORMATIONS WEST OF PARK RANGE,
COLORADO*

No.	Locality	Remarks	Age (m.y.)†	Reference
1	NW¼SE¼ sec. 30, T. 8 N., R. 97 W., Moffat Co., Colo.	30 m above base of Browns Park Formation	24.8 ± 0.8 (B)	Izett and others (1970)
2	NE¼ sec. 10, T. 5 N., R. 96 W., Moffat Co., Colo.	10 m above base of Browns Park Formation	23.3 ± 3.7 (Z)	C. W. Naeser (unpub. data)
3	NE¼ sec. 14, T. 2 S., R. 23 E., Uintah Co., Utah	90 m above base of Browns Park Formation	26.2 ± 0.7 (B)	Damon (1970, p. 52)
4	lat 40°7.7' N., long 106°48' W. Routt Co., Colo.	Lowest part of Browns Park Formation	23.5 ± 2.5 (Z)	C. W. Naeser (unpub. data)
5	SW¼NW¼ sec. 21, T. 9 N., R. 101 W., Moffat Co., Colo.	Upper part of Browns Park Formation	9.1 ± 1.0 (Z)	C. W. Naeser (unpub. data)
			8.2 ± 1.4 (G)	Izett (this paper)
6	Near mouth of Jesse Ewing Canyon, Fig. 4, loc. 5	Upper part of Browns Park Formation	11.8 ± 0.4 (Gl)	Damon (1970, p. 52)
7	Near Yampa, Routt Co., Colo.	Apatite from fragments in diatremes cutting Browns Park Formation	7.5 to 10.0 (A)	C. W. Naeser (unpub. data)
8	Sec. 10, T. 12 N., R. 88 W., Carbon Co., Wyo.	Basalt flow capping about 490 m of Browns Park Formation	10.7 ± 0.5 (WR)	Buffler (1967, p. 175)
9	Sec. 23, T. 12 N., R. 89 W., Moffat Co., Colo.	Basalt flow capping about 150 m of Browns Park Formation	9.5 ± 0.5 (B)	Buffler (1967, p. 174, 175)
10	Sec. 26, T. 11 N., R. 86 W., Routt Co., Colo.	Latite that intrudes Browns Park Formation	7.6 ± 0.4 (B)	Buffler (1967 p. 174)
11	Sec. 34, T. 11 N., R. 88 W., Routt Co., Colo.	Lamprophyre dike cutting Browns Park Formation	11.1 ± 0.5 (WR)	Buffler (1967, p. 174)
12	Yarmony Mountain, Eagle Co., Colo.	Basaltic lava flows below North Park Formation	21.5 ± 1.0 (WR) 24.0 ± 1.0 Best ages for two flow sequences from 12 K-Ar ages	York and others (1971)
13	Flat Tops area	Upper sequence of interbedded basalt flows and sediments	12.4 to 10.1 (WR)	Larson (1968)
		Lower sequence of basalt flows	21.6 to 13.7 (WR)	
14	Grand Mesa, Mesa Co., Colo.	From about middle of sequence of basalts about 240 m thick	9.7 ± 0.5 (WR)	Marvin and others (1966, p. A81)

*Decay constant used for fission-track ages in this paper is $\lambda_F = 6.85 \times 10^{-17}$ yr^{-1}.

†± associated with fission-track ages in this paper are 2σ calculated from number of tracks counted. Fission track ages from A, apatite; G, glass shards; Z, zircon. K-Ar ages from B, biotite; WR, whole rock; Gl, glass shards.

Aphelops ceratorhinus, and a carnivore *Bassariscops willistoni* collected near Greystone, Colorado (Fig. 1) in the type area of the Browns Park, were assigned a Miocene and Pliocene age by Peterson (1928, p. 88). J. G. Honey and I collected fossil mammals from west of Maybell, Colorado (Fig. 4, loc. 7), from the upper part of the formation and determined that they are of late Miocene age. They include a skull of a small Miocene horse, camel jaw fragments and skeletal parts, a ramus of a merycodont and limb bones, and a jaw fragment of a carnivore.

Fossil mammals collected from Arikaree-like sandstone near Yampa, Colorado (Fig. 1), reported to be of early Miocene age, are an erinaceid and the toothless jaw of a merycoidodont (Kucera, 1968, p. 7). From elsewhere in the Yampa area, Kucera (1968, p. 7) reported a tooth of *Hypolagus* from a unit of tan sandstone.

From upper Miocene sedimentary rocks interlayered with radiometrically dated basaltic lavas at Derby Peak in the Flat Tops area of the White River Plateau, Larson (1968) reported fossil mammal remains including a mylagaulid rodent, a rhinoceros, an antelope, two types of camels, and two kinds of horses. One of the horses was identified by Morris Skinner of the American Museum of Natural History as belonging to the genus *Protohippus* of late Miocene age.

Fossil mammal remains collected from the North Park Formation (brown sandstone facies of Buffler, 1967) at Shield Mountain north of Steamboat Springs, Colorado, include camel limb bones and a jaw of an oreodont. The North Park Formation along Piney River is of late Miocene age and has yielded fossil mammals that include a dog *Cynodesmus casei*, erroneously assigned by Wilson (1939) to early Miocene time; "*Mylagaulus*" *laevis*, camel, chalicothere, and prongbuck (Brennan, 1969); and an oreodont *Brachycrus* of late Miocene age found by J. R. Donnell and identified by Edward Lewis. Lewis (1973, written commun.) also has identified fossil carnivore remains of *Amphicyon* cf. *A. major*, a species originally reported from the Sansan (early late Miocene of France).

In the type area of the Browns Park, the lowest part of the formation was K-Ar dated at about 25 m.y. B.P. (Izett and others, 1970) on biotite from a volcanic ash bed that lies about 30 m above the base of the formation northwest of Maybell, Colorado (Fig. 4, loc. 1; Table 1, no. 1). Zircon microphenocrysts from a similar ash bed about 10 m above the base of the formation southwest of Maybell (Fig. 2, loc. 2; Table 1, no. 2) have a fission-track age of 23.3 ± 3.7 m.y. (C. W. Naeser, 1972, written commun.). Biotite from an ash bed about 90 m above the base of the Browns Park Formation from the Diamond Mountain Plateau (Fig. 4, loc. 3; Table 1, no. 3), Uintah County, Utah, was dated (Damon, 1970, p. 52) at 26.2 ± 0.7 m.y. B.P. A fission-track age of 23.5 ± 2.5 m.y. on zircon from an ash bed in the lowest part of the Browns Park Formation near Yampa, Colorado, indicates that the Arikaree-like sandstone in this area is of early Miocene age (C. W. Naeser, 1973, written commun.).

Zircon from an ash bed in the upper part of the Browns Park Formation have a fission-track age of 9.1 ± 1.1 m.y. (C. W. Naeser, 1973, written commun.), and glass shards have a fission-track age of 8.2 ± 1.4 m.y. (Fig. 4, loc. 4; Table 1, no. 5). A K-Ar glass date of 11.8 ± 0.4 m.y. was reported by Damon (1970, p. 52) for an ash in the upper part of the formation near the mouth of Jesse Ewing Canyon in Browns Park, Uintah County, Utah (Fig. 4, loc. 5; Table 1, no. 6).

Radiometric ages of volcanic rocks that are interlayered or that have crosscutting relations with the Browns Park Formation elsewhere in northwestern Colorado bear on the age span of the formation. Basalts interlayered with the Browns Park of the Flat Tops area range in age from 22 to 10 m.y. (Larson, 1968). North of Steamboat Springs, basaltic flows that cap the Browns Park were dated at

9.5 to 11.1 m.y. (Buffler, 1967). Near State Bridge, Colorado, at Yarmony Mountain, two basaltic flow sequences underlie fossil vertebrate-bearing upper Miocene sedimentary rocks and have been K-Ar dated at about 24 and 21.5 m.y. (York and others, 1971). A single whole-rock K-Ar date for a basalt flow sample collected on Grand Mesa indicates that the basalt sequence there is about 10 m.y. old (Marvin and others, 1966, p. 81). Fission-track ages of apatite crystals from rock fragments in diatremes that cut the Browns Park Formation along the Yampa Valley range in age from 7.5 to 10.0 m.y. and average 9.9 m.y. (C. W. Naeser, 1973, written commun.).

LATE CENOZOIC DEFORMATION WEST OF THE PARK RANGE

A classic area of late Cenozoic deformation that has drawn the attention of geologists since the earliest days of the geologic exploration of the West is the eastern Uinta Mountains. Reconstructions of the late Cenozoic geologic history of this area have been given by Powell (1876), Sears (1924a, 1924b, 1962), Bradley (1936, 1964), Hansen (1965, 1969), and Hunt (1969).

A geologic sketch map of the eastern Uinta Mountains, shown in Figure 4, portrays the distribution of the Miocene sedimentary rocks and some of the larger folds and faults of the area. Also shown in Figure 4 is a diagrammatic cross section by Hansen (1966) across the eastern Uinta Mountains from Cold Spring Mountain on the north to near Artesia, Colorado, at Dinosaur National Monument on the south.

In brief, the Cenozoic history of the eastern Uinta Mountains area, taken from the combined accounts of Sears (1924a), Bradley (1936), and Hansen (1965) and modified slightly by me, is as follows:

1. After Laramide uplift, which formed the Uinta arch (Fig. 4), and stripping of the Paleozoic and Mesozoic sedimentary rock cover, the area underwent renewed extensive erosion and epeirogenic uplift in middle Cenozoic time.

2. A major east-trending paleovalley of the ancestral Green River (Fig. 4) was cut along the crestline of the Uinta arch prior to 25 m.y. ago.

3. Deposition of the Browns Park Formation began about 25 m.y. ago and continued until about 9 m.y. ago along the east-trending paleovalley. Miocene sediments deposited away from the paleovalley covered uplands and extended high onto mountain flanks.

4. Collapse of the eastern end of the Uinta arch formed a large east-trending graben. This collapse probably began in Oligocene time, but the downdropping took place chiefly after 9 to 10 m.y. ago. According to Hansen (1973, written commun.), the collapse along the crest of the Uinta arch totaled about 1,500 m. Local unconformities in the Miocene sequence suggest movement along faults during Miocene time. Collapse of the graben took place along the Yampa fault (as much as 1,200 m of combined Laramide and late Cenozoic movement) on the south and on a series of faults and flexures parallel to the Uinta fault on the north. Later tilting of certain blocks of the eastern Uinta Mountains took place after the deposition of the Browns Park Formation, and such collapse or relaxation faulting displays as much as 500-m throw parallel to the Laramide Uinta fault on the north flank of the Uinta Mountains.

5. Present courses of major rivers such as the Green and Yampa have been superposed in post–Browns Park Formation time.

I have chosen two areas in the eastern Uinta Mountains that show specific evidence for post–Browns Park Formation high-angle normal faulting and associated folding. One area is in Browns Park near the mouth of Jesse Ewing Canyon (Fig.

4, loc. 5), a few kilometers west of the Colorado-Utah border. The fault relations in this area were illustrated by Sears (1924a, p. 296) and more recently by Hansen (1965, p. 153). A modified form of Hansen's diagrammatic sketch is given in Figure 5. Near the mouth of Jesse Ewing Canyon, the upper part of the Browns Park Formation (K-Ar dated at about 12 m.y.; see Table 1) is downfaulted relative to the Precambrian rocks of Cold Spring Mountain along the steeply dipping east-trending Mountain Home fault. The amount of throw on the fault is estimated by Hansen (1973, written commun.) to be several hundred meters. The Browns Park Formation is steeply folded adjacent to the fault and is locally overturned. Another fault that lies along the trend of the Mountain Home fault and shows post–Browns Park Formation movement is the Beaver Creek fault (Fig. 4; 457 m vertical and 1,067 m stratigraphic throw; Hansen, 1965, p. 160). Farther east along the north margin of Browns Park, late Cenozoic faults mapped by McKay (1974) and McKay and Bergin (1974) are collinear with the Mountain Home fault and are probably part of the system of faults along which the Uinta arch collapsed.

A second area of late Cenozoic faulting is on the north flank of the Uinta Mountains in the Talamantes Creek area (Fig. 4, loc. 8). According to H. W. Roehler (1973, written commun.), Miocene sedimentary rocks are faulted against Cretaceous rocks (a few hundred meters of vertical movement) along a series of high-angle north-dipping faults that parallel the Sparks Ranch fault (Laramide eastward extension of the Uinta fault). A field sketch made by Roehler (Fig. 6) illustrates his interpretation of late Cenozoic collapse or relaxation faulting in a direction opposite to that of the Sparks Ranch fault. Movements along late Cenozoic faults, which parallel the Uinta fault, probably took place prior to Miocene time and during deposition of the Miocene rocks, because at a few places multiple angular unconformities occur within the Miocene sequence.

Elsewhere in northwest Colorado, late Cenozoic folding, faulting, tilting, and uplift have been reported by many geologists. A few of the better examples of such deformation are described below.

Geologic mapping by Hansen (1965), Sears (1924b), Bergin (1959), McKay and Bergin (1974), McKay (1974), Dyni (1968), and Hancock (1925) has outlined a large east-trending syncline in the Browns Park Formation (Fig. 4). The fold, called the Browns Park syncline, extends eastward from the west end of Browns Park nearly to Craig, Colorado. The syncline has a gentle south limb and a steeper north limb, and it formed, according to Hansen (1965, p. 154) "through depositional rather than tectonic processes," although the north limb was later steepened by post-Browns Park Formation faulting. Dyni (1968) mapped a series of north-north-west-trending normal faults (Figs. 1, 4) that cut the Browns Park Formation on the west flank of Cross Mountain in the Elk Springs quadrangle. Latest movement occurred on the faults in association with uplift of Cross Mountain after the Browns Park Formation seemingly had covered the top of Cross Mountain.

Near Yampa, Colorado (Fig. 1), Kucera (1962, p. 371) described a large fault zone, of Laramide ancestry, that was reactivated in post–Browns Park time. Movement was at least 120 m and possibly as much as 1,200 m. According to Kucera (1962, p. 371), most of the post-Browns Park faults occurred "along Laramide faults but in a direction opposite to that of Laramide movement." Other late Cenozoic deformation in the area is indicated by folding and faulting of Miocene basalt flows and uplift of the Park Range, as indicated by coarse gravel in the Browns Park Formation.

Several faults of late Cenozoic age were mapped by Buffler (1967) in the Elkhead Mountains area northwest of Steamboat Springs (Fig. 1). One of the better documented breaks, called the Steamboat Springs fault (Figs. 1, 7), is coextensive with a large fault mapped by Kucera (1962) and can be traced from near Strawberry Park to a point 50 km to the south. Buffler (1967, p. 113) and G. L. Snyder (1973, written commun.) believed the fault possibly has as much as 300 m of post-Browns Park Formation movement. The Steamboat Springs fault is roughly parallel to a Laramide thrust fault mapped along the west front of the Park Range by Segerstrom and Young (1973) and G. L. Snyder (1973, unpub. data) and has movement in an opposite direction to the Laramide movement. Other faults with movements that postdate the Browns Park Formation occur at Sand Mountain (600 m of movement; Buffler, 1967, p. 114), and along east-trending faults associated with emplacement of the volcanic rocks of the Hahns Peak area (Buffler, 1967; Segerstrom and Young, 1973).

Another area of late Cenozoic deformation (Fig. 1, loc. 1) west of the Park Range is along Piney River near State Bridge, where the Miocene North Park Formation overlies basaltic flows 24 to 21.5 m.y. old (York and others, 1971). Brennan (1969, Pl. 1) mapped a large open fold in the North Park Formation

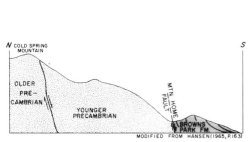

Figure 5. Diagrammatic cross section near locality 5 (Fig. 4), near the north margin of Browns Park, showing steeply folded Browns Park Formation (shaded) downfaulted several hundred meters relative to Precambrian rocks of Cold Spring Mountain block.

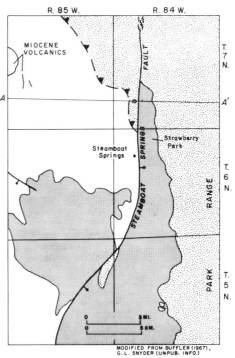

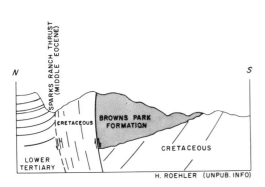

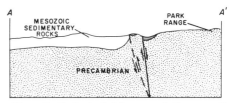

Figure 6. Field sketch for locality 8 (Fig. 4), showing Miocene Browns Park Formation (shaded) downfaulted relative to Cretaceous rocks along late Cenozoic fault parallel to Sparks Ranch fault. Browns Park Formation is no more than 100 m thick at this locality.

Figure 7. Geologic sketch map and diagrammatic section near Steamboat Springs, Colorado, showing Miocene sedimentary rocks (shaded) downfaulted possibly as much as 300 m along late Cenozoic Steamboat Springs fault. Stipple indicates Precambrian rocks.

and underlying basalt flows that he called the Piney River syncline. The fold probably occurred later than 12 to 13 m.y. ago, judging from the age of the fossil mammal assemblage in the North Park Formation of the area.

The White River Plateau north of Glenwood Springs, mapped in part by Bass and Northrop (1963) and Mallory and others (1966), is a large structural block uplifted about 3,600 m by Laramide and later tectonism. It involves 2,400 m of Paleozoic and Mesozoic sedimentary rocks overlain by radiometrically dated basaltic flows (22 to 10 m.y. old) and by interlayered fossil mammal–bearing sedimentary rocks of late Miocene age (Larson, 1968). The basalt flows form the caprock in the Flat Tops area and are the remnants of a once more extensive basaltic volcanic field that may have extended to the east as far as the Park Range and to the south and west as far as Grand Mesa. Hunt (1969, p. 70–71) suggested that the basalt and interlayered sedimentary rocks probably were uplifted about 600 m later than 10 m.y. ago and that the course of the Colorado River has been superposed from the surface of the uplifted basalt flows. He noted that the Miocene sedimentary rocks contain gravels composed of Precambrian rock fragments that, according to Ogden Tweto (in Hunt, 1969), most probably were derived from the Park Range, far to the east.

Another example of late Cenozoic deformation, near Grand Junction in western Colorado, was described by Hunt (1956), Lohman (1965), and Cater (1966). The Uncompahgre Plateau, a monoclinal uplift more than 160 km long and 40 to 50 km wide, was uplifted, according to Cater (1966, p. C86), about 400 to 425 m out of a total uplift of 610 m, beginning in latest Cretaceous and extending to early Pleistocene time. Evidence for this deformation (Cater, 1966) is furnished by the Gunnison River's abandonment of Unaweep Canyon, in response to uplift, leaving a spectacular wind gap about 40 km long. Unaweep Canyon seemingly was abandoned by early Pleistocene time, because the wind gap is locally floored by ancient river gravel—derived from the Gunnison River system—overlain by fanglomerate that Cater (1966) believed to be of early Pleistocene age.

NORTH AND MIDDLE PARKS

Upper Tertiary sedimentary rocks that occur in North and Middle Parks and are chiefly composed of orange-gray siltstone and sandstone are called the North Park and Troublesome Formations (Fig. 3). The rocks of the two formations are very similar in age and general appearance and are akin to the terrestrial upper Cenozoic formations such as the Ogallala of the Great Plains, the Wagontongue (Stark and others, 1949) of South Park, the Dry Union of the Arkansas River valley, and the Tesuque Formation of northern New Mexico. The name Browns Park Formation has been applied by some to an Arikaree-like sandstone sequence that locally underlies the North Park Formation in Big Creek Park (Fig. 8, loc. 2) and near the Pick Ranch along the North Platte River north of the town of Saratoga, Wyoming.

North Park

The North Park Formation (King, 1876; Beekly, 1915; Montagne, 1955; Montagne and Barnes, 1957; Hail, 1965) is as much as 550 m thick, and its major outcrop areas are in its type locality along Peterson and Owl Ridges in the North Park syncline and to the north in Saratoga Valley. These rocks once covered the floor of North Park and extended north along the paleovalley of the ancestral North Platte River, which probably coincided with the axis of thickest deposition of the formation.

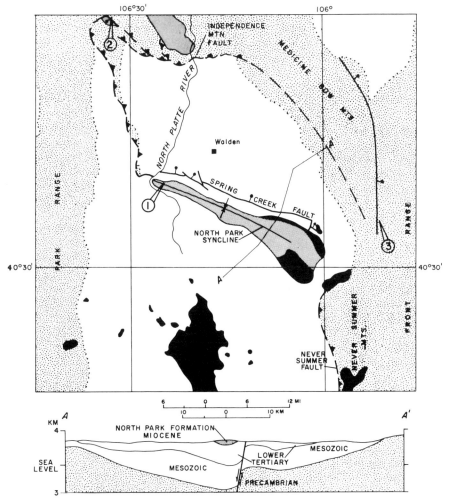

Figure 8. Geologic sketch map of North Park, Colorado, and adjoining areas and diagrammatic cross section showing distribution of Miocene sedimentary rocks (shaded), Oligocene and Miocene volcanic rocks (black), and Precambrian rocks (stippled). Circled numbers as in Figure 1.

The late Miocene age of the North Park Formation is known from its included fossil mammals and its stratigraphic position above the White River Formation of Oligocene age or above rhyolite ash-flow tuffs of Oligocene age K-Ar dated at 31.0 m.y. old (J. D. Obradovich, 1973, written commun.). Mammalian fossils collected from about the middle of the North Park Formation in its type area on Peterson Ridge in North Park, Colorado (Fig. 8, loc. 1) are teeth of a horse *Merychippus* (Hail and Lewis, 1960, p. 260) of late Miocene age. Remains of *Merychippus* and *Brachycrus* of late Miocene age from the North Park Formation in Saratoga Valley, Wyoming, were identified by Lewis (Hail and Lewis, 1960, p. 260). Elsewhere in Saratoga Valley, Montagne and Barnes (1957) found fossil remains of an oreodont *Ustatochoerus*, an antilocaprid *Merycodus*, a beaver *Eucastor*, a horse *Merychippus*, and a camel similar to *Pliauchenia*. P. O. McGrew (1951) identified a gomphotherid mastodon, a horse *Nannippus*, a camel *Procamelus*, and antilocaprids *Merycodus* and *Paracosoryx* in the North Park Formation of

Saratoga Valley. The fossil mammals listed above all indicate that the North Park Formation is of late Miocene age.

Oreodont remains collected in Big Creek Park (Fig. 8, loc. 2) near the Colorado-Wyoming border were identified by C. H. Falkenbach (Montagne and Barnes, 1957, p. 59) as being close to *Merychyus arenarum*, an index fossil of early Hemingford (middle Miocene) age (Schultz and Stout, 1961, p. 13). The fossils were taken from an Arikaree-like sandstone that Montagne and Barnes (1957, Pl. 1) assigned to the Browns Park Formation.

One of the more conspicuous structural features in North Park is the North Park syncline, which has been mapped by Hail (1965), Kinney (1970a), and Kinney and Hail (1970b) and studied geophysically by Behrendt and others (1969). A geologic sketch map (Fig. 8) adapted from the maps of Hail (1965, 1968), Kinney (1970a, 1970b, 1970c), Kinney and Hail (1970a, 1970b), and Kinney and others (1970) shows the distribution of Miocene sedimentary rocks in the North Park syncline and some of the other important structural features of North Park. The North Park syncline is a narrow west-northwest–trending fold that extends from near the North Platte River to about 32 km to the east-southeast. This syncline is south of the large Spring Creek fault which, according to Behrendt and others (1969), has about 1,500 m of throw. The part of the total throw that is of late Cenozoic age cannot be determined, but presumably it is significant, as suggested in Figure 8 (the cross section in Fig. 8, through the North Park syncline, is based on the work of Behrendt and others, 1969). The surface rocks involved in the folding are the Coalmont Formation of Paleocene and early Eocene age, rhyolitic ash-flow tuff of Oligocene age, and the North Park Formation of Miocene age.

Other localities of late Cenozoic deformation in the North Park area have been described by Montagne (1957) and Montagne and Barnes (1957) in Big Creek and Cunningham Parks near the Colorado-Wyoming border and farther north in Saratoga Valley (Montagne, 1955). In Big Creek and Cunningham Parks (Fig. 8, loc. 2), Miocene rocks occur in wedge-shaped bodies bounded on the east by northwest-trending late Cenozoic faults that parallel the Independence Mountain fault of Laramide age (Fig. 8). These late Cenozoic faults seemingly are part of a post–North Park Formation structural reactivation along the Laramide Independence Mountain thrust fault.

Another area of suspected late Cenozoic faulting occurs in the Chambers Lake area (Fig. 8, loc. 3). Outcrops around the shores of Chambers Lake and to the east are part of a fairly large area underlain by Cenozoic rocks I interpret to be Oligocene in age. According to M. E. McCallum (1973, written commun.), these sedimentary rocks were downdropped in late Cenozoic time along a large north-trending fault (Fig. 8) that lies along the Laramie River valley and probably has a long history of movement beginning in Precambrian time. It is noteworthy that a segment of the upper Colorado River valley south of Cameron Pass is collinear with the Laramie River fault and probably is a late Cenozoic fault valley.

A brief summary of the geologic events in North Park as interpreted from geologic and geophysical maps is as follows:

1. Laramide orogenesis formed the North Park–Middle Park basin, and the basin was filled with considerable coarse detritus (Middle Park and Coalmont Formations); post–Coalmont Formation faulting and folding followed.

2. Erosion in late Eocene time was followed by deposition of the White River Formation of Oligocene age; nearly all of the White River Formation then was stripped from North Park.

3. Volcanism occurred in the Rabbit Ears Range to the south and in the Never Summer Range to the east in Oligocene and Miocene time. Tongues of rhyolitic

ash flows derived from the Specimen Mountain area spread locally into eastern North Park. Pumice associated with rhyolitic ash flows about 28 m.y. old (Corbett, 1968) in the Specimen Mountain area possibly swept down the ancestral North Platte and Laramie River valleys and was deposited in the lowest part of the Gering Sandstone near Scotts Bluff, Nebraska, where rhyolite pumice fragments about 28 m.y. old (Obradovich and others, 1973) occur.

4. The North Park Formation, deposited in Miocene time, covered the floor of North Park and extended north into Saratoga Valley.

5. The North Park Formation was folded into the North Park syncline south of the Spring Creek fault. This folding and faulting is probably younger than 12 to 13 m.y., judging from the kinds of fossil mammals found in the North Park Formation.

Middle Park

Tuffaceous sedimentary rocks of Miocene age, which are as much as 450 m thick, compose the Troublesome Formation in Middle Park (Richards, 1941; Izett, 1968). These rocks occur in three intermontane basins (Fig. 1), one near Kremmling, a second near Granby, and a third near Fraser. Deposition of the Troublesome began in paleovalleys and in gently subsiding basins in response to influx of locally derived material and also enormous quantities of airborne ash derived from volcanic sources in the Great Basin. The formation is locally divisible into two main lithologic types—a lower unit of Arikaree-like sandstone (occurring mainly north of Granby but also near Kremmling) and an upper unit of orange-gray siltstone and sandstone. Many other rock types can be found in the formation, including conglomerate composed of clasts from the Rabbit Ears Volcanics (Oligocene and Miocene?) and Precambrian rocks, and also earthy limestones, varicolored sandy mudstones, and many volcanic ash beds.

The Troublesome Formation is underlain in some places by the Rabbit Ears Volcanics (Fig. 9) of Oligocene and Miocene(?) age, and locally basalt flows are interlayered with the Troublesome north of Granby (Fig. 10). East of Kremmling, the Troublesome is locally interlayered with basalt flows (Grouse Mountain Basalt) assigned a Pliocene(?) age by Izett (1968), but these basalts are here considered Miocene in age as a result of a recent whole-rock K-Ar age determination (DKA2847) of 20.1 ± 0.4 m.y. made by J. D. Obradovich (1974, written commun.).

Sedimentary rocks included in the Troublesome Formation in Middle Park are here interpreted to be early to late Miocene in age (Izett and Lewis, 1963; Izett, 1968; Lewis, 1969; Izett and Barclay, 1972). The lowest part of the Troublesome (Arikaree-like sandstone) near Kremmling (Fig. 9, loc. 1) is latest early Miocene in age, judging by the occurrence of the fossil oreodont *Merycochoerus matthewi*, an index fossil (Schultz and Stout, 1961, p. 13) of the lower part of the Marsland Formation of Nebraska (late early Miocene of this report). An even older Arikaree-like sandstone part of the Troublesome Formation lies north of Granby (Fig. 10), where a volcanic ash bed (Fig. 10, loc. 4) was fission-track dated at about 23 m.y. (Table 2, no. 4). A basalt flow (Fig. 10, loc. 3) interbedded with the Troublesome above the fission-track-dated ash bed is about 26 m.y. old (DKA2682) (Table 2, no. 5; J. D. Obradovich, 1973, written commun.). Although the radiometric ages conflict, they suggest that the rocks in the area are certainly of early Miocene age. Fossil mammals reported by Lovering (1930) were collected from Miocene rocks, most probably from near locality 2 (Fig. 10). Peter Robinson of the University of Colorado found a *Gregorymys*-like rodent of early Miocene age from near locality 1 (Fig. 10). Fossil mammals from the lower part of the orange-gray siltstone and sandstone unit of the Troublesome near Kremmling (Fig. 9, loc. 2) are the oreodont

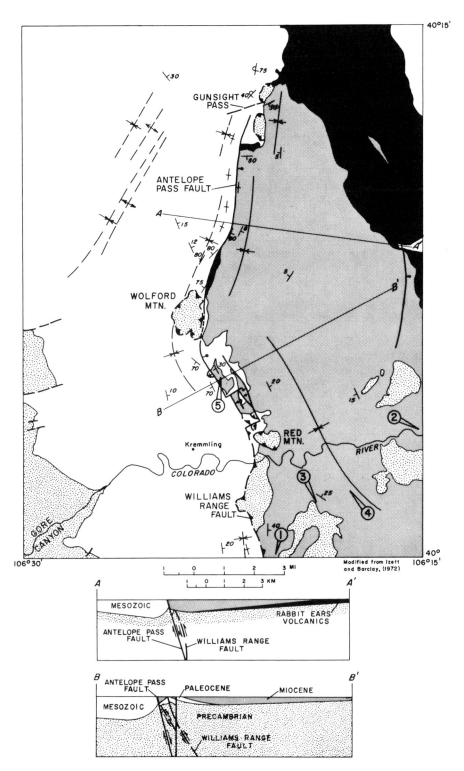

Figure 9. Geologic sketch map and diagrammatic cross sections of Kremmling 15′ quadrangle, Grand County, Colorado, showing distribution of Precambrian rocks (stippled), Oligocene and Miocene extrusive volcanic rocks (black), and Miocene sedimentary rocks (shaded).

Merycochoerus proprius and a rodent similar to *Mesogaulus paniensis* (Izett, 1968; Lewis, 1969), which typically are found in rocks of late early Miocene age. A zircon and sphene fission-track age determination on a sample from an ash bed presumably at about the stratigraphic position of these fossils is about 18 to 20 m.y. (Fig. 9, loc. 3; Table 2, no. 3).

Fossil mammal remains that clearly date the upper part of the Troublesome Formation (orange-gray siltstone and sandstone) as late Miocene (Fig. 9, loc. 4) are an oreodont *Brachycrus*, several kinds of horses, including *Parahippus*, *Hypohippus*, and *Merychippus*, a rodent *Mesogaulus laevis*, a camel *Protolabis angustidens*, and a rhinoceros *Aphelops* (Izett and Lewis, 1963; Lewis, 1969). Some of these fossils typically occur in the Sheep Creek Formation (late Hemingfordian of Lugn, 1939) of western Nebraska. A zircon fission-track age determination on a volcanic ash bed at the stratigraphic level of these fossils is about 12.8 ± 2.0 m.y. (C. W. Naeser, 1973, written commun.), and a glass fission-track age determination made by me is 10.9 ± 1.4 m.y. (Table 2, no. 2). Glass fission-track ages are thought to be minimum ages, because of the probability of track fading or annealing through geologic time (Lakatos and Miller, 1972). A younger part of the Troublesome is dated by the presence of a milk molar of a mastodon in the formation north of Kremmling near locality 5 (Fig. 9). A volcanic ash at about the stratigraphic level of the fossil was fission-track dated at 11.2 ± 1.8 m.y. (Table 2, no. 1) on zircons.

Fossil mammal assemblages and radiometric age determinations (Table 2) indicate clearly that the Troublesome Formation in Middle Park ranges in age from early to late Miocene and is perhaps as old as 26 m.y. and as young as 11 m.y.

Effects of late Cenozoic faulting and associated folding, visible at several places in Middle Park, are particularly well illustrated at localities near Kremmling, north of Granby, and along the Blue River between Kremmling and Breckenridge (Fig. 1).

East of Kremmling, the Troublesome Formation was deposited in the small intermontane Troublesome basin (Fig. 9) that formed in Miocene time and that was filled with sediments stripped from the adjoining highlands after an episode

TABLE 2. RADIOMETRIC AGE DETERMINATIONS FROM INTERBEDDED TUFF AND LAVA BEDS OF TROUBLESOME FORMATION, MIDDLE PARK, COLORADO

No.	Locality	Remarks	Age (m.y.)*	Reference
1	Fig. 9, loc. 5, SW¼SE¼ sec. 32, T. 2 N., R. 80 W.	Uppermost part of Troublesome Formation	11.2 ± 1.8 (Z) 8.3 ± 1.2 (G)	Izett (this paper)
2	Fig. 9, loc. 4, NE¼ SE¼ sec. 24, T. 1 N., R. 80 W.	Upper part of Troublesome Formation	12.8 ± 2.0 (Z)	C. W. Naeser (unpub. data)
			10.9 ± 1.4 (G)	Izett (this paper)
3	Fig. 9, loc. 3, SW¼SW¼ sec. 24, T. 1 N., R. 80 W.	Lower part of Troublesome Formation	18.0 ± 2.0 (Z) 20.0 ± 2.3 (S)	C. W. Naeser (unpub. data)
4	Fig. 10, loc. 4, NW¼SW¼ sec. 18, T. 2 N., R. 76 W.	Lowest part of Troublesome Formation	22.8 ± 3.1 (Z)	Izett (1973)
5	Fig. 10, loc. 3, SW¼SE¼ sec. 7, T. 2 N., R. 76 W.	Lowest part of Troublesome Formation	25.9 ± 0.4 (WR)	Izett (1973)

*Fission-track ages from Z, zircon; S, sphene; G, glass shards. WR, whole rock K-Ar age.

of middle Cenozoic volcanism (Rabbit Ears Volcanics). The Troublesome basin is about 48 km long, and its axis lies west of Troublesome Creek and along the Williams Fork River.

North of the Colorado River along the west margin of outcrops of the Troublesome Formation, a series of faults is associated with late Cenozoic collapse along the upthrown block of the Williams Range thrust fault, one of the larger Laramide faults of northern Colorado. These late Cenozoic faults, most of which are high-angle normal faults, parallel the partly buried trace of the Williams Range fault. Late Cenozoic normal faulting is illustrated by generalized cross sections A-A' and B-B' of Figure 9 and a diagrammatic cross section (Fig. 11). In the area of cross section A-A', a post-Troublesome Formation fault (Antelope Pass fault of Fig. 9) separates steeply west-dipping to locally overturned Cretaceous shale and sandstone of the Pierre Shale from steeply east-dipping Miocene siltstone of the Troublesome Formation. The steep dips in the Cretaceous rocks are along the east limb of a large syncline that formed in the Cretaceous rocks in front of the upthrown Precambrian rocks of the Williams Range fault in Laramide time. The trace of the Williams Range thrust fault lies buried below the Troublesome Formation probably less than 1.2 km east of the Antelope Pass fault. The steep dips in the Miocene Troublesome resulted from drag along the Antelope Pass fault, whose throw may be as much as 300 m.

In the area of cross section B-B' (Fig. 9), a post-Troublesome Formation fault with several hundred meters of throw lies along the trend of the Antelope Pass fault. This fault separates steeply dipping to locally overturned Cretaceous shale west of the fault from a rotated block of Miocene Troublesome Formation that dips about 30° into the fault plane. A diagrammatic cross section of the area is shown in Figure 11. The Troublesome Formation in the area overlies arkosic grit of the lower Tertiary Middle Park Formation, which in turn overlies the eroded upthrown block of the Williams Range thrust fault.

The Antelope Pass fault probably was active in Miocene time, as suggested by outcrops east of the fault trace south of Wolford Mountain and north of Red Mountain (Fig. 12). These outcrops consist of a large mélangelike landslide block composed of an intricate mixture of lenses of Miocene (Troublesome Formation) and Paleocene (Middle Park Formation) sandstone and conglomerate enclosed by marine Cretaceous shale (Pierre Shale). Seemingly the landslide block slid in Miocene time from an eastern area where Pierre Shale was overlain by Troublesome Formation toward a western area along the active trace of the Antelope Pass fault. Latest movement on the Antelope Pass fault occurred later than about 11 m.y. ago, because the sequence of beds that compose the Troublesome Formation includes a vitric rhyolitic ash bed fission-track dated at 11.2 m.y. (Table 2, no. 1; Fig. 11).

A third area of late Cenozoic deformation in Middle Park lies north of U.S. Highway 40 east of Kremmling at Red Mountain. Red Mountain (Fig. 9) is seemingly a klippe of Precambrian rocks that lies along the gently east-dipping trace of the Williams Range fault. Outcrops on the south side of the mountain show an inverted sequence that includes in ascending order (1) Troublesome Formation (Miocene), overlain in fault contact by (2) thin remnants of the Middle Park Formation (early Tertiary) and Dakota Sandstone (Cretaceous), overlain by the klippe of Precambrian crystalline rocks. The klippe of Precambrian rock at Red Mountain (formed in Laramide time) moved in Miocene time as a gravity slide block in response to reactivation along the Laramide Williams Range fault.

Evidence that the Gore Range (Fig. 1) was uplifted in late Cenozoic time was given by Tweto and others (1970), who mapped a fault block (Fig. 1, loc. 2) of upper Cenozoic sedimentary rocks with minor interbedded basalt flows in the

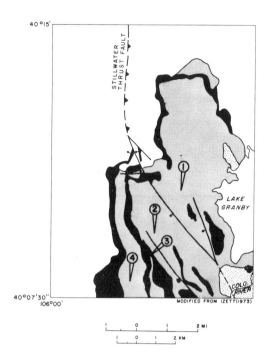

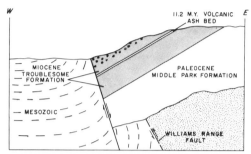

←Figure 10. Geologic sketch map of Trail Mountain 7½′ quadrangle, Grand County, Colorado, showing distribution of Precambrian rocks (stippled), Miocene basalt flows (black), and Miocene sedimentary rocks (shaded). Circled numbers as in Figure 1.

Figure 11. Diagrammatic east-west section across Antelope Pass fault near locality 5 (Fig. 9) in Kremmling 15′ quadrangle, Grand County, Colorado, showing Miocene Troublesome Formation (shaded) overlying Paleocene Middle Park Formation. Troublesome Formation is about 100 m thick at this locality. Middle Park Formation was deposited on eroded upthrown block of Laramide Williams Range thrust fault. Stipple indicates Precambrian rocks.

Brush and Squaw Creeks area on the east flank of the Gore Range. The upper Cenozoic rocks in the fault block (Tweto and others, 1970, p. C19) "are possibly, but not certainly a facies of the Troublesome Formation." The fault block has been downfaulted relative to the Gore Range block, and this relationship suggests that the upper Cenozoic rocks assigned to the Miocene Troublesome Formation were, according to Tweto and others (1970, p. C19), "deposited after the first elevation and subsequent erosion of the modern Gore Range but before rejuvenation of the range in its present heights." Further evidence of the late Cenozoic rejuvenation of the Gore Range is found (Fig. 1, loc. 3) in the Breckenridge area, where Tweto (1973, unpub. map) identified fault blocks of the upper Cenozoic Dry Union Formation. These blocks are downfaulted relative to the Gore Range along a major fault zone called the Frontal fault by Tweto and others (1970, p. C26).

In the Granby area, late Cenozoic sedimentation began in early Miocene time in a small intermontane basin that extends from near Grand Lake on the north to a few miles south of Granby (Fig. 1); farther south the Miocene rocks of the Troublesome Formation can be traced across a bedrock high into the Fraser basin. North of Granby in the Trail Mountain quadrangle (Izett, 1973), the Troublesome Formation is interlayered with basalt flows that outline an embayment of the Granby basin (Fig. 10). Mapping of the basalt flows coupled with paleomagnetic work suggests that the flows are in east-dipping tilted fault blocks. The associated post–Troublesome Formation faults suggest renewed movements along Laramide faults as they are parallel to and dip toward the trace of the large Laramide Stillwater thrust (Fig. 10).

A brief summary of the geologic history of Middle Park is as follows:

1. Retreat of Cretaceous sea about 67 m.y. ago (Obradovich and Cobban, 1974) was followed by Laramide orogenesis, during which the main structural framework of Middle Park and the surrounding mountain ranges was formed. Mesozoic sedimentary rocks were folded, faulted, and overridden by Precambrian rocks along

the Williams Range fault. Some of the Laramide faults may have a Precambrian ancestry. Precambrian cores of uplifted mountain ranges were exposed by about 65 m.y. ago.

2. An episode of andesitic volcanism (Windy Gap Member of Middle Park Formation, possibly in Late Cretaceous time) and deposition of arkosic basin-fill deposits of the Middle Park Formation (Paleocene) occurred. The Middle Park Formation locally truncates Mesozoic sedimentary rocks and laps onto the Precambrian rocks.

3. A late phase of Laramide folding and faulting took place, probably during Eocene time, after and possibly during deposition of the Middle Park Formation at several places, including the Mount Bross fault at Hot Sulphur Springs (Izett, 1968).

4. Erosion occurred during middle and late Eocene time.

5. Volcanism started about 30 m.y. ago in the Rabbit Ears Range and Never Summer Mountains.

6. Deposition of the Troublesome Formation in paleovalleys and structural basins started in early Miocene time (possibly as early as 26 m.y. ago) and continued until about 11 m.y. ago. Basaltic volcanism accompanied deposition of the Troublesome Formation. Late Cenozoic faulting and uplift folding of Miocene rocks took place principally along pre-existing fault zones of the Gore and Park Ranges after about 11 m.y. ago.

NORTHEAST COLORADO

Upper Cenozoic rocks of northeast Colorado here assigned to the Miocene Series (Fig. 2) include the Arikaree and Ogallala Formations. The Arikaree Formation as used in this report is the lower part of the Pawnee Creek Formation of Galbreath (1953), and the Ogallala Formation of this report is the upper part of the Pawnee Creek Formation of Galbreath (1953) and younger Tertiary rocks. The Arikaree and Ogallala are continuous to the north with the beds in the classical areas of upper Cenozoic rocks of western Nebraska and eastern Wyoming. The Miocene formations now underlie large areas north and south of the South Platte River, and they must have covered much larger areas prior to latest Cenozoic erosion.

Arikaree Formation

The Arikaree Formation of Miocene age consists of very fine grained, well-sorted, buff to gray sandstone characterized by its high content of volcanic plagioclase, glass shards, and magnetite grains, and by a heavy mineral assemblage mainly composed of volcanically derived minerals such as green-brown hornblende and augite (Denson, 1969, p. 31). The volcanically derived material in the Arikaree resulted from the influx of large amounts of ash derived from volcanic centers far to the west. In northeast Colorado, the formation is not nearly as widespread as might be inferred from inspection of the Geologic Map of Colorado (Burbank and others, 1935), because much of what is shown on the map as Arikaree is either White River Formation or Ogallala Formation (N. M. Denson, 1973, unpub. maps). In northeast Colorado there are only scattered remnants of the Arikaree (lower part of the Pawnee Creek Formation of Galbreath, 1953), less than 23 m thick, such as those at Martin Canyon north of Sterling, Colorado. The Arikaree is much thicker (about 180 m) and more widespread along the North Platte River in Nebraska, where it is of group rank and has been divided into formations. In northeast Colorado, either most of the Arikaree was eroded prior to deposition of the overlying Ogallala, or the Arikaree thins by onlap from a northern area along the paleovalley of the North Platte River to a southern area in northeast Colorado. It seems more likely to me that the Arikaree thins chiefly by onlap, because, had it been deposited uniformly and then removed by pre-Ogallala erosion, only the lowermost part of the Arikaree Group (Gering and Monroe Creek Formations) would be preserved. Instead, only the uppermost part of the Arikaree (Marsland Formation age) is known in northeast Colorado (N. M. Denson, 1973, unpub. map), supporting the interpretation of onlap with minor pre-Ogallala erosion.

The late early Miocene age assignment of the beds here included in the Arikaree Formation (lower part of the Pawnee Creek Formation of Galbreath, 1953) in northeast Colorado is based on their stratigraphic position above the Oligocene White River Formation and below beds here included in the Ogallala Formation (upper part of the Pawnee Creek Formation of Galbreath, 1953). The Arikaree contains the oreodont *Merycochoerus proprius magnus* (Galbreath, 1953, p. 33) at Martin Canyon. This oreodont characteristically occurs in the upper part of the Marsland Formation in Nebraska (Schultz and Stout, 1961, p. 13).

Ogallala Formation

The Ogallala Formation (upper Miocene of this report) in northeast Colorado consists of poorly sorted medium- to coarse-grained sandstone and minor beds of conglomerate, volcanic ash, and impure limestone that may be as much as 165 m thick (Denson and Bergendahl, 1961, p. C168). The amount of conglomerate and coarse-grained sandstone in the Ogallala in northeast Colorado increases from east to west. The Ogallala lithology contrasts markedly with that of the underlying Arikaree Formation in its coarser grain size, poorer sorting, paucity of magnetite grains, and a heavy mineral assemblage mainly composed of minerals derived from plutonic granitic and metamorphic source rocks of Precambrian age (Denson, 1969). On the basis of the presence of fossil mammals and very minor lithologic differences, the Ogallala has been divided into several subordinate rock units in the Great Plains, but these subordinate units are nearly impossible to map regionally and can only be recognized by their different faunal contents (Denson, 1969, p. 28).

The Ogallala, which overlies all the older rocks of northeast Colorado and adjoining areas, is the caprock formation that can be traced from east to west along the northern boundary of Colorado. From east to west along this line, the Ogallala successively truncates the older Tertiary sedimentary rocks. Near the Front Range it crosses the upturned edges of the Paleozoic and Mesozoic sedimentary rocks and laps onto the Precambrian rocks.

The age of the Ogallala Formation in northeast Colorado is here interpreted to be late Miocene on the basis of its content of fossil mammal remains. Fossil mammals have been collected from the formation since the turn of the century (Matthew, 1901), and more recently, Galbreath (1953) compiled a list of fossil mammals taken from Ogallala-age rocks (upper part of Pawnee Creek Formation of Galbreath, 1953).

No radiometric age determinations are available on volcanic ashes in the Ogallala in northeast Colorado as yet, but a K-Ar age determination of 14.7 m.y. (Evernden and others, 1964, p. 184) was made on hydrated glass shards from a volcanic ash bed about 50 m above the base of the Ogallala Formation at the head of Merychippus Draw in southwest Nebraska (Sheep Creek Formation of Lugn, 1939). Because the age determination was made on hydrated glass shards, which frequently give spurious ages, an attempt was made to date the ash using the fission-track method. C. W. Naeser (1973, written commun.) determined the age of the ash at 16.1 ± 3.7 m.y. using zircon microphenocrysts. The best age for this ash (zircon) in the lower part of the Ogallala is perhaps about 16 to 17 m.y.

An attempt was made to date several volcanic ash beds in the uppermost part of the Ogallala over a large area of the Great Plains. An ash in the Kimball Member about 5 km north of Dalton, Morrill County, Nebraska (Stout, 1971, p. 73), was dated using the K-Ar method on glass shards and the fission-track method on zircon microphenocrysts. J. D. Obradovich determined the age of the ash (DKA2746) to be 8.2 ± 0.5 m.y., but this age is maximum because it was made on hydrated glass shards that probably contain excess argon trapped in the glass at its time of quenching. A fission-track age determination using zircons gave an age of 4.6 ± 1.0 m.y. for the ash in the Kimball Member, which is thought to be the best age for the uppermost part of the Ogallala. A part of the Ogallala Formation older than the Kimball was radiometrically dated by me, using the fission-track method on zircon and hydrated glass shards from an ash bed directly overlying an important Hemphillian land mammal assemblage on the Coffee Ranch in Hemphill County, Texas (Reed and Longnecker, 1932, loc. 20). Glass-mantled zircon microphenocrysts yielded an age of 6.6 ± 0.8 m.y., and glass shards gave an age of 4.7 ± 0.8 m.y. Yet an older part of the Ogallala was dated using the fission-track method on a volcanic ash bed that lies about 15 m above the Burge fossil mammal quarry in the NE$\frac{1}{4}$SE$\frac{1}{4}$ sec. 15, T. 32 N., R. 70 W., Cherry County, Nebraska. Zircon from the ash yielded an age of 9.7 ± 1.2 m.y., and glass shards an age of 7.5 ± 2.2 m.y.

In summary, the age of the Ogallala in the Great Plains region is late Miocene and ranges from about 16 to 17 m.y. to about 5 m.y.

The upper Cenozoic sedimentary rocks of northeast Colorado are seemingly undeformed, and they are nearly on their original depositional slopes. Though the rocks are not deformed, there are indications from the increases in grain size and amount of conglomerate, in passing from the Arikaree to the Ogallala, that the Front Range was uplifted several hundred meters at the beginning of Ogallala time. Accumulation of the Ogallala Group ended with deposition of the Kimball Formation, about 5 m.y. ago. Although the Miocene rocks are not deformed in northeast Colorado, to the north in southeast Wyoming several large faults cut the Miocene sequence—the Wheatland fault system (L. W. McGrew, 1963), Rawhide fault (Denson and Botinelly, 1949), and the Whalen fault system (Schlaikjer, 1935, p. 121; L. W. McGrew, 1963). These post-Miocene faults have a minimum throw of about 200 m.

ANCIENT RIVER COURSES

This paper cannot relate in any detail the antiquity of the larger rivers of the Southern Rocky Mountains and adjoining areas, but it is pertinent to note that scraps of evidence gained from recent studies of Cenozoic rocks, particularly the Miocene ones, bear on this problem. It seems fairly well established that the major rivers of northern Colorado, including the Colorado, Laramie, North and South Platte, and Yampa, have been in about their present courses for the past 10 to 12 m.y. (Hunt, 1969, p. 127; Larson and others, 1975). Well-rounded stream gravel in the upper part of the Troublesome Formation in the area north of Kremmling (Fig. 11) appears to be one important clue indicating such an age for the Colorado River in the region discussed here.

Evidence that the present rivers existed prior to 10 to 12 m.y. ago is scant but nevertheless present in some places (compare Denson and Chisholm, 1971, p. C125). In early Miocene time, the ancestral Colorado River possibly flowed west from Middle Park, Colorado, across the Park Range to near Yarmony Mountain, following a course much the same as today's. Rounded fragments of intermediate to silicic volcanic rocks found by E. E. Larson (1973, written commun.) occur between lower Miocene basaltic lava flows at Yarmony Mountain north of State Bridge, Colorado. The source for the rounded volcanic rocks may have been certain rocks of the Rabbit Ears Volcanics (Oligocene and Miocene?) of Middle Park, fragments of which could have been carried downstream by the ancestral Colorado River. Other rounded cobbles in the conglomerates at Yarmony Mountain are identical to intrusive masses of quartz porphyry mapped east of Monarch Lake by Robert Pearson (unpub. map) in tributaries to the Colorado River.

Newly found evidence suggests that the ancestral Yampa River flowed west from the Park Range east of Steamboat Springs to at least as far west as Juniper Mountain (Fig. 1) in early Miocene time. The evidence consists of rounded fragments, as large as 10 cm in diameter, of granitic and metamorphic rocks of Precambrian age that occur at the base of the Browns Park Formation just north of Juniper Hot Springs, Colorado, near Juniper Mountain (Fig. 13). The pebbles and cobbles of Precambrian rocks are very similar to Precambrian rocks of the Park Range and also to Precambrian pebbles and cobbles in Pleistocene terrace deposits along the Yampa River near Juniper Mountain.

The Gering Sandstone (lower Miocene) near Scotts Bluff, Nebraska, contains a pumice lapilli bed K-Ar dated by Obradovich and others (1973) at 28 m.y. The character of the pumice bed suggests that the pumice was washed down an early Miocene river (ancestral Laramie River) as a slurry flood. A possible source of the pumice that has compatible age, composition, and availability lies in the headwaters of the North Platte and Laramie Rivers in the Never Summer Mountains near Specimen Mountain.

The Castle Rock Conglomerate (lower Oligocene) near Castle Rock, Colorado, contains rounded fragments of Tertiary volcanic rocks of the Thirtynine Mile volcanic field of South Park, indicating that in early Oligocene time the South Platte River probably flowed from South Park to the east into the Denver basin.

Figure 12. Cow Gulch south of Wolford Mountain in Kremmling quadrangle, Grand County, Colorado (Fig. 9), showing outcrops of faulted and folded Troublesome Formation (Miocene) separated from overlying Pierre Shale (Cretaceous) by nearly flat-lying fault. Rocks were probably deformed in Miocene time in area just east of active Antelope Pass fault.

Figure 13. Conglomeratic sandstone outcrop in lower part of Browns Park Formation. Well-rounded pebbles and cobbles of Precambrian granitic and metamorphic rocks probably were derived from Park Range about 96 km east of Juniper Hot Springs.

GENERAL SUMMARY

In northern Colorado and adjoining areas, deposition of Miocene sedimentary rocks began about 26 m.y. ago and locally continued until about 5 m.y. ago. Deposition occurred in subsiding basins and along Oligocene paleovalleys and extended high onto mountain flanks. For the most part, the Miocene sedimentary rocks are fine-grained, nonorogenic sandstone and siltstone containing large amounts of volcanic minerals and glass shards as well as many rhyolitic volcanic ash beds. Conglomerate and coarse-grained sandstone are locally present in upper Miocene rocks, especially near mountain fronts. Miocene volcanic rocks, which consist chiefly of basaltic lava flows and intermediate to silicic intrusive rocks, are remnants of large volcanic fields in the White River Plateau, Elkhead Mountains, State Bridge area, Rabbit Ears Range, and Never Summer Mountains. These volcanic rocks, interbedded with or underlying upper Miocene sedimentary rocks, are as old as 26 m.y. and perhaps as young as 8 m.y. (Larson and others, 1975).

Late Cenozoic normal faulting, folding, collapse, and uplift in northern Colorado are part of widespread regional tectonism and volcanism in the Western Cordillera. Evidence of this widespread tectonism is especially well shown by Miocene rocks of many areas in northern Colorado, including the eastern Uinta Mountains, the west flank of the Park Range, the State Bridge area, and the North and Middle Park areas. Other parts of northern Colorado probably were also affected by late Cenozoic tectonism, but because large areas are devoid of upper Cenozoic rocks, direct evidence of such late deformation is generally lacking.

Folding of Miocene rocks is apparent chiefly adjacent to large late Cenozoic faults such as the faults that bound the north margin of Browns Park in the eastern Uinta Mountains, the west margin of the Troublesome basin in Middle Park, and the north margin of the North Park syncline in North Park.

Miocene rocks are faulted along zones of older thrust faults, including the Sparks Ranch fault on the north flank of the Uinta Mountains, thrust faults on the west flank of the Park Range, the Williams Range fault in Middle Park, the Stillwater fault in North and Middle Parks, and the Independence Mountain fault in North Park. These late Cenozoic faults probably have a middle Tertiary ancestry and reflect collapse of blocks previously upthrown on Laramide faults.

Tilting or rotation of blocks of Miocene rocks in late Cenozoic time is documented in the eastern Uinta Mountains and in Middle Park. Uplifted mountain blocks include the Uncompahgre and White River Plateaus and the Gore and Park Ranges.

The youngest rocks deformed by the tectonism are about 9 m.y. old, as inferred from fossil mammal remains and radiometric dating of volcanic rocks that are interlayered with or cut Miocene rocks. Although the time of last movement on many of the normal faults is younger than 9 m.y., many of the faults were probably active throughout Cenozoic time, and some seemingly have a very long history of movement beginning in Precambrian time. The magnitude of late Cenozoic tectonism in northern Colorado is not nearly as large as the magnitude in southern Colorado. In northern Colorado, displacements by late Cenozoic faulting and uplift measure generally less than 1,000 m, whereas in the San Luis Valley, Wet Mountains, and Arkansas Valley (Taylor, 1975a), they locally measure thousands of meters.

ACKNOWLEDGMENTS

I am indebted to several colleagues, including N. M. Denson, W. R. Hansen, H. W. Roehler, G. L. Snyder, and Ogden Tweto for discussion of the evolution and the late Cenozoic geologic history of the Southern Rocky Mountains and for making available unpublished maps and drawings. I thank especially C. W. Naeser and J. D. Obradovich for radiometric age determinations of volcanic ash beds and rocks. I also thank M. E. McCallum for information on the fault relations in the Medicine Bow Mountains, shown in Figures 1 and 8, and Gerald T. Cebula and Jack W. Groen, who made mineral separations used for some of the radiometric age determinations.

REFERENCES CITED

Bass, N. W., and Northrop, S. A., 1963, Geology of Glenwood Springs quadrangle and vicinity, northwestern Colorado: U.S. Geol. Survey Bull. 1142-J, p. J1-J74.

Beekly, A. L., 1915, Geology and coal resources of North Park, Colorado: U.S. Geol. Survey Bull. 596, 122 p.

Behrendt, J. C., Popenoe, Peter, and Mattick, R. E., 1969, A geophysical study of North Park and the surrounding ranges, Colorado: Geol. Soc. America Bull., v. 80, no. 10, p. 1523-1538.

Berggren, W. A., 1969, Cenozoic chronostratigraphy, planktonic foraminiferal zonation and the radiometric time scale: Nature, v. 224, p. 1072-1075.

——1972, A Cenozoic time-scale—some implications for regional geology and paleobiogeography: Lethaia, v. 5, p. 195-215.

Bergin, M. J., 1959, Preliminary geologic map of the Maybell-Lay area, Moffat County, Colorado: U.S. Geol. Survey Open-File Map.

Blackstone, D. L., Jr., 1953, Notes on the tectonic map of a portion of southern Wyoming and northern Colorado, in Wyoming Geol. Assoc. Guidebook 8th Ann. Field Conf., Laramie Basin, Wyoming, and North Park, Colorado, 1953: p. 85-86.

Bradley, W. H., 1936, Geomorphology of the north flank of the Uinta Mountains: U.S. Geol. Survey Prof. Paper 185-I, p. 163-199.

——1964, Geology of the Green River Formation and associated Eocene rocks in southwestern Wyoming and adjacent parts of Colorado and Utah: U.S. Geol. Survey Prof. Paper 496-A, 86 p.

Brennan, W. J., 1969, Structural and surficial geology of the west flank of the Gore Range, Colorado [Ph.D. thesis]: Boulder, Univ. Colorado.

Bryan, Kirk, 1938, Geology and ground-water conditions of the Rio Grande depression in Colorado and New Mexico, in Regional planning, Pt. 6, Upper Rio Grande: Washington, D.C., Nat. Resources Comm., p. 197-225.

Buffler, R. T., 1967, The Browns Park Formation and its relationship to the late Tertiary geologic history of the Elkhead region, northwestern Colorado-south-central Wyoming [Ph.D. thesis]: Berkeley, Univ. California, Berkeley.

Burbank, W. S., Lovering, T. S., Goddard, E. M., and Eckel, E. B., 1935, Geologic map of Colorado: U.S. Geol. Survey, scale 1:500,000.

Cater, F. W., 1966, Age of the Uncompahgre uplift and Unaweep Canyon, west-central Colorado, in Geological Survey research 1966: U.S. Geol. Survey Prof. Paper 550-C, p. 86-92.

Corbett, M. K., 1965, Tertiary igneous petrology of the Mount Richthofen-Iron Mountain area, north-central Colorado, in Abstracts for 1964: Geol. Soc. America Spec. Paper 82, p. 324-325.

——1968, Tertiary volcanism of the Specimen-Lulu-Iron Mountain area, north-central Colorado: Colorado School Mines Quart., v. 63, no. 3, p. 1-37.

Damon, P. E., 1970, Correlation and chronology of ore deposits and volcanic rocks: Arizona Univ. Geochronology Dept. Ann. Prog. Rept. C00-689-130 to U.S. Atomic Energy Comm.

Denson, N. M., 1969, Distribution of nonopaque heavy minerals in Miocene and Pliocene rocks of central Wyoming and parts of adjacent states, in Geological Survey research 1969: U.S. Geol. Survey Prof. Paper 650-C, p. C25-C32.

Denson, N. M., and Bergendahl, M. H., 1961, Middle and upper Tertiary rocks of southeastern Wyoming and adjoining areas, in Geological Survey research 1961: U.S. Geol. Survey Prof. Paper 424-C, p. C168-C172.

Denson, N. M., and Botinelly, Theodore, 1949, Geology of the Hartville uplift, eastern Wyoming: U.S. Geol. Survey Oil and Gas Inv. Prelim. Map 102, 2 sheets.

Denson, N. M., and Chisholm, W. A., 1971, Summary of mineralogic and lithologic characteristics of Tertiary sedimentary rocks in the middle Rocky Mountains and the northern Great Plains, in Geological Survey research 1971: U.S. Geol. Survey Prof. Paper 750-C, p. C117-C126.

Donner, H. F., 1949, Geology of the McCoy area, Eagle and Routt Counties, Colorado: Geol. Soc. America Bull., v. 60, p. 1215-1248.

Dyni, J. R., 1968, Geologic map of the Elk Springs quadrangle, Moffat County, Colorado: U.S. Geol. Survey Geol. Quad. Map GQ-702.

Evernden, J. F., Savage, D. E., Curtis, G. H., and James, G. T., 1964, Potassium-argon dates and the Cenozoic mammalian chronology of North America: Am. Jour. Sci., v. 262, no. 2, p. 148-198.

Funnell, B. M., 1964, The Tertiary period, in Harland, W. B., Smith, A. G., and Wilcock, Bruce, eds., The Phanerozoic time-scale—A symposium: Geol. Soc. London Quart. Jour., v. 120 S, p. 179-191.

Galbreath, E. C., 1953, A contribution to the Tertiary geology and paleontology of northeastern Colorado: Kansas Univ. Paleont. Contr. 13, Vertebrata, art. 4, 119 p.

Gill, J. B., and McDougall, Ian, 1973, Biostratigraphic and geological significance of Miocene-Pliocene volcanism in Fiji: Nature, v. 241, p. 176-180.

Gorton, K. A., 1953, Geology of the Cameron Pass area, Grand, Jackson, and Larimer Counties, Colorado, in Wyoming Geol. Assoc. Guidebook 8th Ann. Field Conf., Laramie Basin, Wyoming, and North Park, Colorado, 1953: p. 87-99.

Hail, W. J., Jr., 1965, Geology of northwestern North Park, Colorado: U.S. Geol. Survey Bull. 1188, 133 p.

——1968, Geology of southwestern North Park and vicinity, Colorado: U.S. Geol. Survey Bull. 1257, 119 p.

Hail, W. J., Jr., and Lewis, G. E., 1960, Probable late Miocene age of the North Park Formation in the North Park area, Colorado, in Geological Survey research 1960: U.S. Geol. Survey Prof. Paper 400-B, p. B259-B260.

Hancock, E. T., 1925, Geology and coal resources of the Axial and Monument Butte quadrangles, Moffat County, Colorado: U.S. Geol. Survey Bull. 757, 134 p.

Hansen, W. R., 1965, Geology of the Flaming Gorge area, Utah-Colorado-Wyoming: U.S. Geol. Survey Prof. Paper 490, 196 p.

——1966, Dinosaur National Monument: U.S. Geol. Survey Spec. Map.

——1969, The geologic story of the Uinta Mountains: U.S. Geol. Survey Bull. 1291, 144 p.

Hansen, W. R., Kinney, D. M., and Good, J. M., 1960, Distribution and physiographic significance of the Browns Park Formation, Flaming Gorge and Red Canyon areas, Utah-Colorado: U.S. Geol. Survey Prof. Paper 400-B, p. B257-B259.

Houston, R. S., 1968, Geologic map of the Medicine Bow Mountains, Albany and Carbon Counties, Wyoming: Wyoming Geol. Survey Mem. 1, Pl. 1.

Hunt, C. B., 1956, Geology of the Taylor site, Unaweep Canyon, Colorado, in Archaeological Investigations of Uncompahgre Plateau in west-central Colorado: Denver Mus. Nat. History Proc., no. 2, p. 64-69.

——1969, Geologic history of the Colorado River: U.S. Geol. Survey Prof. Paper 669-C, p. C59-C130.

Izett, G. A., 1968, Geology of the Hot Sulphur Springs quadrangle, Grand County, Colorado: U.S. Geol. Survey Prof. Paper 586, 79 p.

——1973, Geology of the Trail Mountain quadrangle, Grand County, Colorado: U.S. Geol. Survey Geol. Quad. Map GQ-1156.

Izett, G. A., and Barclay, C.S.V., 1972, Preliminary geologic map of the Kremmling quadrangle, Grand County, Colorado: U.S. Geol. Survey Open-File Map.

Izett, G. A., and Lewis, G. E., 1963, Miocene vertebrates from Middle Park, Colorado, in Short papers in geology and hydrology: U.S. Geol. Survey Prof. Paper 475-B, p. B120-B122.

Izett, G. A., Denson, N. M., and Obradovich, J. D., 1970, K-Ar age of the lower part of the Browns Park Formation, northwestern Colorado: U.S. Geol. Survey Prof. Paper 700-C, p. C150-C152.

King, Clarence, 1876, Geological and topographical atlas: U.S. Geol. Expl. 40th Parallel.

Kinney, D. M., 1970a, Preliminary geologic map of the Gould quadrangle, North Park, Jackson County, Colorado: U.S. Geol. Survey Open-File Map.

——1970b, Preliminary geologic map of the Rand quadrangle, North and Middle Parks, Jackson and Grand Counties, Colorado: U.S. Geol. Survey Open-File Map.

——1970c, Preliminary geologic map of southwest third of Kings Canyon quadrangle, North Park, Jackson County, Colorado: U.S. Geol. Survey Open-File Map.

Kinney, D. M., and Hail, W. J., Jr., 1970a, Preliminary geologic map of the Hyannis Peak quadrangle, North and Middle Parks, Jackson and Grand Counties, Colorado: U.S. Geol. Survey Open-File Map.

——1970b, Preliminary geologic map of the Walden quadrangle, North Park, Jackson County, Colorado: U.S. Geol. Survey Open-File Map.

Kinney, D. M., Hail, W. J., Steven, T. A., and others, 1970, Preliminary geologic map of the Cowdrey quadrangle, North Park, Jackson County, Colorado: U.S. Geol. Survey Open-File Map.

Kucera, R. E., 1962, Geology of the Yampa district, northwest Colorado [Ph.D. thesis]: Boulder, Univ. Colorado.

——1968, Geomorphic relationship of Miocene deposits in the Yampa district, northwest Colorado, in Field Conf. Guidebook for high altitude and mountain basin deposits of Miocene age in Wyoming and Colorado, 1968: Boulder, Univ. Colorado Museum.

Lakatos, Stephen, and Miller, D. S., 1972, Fission-track stability in volcanic glass of different water contents: Jour. Geophys. Research, v. 77, no. 35, p. 6990-6993.

Larson, E. E., 1968, Miocene and Pliocene rocks in the Flat Tops primitive area, in Field Conf. Guidebook for high altitude and mountain basin deposits of Miocene age in Wyoming and Colorado, 1968: Boulder, Univ. Colorado Museum.

Larson, E. E., Ozima, Minoru, and Bradley, W. C., 1975, Late Cenozoic basic volcanism in northwestern Colorado and its implications concerning tectonism and the origin of the Colorado River system, in Curtis, Bruce, ed., Cenozoic history of the Southern Rocky Mountains: Geol. Soc. America Mem. 144, p. 155-178.

Lewis, G. E., 1968, Stratigraphic paleontology of the Barstow Formation in the Alvord Mountain area, San Bernardino County, California, in Geological Survey research 1968: U.S. Geol. Survey Prof. Paper 600-C, p. C75-C79.

——1969, Larger fossil mammals and mylagaulid rodents from the Troublesome Formation (Miocene) of Colorado, in Geological Survey research 1969: U.S. Geol. Survey Prof. Paper 650-B, p. 53-56.

Lipman, P. W., and Mehnert, H. W., 1975, Late Cenozoic basaltic volcanism and development of the Rio Grande depression in the Southern Rocky Mountains, in Curtis, Bruce, ed., Cenozoic history of the Southern Rocky Mountains: Geol. Soc. America Mem. 144, p. 119-154.

Lohman, S. W., 1965, Geology and artesian water supply of the Grand Junction area, Colorado: U.S. Geol. Survey Prof. Paper 451, 149 p.

Love, J. D., Weitz, J. L., and Hose, R. K., 1955, Geologic map of Wyoming: U.S. Geol. Survey, scale 1:500,000.

Lovering, T. S., 1930, The Granby anticline, Grand County, Colorado, in Contributions to economic geology, 1930: U.S. Geol. Survey Bull. 822, p. 71-76.

Lugn, A. L., 1939, Classification of the Tertiary system in Nebraska: Geol. Soc. America Bull., v. 50, p. 1245-1276.

Mallory, W. W., Post, E. V., Ruane, P. J., and Lehmbeck, W. L., 1966, Mineral resources of the Flat Tops primitive area, Colorado: U.S. Geol. Survey Bull. 1230-C, 30 p.

Marvin, R. F., Mehnert, H. H., and Mountjoy, W. M., 1966, Age of basalt cap on Grand Mesa, in Geological Survey research 1966: U.S. Geol. Survey Prof. Paper 550-A, p. A81.

Matthew, W. D., 1901, Fossil mammals of the Tertiary of northeastern Colorado: Am. Mus. Nat. History Mem., v. 1, pt. 7, p. 355-447.

McGrew, L. W., 1963, Geology of the Fort Laramie area, Platte and Goshen Counties, Wyoming: U.S. Geol. Survey Bull. 1141-F, 39 p.

McGrew, P. O., 1951, Tertiary stratigraphy and paleontology of south-central Wyoming, *in* Wyoming Geol. Assoc. Guidebook 6th Ann. Field Conf., south-central Wyoming, 1951: p. 54-57.

McKay, E. J., 1974, Geologic map of the Lone Mountain quadrangle, Moffat County, Colorado: U.S. Geol. Survey Geol. Quad. Map GQ-1144 (in press).

McKay, E. J., and Bergin, M. J., 1974, Geologic map of the Maybell quadrangle, Moffat County, Colorado: U.S. Geol. Survey Geol. Quad. Map GQ-1145 (in press).

Montagne, John de la, 1955, Cenozoic history of the Saratoga Valley area, Wyoming and Colorado [Ph.D. thesis]: Laramie, Univ. Wyoming (Ann Arbor, Mich., Univ. Microfilms, Inc.).

——1957, Cenozoic structural and geomorphic history of northern North Park and Saratoga Valley, Colorado and Wyoming, *in* Rocky Mtn. Assoc. Geologists Guidebook 9th Ann. Field Conf., North and Middle Park Basins, Colorado, 1957: p. 36-41.

Montagne, John de la, and Barnes, W. C., 1957, Stratigraphy of the North Park Formation in the North Park area, Colorado, *in* Rocky Mtn. Assoc. Geologists Guidebook 9th Ann. Field Conf., North Park and Middle Park Basins, Colorado, 1957: p. 55-60.

Obradovich, J. D., and Cobban, W. A., 1974, A time scale for the late Cretaceous of the Western Interior of North America: Canada Geol. Soc. Spec. Paper (in press).

Obradovich, J. D., Izett, G. A., and Naeser, C. W., 1973, Radiometric ages of volcanic ash and pumice beds in the Gering Sandstone (earliest Miocene) of the Arikaree Group, southwestern Nebraska [abs.]: Geol. Soc. America, Abs. with Programs (Rocky Mountain Sec.), v. 5, no. 6, p. 499-500.

Peterson, O. A., 1928, The Browns Park Formation: Carnegie Mus. Mem., v. 11, no. 2, 121 p.

Powell, J. W., 1876, Report on the geology of the eastern portion of the Uinta Mountains and a region of country adjacent thereto: U.S. Geol. and Geog. Survey Terr., v. 7, 218 p.

Reed, L. C., and Longnecker, O. M., Jr., 1932, The geology of Hemphill County, Texas: Texas Univ. Bull. 3231, 98 p.

Richards, Arthur, 1941, Geology of the Kremmling area, Grand County, Colorado [Ph.D. thesis]: Ann Arbor, Univ. Michigan (Ann Arbor, Mich., Univ. Microfilms, Inc.).

Sato, Yoshiaki, and Denson, N. M., 1967, Volcanism and tectonism as reflected by the distribution of nonopaque heavy minerals in some Tertiary rocks of Wyoming and adjacent states, *in* Geological Survey research 1967: U.S. Geol. Survey Prof. Paper 575-C, p. C42-C54.

Schlaikjer, E. M., 1935, Contributions to the stratigraphy and paleontology of the Goshen Hole area, Wyoming: Harvard College Mus. Comp. Zoology Bull., v. 76, no. 2-4, p. 33-189.

Schultz, C. B., and Stout, T. M., 1961, Field conference on the Tertiary and Pleistocene of western Nebraska: Nebraska Univ. State Mus. Spec. Pub. 2, 55 p.

Sears, J. D., 1924a, Relations of the Browns Park Formation and the Bishop Conglomerate, and their role in the origin of Green and Yampa Rivers: Geol. Soc. America Bull., v. 35, p. 279-304.

——1924b, Geology and oil and gas prospects of part of Moffat County, Colorado, and southern Sweetwater County, Wyoming: U.S. Geol. Survey Bull. 751-G, 319 p.

——1962, Yampa Canyon in the Uinta Mountains, Colorado: U.S. Geol. Survey Prof. Paper 374-I, 31 p.

Segerstrom, K. G., and Young, E. J., 1973, General geology of the Hahns Peak and Farwell Mountain quadrangles, Routt County, Colorado: U.S. Geol. Survey Bull. 1349, 60 p.

Stark, J. T., Johnson, J. H., Behre, C. H., Jr., Powers, W. E., Hewland, A. L., Gould, D. B., and others, 1949, Geology and origin of South Park, Colorado: Geol. Soc. America Mem. 33, p. 68-69.

Steven, T. A., 1975, Middle Tertiary volcanic field in the Southern Rocky Mountains, *in* Curtis, Bruce, ed., Cenozoic history of the Southern Rocky Mountains: Geol. Soc. America Mem. 144, p. 75-94.

Stokes, W. L., and Madsen, J. H., Jr., compilers, 1961, Geologic map of Utah—Northeast quarter: Utah Geol. and Mineralog. Survey, scale 1:250,000.

Stout, T. M., 1971, Geologic sections, in Rocks, real estate, and resources: Geol. Soc. America North Central Sec. Guidebook 5th Ann. Mtg., Lincoln, Nebraska: p. 29-80.

Taylor, R. B., 1975a, Geologic map of the Bottle Pass quadrangle, Grand County, Colorado: U.S. Geol. Survey Geol. Quad. Map GQ-1224 (in press).

——1975b, Neogene tectonism in south-central Colorado, in Curtis, Bruce, ed., Cenozoic history of the Southern Rocky Mountains: Geol. Soc. America Mem. 144, p. 211-226.

Theobald, P. K., 1965, Preliminary geologic map of the Berthoud Pass quadrangle, Clear Creek and Grand Counties, Colorado: U.S. Geol. Survey Misc. Geol. Inv. Map I-443.

Tongiorgi, E., and Tongiorgi, M., 1964, Age of the Miocene-Pliocene limit in Italy: Nature, v. 201, p. 365-367.

Tweto, Ogden, 1975, Laramide (Late Cretaceous-early Tertiary) orogeny in the Southern Rocky Mountains, in Curtis, Bruce, ed., Cenozoic history of the Southern Rocky Mountains: Geol. Soc. America Mem. 144, p. 1-44.

Tweto, Ogden, Bryant, Bruce, and Williams, F. E., 1970, Mineral resources of the Gore Range-Eagles Nest primitive area and vicinity, Summit and Eagle Counties, Colorado: U.S. Geol. Survey Bull. 1319-C, 127 p.

Van Couvering, J. A., 1972, Radiometric calibration of the European Neogene, in Bishop, W. W., and Miller, J. A., eds., Calibration of hominoid evolution: Edinburgh, Scottish Academic Press, p. 247-271.

Wilson, J. A., 1939, A new species of dog from the Miocene of Colorado: Michigan Univ. Mus. Paleont. Contr., v. 5, no. 12, p. 315-318.

Wood, H. E., and others, 1941, Nomenclature and correlation of the North American continental Tertiary: Geol. Soc. America Bull., v. 52, no. 1, p. 1-48.

York, D., Strangway, D. W., and Larson, E. E., 1971, Preliminary study of a Tertiary magnetic transition in Colorado: Earth and Planetary Sci. Letters, no. 11, p. 333-338.

MANUSCRIPT RECEIVED BY THE SOCIETY MAY 8, 1974

Printed in U.S.A.

The Geological Society of America
Memoir 144
© 1975

Neogene Tectonism in South-Central Colorado

RICHARD B. TAYLOR

U.S. Geological Survey
Denver Federal Center
Denver, Colorado 80225

ABSTRACT

Miocene-Pliocene history is recorded in south-central Colorado by sediments deposited in subsiding basins bounded by fault-block mountains and by faulted sedimentary and volcanic deposits lying on a channeled late Eocene erosion surface of regional extent. The San Luis Valley and upper Arkansas Valley are en echelon segments of the Rio Grande trough that are constricted and faulted south of Salida at the northern end of the Sangre de Cristo Range. Great movements on the bounding faults during Neogene time are indicated by clastic and volcanic trough fill, which may be 10,000 m thick near Alamosa and 1,500 m thick near Salida, and by adjoining mountains which stand as much as 1,500 m above the valley floors. The Sangre de Cristo Range owes most of its present elevation to Neogene faulting that cut the Miocene-Pliocene Dry Union Formation and some volcanic deposits in Oligocene-Miocene paleovalleys. Valley fill in the Wet Mountain Valley graben is as much as 1,500 m thick, and at least the upper 300 m of it consists of pinkish beds correlated with the Santa Fe(?) Formation. The crest of the Wet Mountains was uplifted about 400 m above the valley to the west of the mountains and nearly 1,200 m above the high plains which lie to the east. To the north, Neogene faulting elevated the Rampart Range at the front of the Rocky Mountains, dropped the Fourmile Creek graben nearly 400 m, formed complex fault-bounded basins at the southern margin of South Park, and segmented volcanic deposits in paleovalleys that once crossed the upper Arkansas Valley.

INTRODUCTION

Neogene block faulting was recognized as a locally important geologic feature in south-central Colorado early in the geologic study of the state (Hills, 1888, 1900), but its regional significance has generally been neglected. This paper proposes that Miocene-Pliocene block faulting disrupted an earlier landscape of relatively gentle relief and formed the major modern mountains and basins. Thus, the present configuration of the Southern Rocky Mountains is largely of Neogene age.

The specific effects of late deformation in south-central Colorado have been difficult to identify, because the faulting and folding in the area are the cumulative results of many episodes of deformation. Most major faults in the area originated in Precambrian time and have been reactivated periodically. Much of the folding of the sedimentary rocks above the crystalline basement has also been incremental. Historically, geologists have commonly assumed Laramide (Late Cretaceous-early Tertiary) ages for most of the structural features, hypabyssal igneous rocks, and associated ore deposits in Colorado; these poorly founded assumptions have largely ignored younger Tertiary events.

Isotopic dating techniques now indicate that the igneous sequences and related mineralizations in south-central Colorado represent a span from Late Cretaceous to late Miocene. Times of tectonism can be established where isotopically dated igneous rocks—particularly ash-flow tuffs of great regional extent—and paleontologically dated sedimentary rocks are faulted and folded. Further evidence of Neogene block faulting comes from reconstruction of paleotopography, including the evolution of mountain ranges, basins, and drainage systems.

REGIONAL DISCUSSION

In south-central Colorado (Fig. 1) a regional surface of low relief was cut across most of the Southern Rocky Mountains area by the end of the Eocene (Epis and Chapin, 1975; Scott, 1975). The relative tectonic stability that permitted this surface to be cut persisted through the Oligocene, most importantly through the period of ash-flow eruptions from sources in the central San Juan Mountains and the southern Sawatch Range (Steven, 1975). These eruptions ended about 26 m.y. ago. Ash-flow units spread across hundreds of square kilometers, with the deepest accumulations localized by broad, shallow paleochannels. Block faulting followed the major ash-flow eruptions and was broadly concurrent with deposition of sedimentary rocks in local basins during the Miocene and Pliocene. This paper focuses on a number of localities where the nature, extent, and time of Neogene deformation can be documented.

Wet Mountains

The front of the Southern Rocky Mountains between the Arkansas River and Huerfano Park is formed by the Wet Mountains. This flat-crested range is about 75 km long and 8 to 16 km wide, and it trends north-northwest. The summits of the range vary in altitude from a little more than 2,700 m at the north end to more than 3,600 m at Greenhorn Mountain at the extreme southern end.

The eastern segment of the range is a narrow, flat-topped block of Precambrian rock elevated between the Ilse fault on the west and the Wet Mountains fault on the east (see Figs. 1, 5). The northern border of this uplifted segment is a topographic escarpment that follows a fault along the south side of a graben whose northern part contains Mesozoic sedimentary rocks. The southern part of the segment is less elevated than the northern, and displacements on the bounding faults die out gradually in a series of step faults.

Toward the south the Wet Mountains extend across the Ilse fault, where their western segment is defined by the Ilse fault to the east and by the mountain-margin Westcliffe fault to the west. This segment consists largely of Precambrian rocks overlain locally by Oligocene volcanic rocks. The flat crest of this segment is probably an exhumed provolcanic surface that was cut in late Eocene time and is believed to correlate with the flat top that caps the narrow mountain block east of the Ilse fault. To the north, the western segment of the Wet Mountains

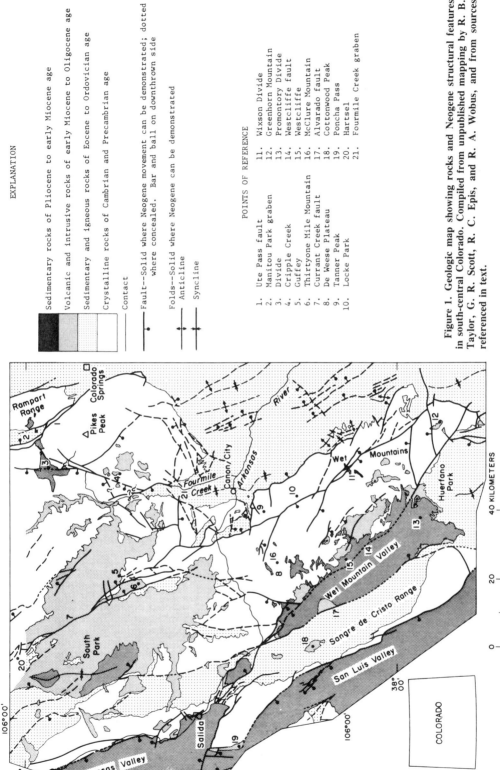

Sedimentary rocks of Pliocene to early Miocene age

Volcanic and intrusive rocks of early Miocene to Oligocene age

Sedimentary and igneous rocks of Eocene to Ordovician age

Crystalline rocks of Cambrian and Precambrian age

———— Contact

———— Fault--Solid where Neogene movement can be demonstrated; dotted
where concealed. Bar and ball on downthrown side

Folds--Solid where Neogene can be demonstrated
———— Anticline
———— Syncline

POINTS OF REFERENCE

1. Ute Pass fault
2. Manitou Park graben
3. Divide
4. Cripple Creek
5. Guffey
6. Thirtyone Mile Mountain
7. Currant Creek fault
8. De Weese Plateau
9. Tanner Peak
10. Locke Park
11. Wixson Divide
12. Greenhorn Mountain
13. Promontory Divide
14. Westcliffe fault
15. Westcliffe
16. McClure Mountain
17. Alvarado fault
18. Cottonwood Peak
19. Poncha Pass
20. Hartsel
21. Fourmile Creek graben

Figure 1. Geologic map showing rocks and Neogene structural features
in south-central Colorado. Compiled from unpublished mapping by R. B.
Taylor, G. R. Scott, R. C. Epis, and R. A. Wobus, and from sources
referenced in text.

diminishes in altitude and passes gradually into the De Weese Plateau just north of the latitude of Westcliffe.

The Ilse fault, where it divides the southern part of the Wet Mountains, is only partly responsible for the uplift there; major uplift also took place on mountain-margin faults to the west, south, and east. To the north, however, the Ilse fault is the dominant structural feature separating the range from the lower terrain to the west. It forms a prominent fault-line scarp more than 20 km long and 300 m high. In describing this fault, Singewald (1966) properly emphasized its Precambrian ancestry, its successive periods of movement, and its regional importance, but he failed to recognize its role in the Neogene uplift of the Wet Mountains.

Recurrent movement along the Ilse fault during Precambrian (and possibly later) time is indicated by the following relations: (1) no rock types or structures in the Precambrian gneiss can be correlated across the Ilse fault in the nearly 20 km where this rock forms its walls; (2) steep contacts of a granitic pluton of Boulder Creek age (1.72 b.y.) with older gneiss are apparently displaced about 3 km in a right-lateral sense near the Arkansas River; and (3) steep contacts of the San Isabel granitic pluton (1.4 b.y.; Boyer, 1962) in the southern Wet Mountains are displaced a total of about 610 m in a left-lateral sense on two strands of the fault.

Details of the Paleozoic, Mesozoic, and early Tertiary movements on the Ilse fault are unknown. Rocks were beveled on both sides of the Ilse fault during late Eocene time. An Oligocene-Miocene drainage pattern was established across the fault, and streams of low gradient flowed eastward across a terrain of low relief. The flat summit areas on many parts of the present range are little-modified segments of this early Tertiary erosion surface. At three localities, offset Oligocene-Miocene deposits and disrupted paleodrainages show major late Tertiary displacement along the Ilse fault and indicate that uplift of the modern Wet Mountains took place in Neogene time.

At Tanner Peak (Fig. 2), the northernmost summit of the Wet Mountains, a major Miocene stream crossed the area of the mountain block. This stream drained the area to the west, and remnants of its deposits can be identified easily by clasts of Cambrian syenite (McClure Mountain Complex) from the pluton at McClure Mountain (Parker and Hildebrand, 1963; Shawe and Parker, 1967) west of the Ilse fault. Oligocene ash-flow tuffs once widely covered the late Eocene surface in this area, and a few remnants of the Gribbles Park Tuff (ash flow VII of Epis and Chapin, 1968) can still be identified. This rock is dated isotopically as about 29 m.y. old and is sufficiently distinctive that cobbles can be recognized in Tertiary alluvial deposits. West of Tanner Peak, Tertiary gravel deposits containing fragments of syenite and Gribbles Park Tuff lie in broad channels cut a few meters to a few tens of meters below the main flat surface. These deposits are almost certainly Miocene in age because they contain fragments of the 29-m.y.-old rock, and yet they are dissected by canyons cut during Pliocene and Pleistocene time. A single remnant of boulder alluvium containing fragments of syenite and Gribbles Park Tuff makes up the top of Tanner Peak. It closely resembles the gravels in channels to the west, and its contained materials clearly came from that direction. The alluvium on Tanner Peak is about 500 m higher than the gravels upstream in the paleodrainage; this separation provides a measure of minimum Neogene displacement on the Ilse fault in this vicinity.

Although no Miocene sedimentary deposits remain on the high plains east of the northern Wet Mountains, two remnants of volcanic material in Wolf Park southwest of Canon City (Fig. 2) furnish critical evidence of topographic levels at the end of the Oligocene. In a small butte topped by a large water tank 1,100 m south of the abandoned New Jersey Zinc Company smelter, Pierre Shale is overlain successively by a few meters of volcanic mudflows, by about 10 m of Gribbles Park Tuff, and by additional conglomeratic mudflows. These rocks are intruded by a small plug of resistant olivine basalt which protected the whole assemblage from Pleistocene pediment cutting. The mudflows contain abundant volcanic detritus, fragments of Precambrian gneiss and granodiorite, and conspicuous cobbles of syenite from the McClure Mountain area. The syenite fragments clearly came from the same source area that supplied the Miocene(?) gravels at the top of Tanner Peak. A second exposure of Oligocene mudflow crops out at almost the same altitude on the east side of a small butte about 1.6 km south of the Canon City dump. This mudflow also contains cobbles of syenite and Gribbles Park Tuff, together with more abundant Precambrian and volcanic detritus.

These Oligocene volcanic and alluvial deposits south of Canon City are about 1,130 m below the base of the Miocene(?) gravels capping Tanner Peak. Because latest Oligocene and early Miocene topographic levels elsewhere in these mountains commonly differ by only a few tens of meters, this figure can be taken as a

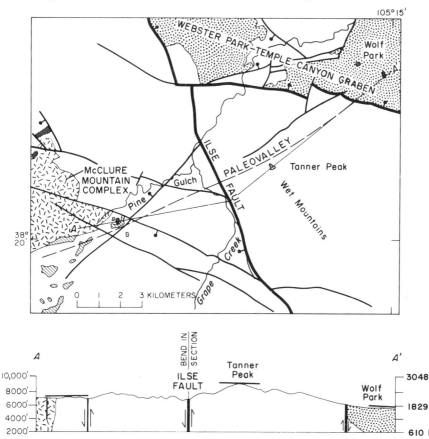

Figure 2. Geologic sketch map of Tanner Peak area. Miocene(?) boulder alluvium shown by diagonal line pattern, Oligocene volcanic rocks and mudflows shown by fine stipple, Cretaceous to Ordovician sedimentary rocks shown by coarse stipple, Cambrian syenite (McClure Mountain Complex) shown by broken line pattern. Heavy lines on cross section above the syenite, on Tanner Peak, and at Wolf Park show Oligocene and Miocene levels on paleodrainage offset by Neogene block faults outlining the Wet Mountains.

general indication of the Neogene offset on the Wet Mountains fault that forms the eastern margin of the Wet Mountains.

At Locke Park, about 12.8 km south of Tanner Peak, a second eastward-flowing paleodrainage crossed the Ilse fault and the Wet Mountains area. Gravel deposits in this drainage can be identified by contained fragments of quartz-bearing syenite derived from a small pluton exposed 8 km west of the fault (Christman and others, 1959). The gravel deposits at the extreme western tip of Locke Park lie 370 m above correlative deposits on the area west of the fault. The fault-line scarp west of Locke Park strikingly displays the Neogene movement on the Ilse fault (Fig. 3). Although no middle and upper Tertiary deposits have been found in the Locke Park paleodrainage east of the Wet Mountains, vertical displacement on the Wet Mountains fault nearly comparable to that at Tanner Peak is inferred from the modern elevation of the flat crest of the Wet Mountains above the adjacent high plains.

At Wixson Divide, 32 km south of Tanner Peak, two east-directed paleochannels are truncated by the Ilse fault (Fig. 4). The fault here is a braided zone more than 1.6 km wide. The southern channel is just south of Wixson Divide; it is about 800 m wide and was cut into Precambrian rocks. Gravels in this channel contain volcanic rock fragments derived from dissection of an Oligocene-Miocene volcano east of Deer Peak, about 9.6 km to the south. The northern channel is about 2 km north of Wixson Divide and lies at about the same level as the southern channel. This channel drained a source area different from that of the southern channel, and the contained gravels have no volcanic debris. Gravels in

Figure 3. Ilse fault escarpment near Locke Park. Fault-line scarp along the Ilse fault divides segments of late Eocene surface on the Wet Mountains (skyline) from equivalent surface on the De Weese Plateau (middle distance). Modern drainage of Grape Creek (foreground) cuts deep, steep-walled canyon below late Eocene surface. Timber-covered ridges on the De Weese Plateau are upheld by resistant layers of metamorphic rock and stand above general surface level.

the northern channel are offset by three different branches of the Ilse fault, with displacement totaling nearly 370 m. More than half this displacement is on a southeast-trending subsidiary fault that leaves the main zone 1.6 km north of the channel. East of Wixson Divide no accurate estimate can be made of Neogene displacement on the Wet Mountains fault to the east. Only the present elevation of the mountain block above the plains indicates that extensive recent uplift occurred here, but such displacement on the fault steadily decreases to the south.

The southern end of the Wet Mountains lies west of the Ilse fault and forms a structural block greatly elevated above the plains to the east and above Huerfano Park to the south. The Precambrian rocks of the mountains are beveled by a flat summit area that reflects a largely exhumed late Eocene (prevolcanic) surface, covered at Greenhorn Mountain (Boyer, 1962) by poorly consolidated water-laid tuffs overlain by andesitic flows. Scarps more than 1.8 km high extend from this peak to the lowlands below. According to Burbank and Goddard (1937, p. 973), Greenhorn Mountain has been elevated between 850 and 1,500 m since early Bridger (late Eocene) time. More recent geologic mapping indicates that the Neogene displacements on the boundary faults to the south and east of the peak are close to the maximum suggested by Burbank and Goddard.

De Weese Plateau

West of the Ilse fault and the northern Wet Mountains, and between Westcliffe and the Arkansas River, well-preserved remnants of the late Eocene (prevolcanic) surface form an area of remarkably gentle relief called the De Weese Plateau. The area east of Westcliffe was covered by volcanic rocks from southern sources in Oligocene time, whereas the northern part of the plateau was partly covered by ash flows from distant western or northwestern sources, including the 29-m.y.-old Gribbles Park Tuff.

The volcanic rocks were widely eroded from the De Weese Plateau, and Miocene(?) alluvial deposits provisionally correlated with the Santa Fe(?) Formation were laid down on the exhumed Precambrian rocks. Miocene deposits in the more rugged southern part of the area are principally boulder alluvium which accumulated in paleovalleys bounded by ridges of resistant Precambrian gneiss. To the north, similar boulder alluvium in shallow paleochannels is widely covered by pinkish to light-brown or gray alluvial deposits composed of crudely stratified and poorly sorted muddy sandstones with lenses of boulders and cobbles. In these deposits, soft beds alternate with layers and lenses well cemented by carbonate. Contained volcanic clasts including fragments of Gribbles Park Tuff show that these alluvial deposits can be no older than latest Ologocene, and dissection by Pliocene and younger canyon cutting sets a minimum age. The deposits thus were probably laid down in Miocene time and are comparable in age and lithology to the Dry Union Formation (correlative of the Santa Fe Group) in the upper Arkansas Valley to the northwest.

Neogene displacement on minor faults in this area can be recognized by offset remnants of Gribbles Park Tuff or other volcanic rocks, by offset paleodrainages on the exhumed provolcanic surface, and by Miocene beds faulted against Precambrian gneiss. Faults cannot be traced through the poorly consolidated Miocene beds where relations are commonly obscured by a thick covering of lag gravels left by sheet-wash erosion of the finer detritus.

Wet Mountain Valley

The Wet Mountain Valley is a low area marking a Neogene graben that lies between the Westcliffe and Alvarado fault zones (Fig. 1). It is bounded by the Sangre de Cristo Range to the west and by the Precambrian rocks of the elevated

De Weese Plateau to the east. This graben is a structural continuation of Huerfano Park to the south, and like this depositional basin, it probably started to form during the Laramide tectonism but continued to subside during later Cenozoic time as indicated by a thick fill of Miocene(?) alluvial deposits. The valley is largely covered by Pleistocene glacial deposits, but significant exposures of Oligocene and Miocene fill crop out at the northern and southern ends.

Near Promontory Divide, at the south end of the Wet Mountain Valley (Fig. 1), Oligocene deposits are composed largely of clastics derived from nearby volcanic centers. Above these are pink, orange-tan, and gray beds correlated with the Santa Fe Group on the basis of lithology and stratigraphic position. The younger beds are poorly sorted water-laid muddy sandstone and mudstone with conglomeratic lenses. Carbonate-cemented lenses alternate with soft beds. These deposits extend continuously up the axis of the valley for a distance of almost 65 km, and toward the north they locally cover Oligocene volcanic rocks, including the 29-m.y.-old Gribbles Park Tuff. The thickness of the Santa Fe(?) beds ranges from a few hundred meters or less outside the main graben to more than 400 m penetrated by water wells in the graben southwest of Westcliffe.

San Luis Valley

The San Luis Valley has long been recognized as a part of the Rio Grande trough, the dominant Neogene structural feature in Colorado. As documented by

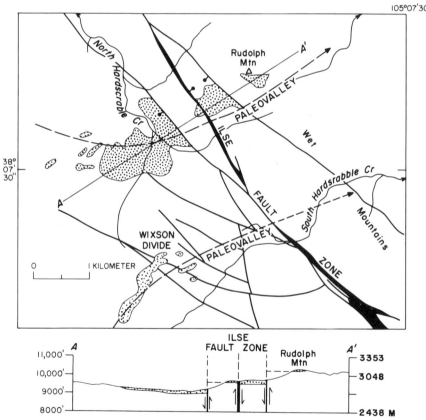

Figure 4. Geologic sketch map of Wixson Divide area. Oligocene(?) or Miocene(?) boulder alluvium of two paleovalleys shown by stippled patterns. Deposits of north paleovalley are offset by branches of the Ilse fault, with cumulative Neogene displacement of about 370 m.

Lipman and Mehnert (1975), the beginning of the downdropping of the west side of this trough took place less than 27 m.y. ago. Faulting has continued on the eastern side of the valley at least into Pleistocene (Scott, 1970) and perhaps into Holocene time. To the south, near Alamosa, the clastic and volcanic trough fill may exceed 10,000 m in thickness (Gaca and Karig, 1966). In the area considered here, the timing and magnitude of tectonism cannot be determined. This major structural feature merits consideration far beyond the limited scope of this paper.

Basin-Range Structure South of the Arkansas River

The major block fault features of Neogene age in the southern part of the area being considered are shown in Figure 5. The faults are drawn as nearly vertical, expressing my conviction that high-angle faulting was dominant. The modern topography is obviously structurally controlled, and similarities to basin-range structures are evident. The late Eocene erosion (prevolcanic) surface is commonly buried by Miocene(?) valley fill in the basins and is widely exhumed between the Ilse and Westcliffe faults and on the Wet Mountains but is eroded entirely from the Sangre de Cristo Range and from the high plains (Fig. 5).

Neogene movements on the Westcliffe and Alvarado faults are documented by offsets on the 29-m.y.-old Gribbles Park Tuff and by the thick Miocene(?) valley fill of the Santa Fe(?) beds. A more than 400-m Neogene subsidence of the Wet Mountain Valley graben is proved by drilling, and preliminary geophysical data suggest that the valley fill may be three times this thick.

Sangre de Cristo Range

The Sangre de Cristo Range is the northern segment of the Sangre de Cristo Mountains, which extend southward more than 160 km into New Mexico. These mountains tower more than 1.6 km over the flanking Wet Mountain Valley to the east and the San Luis Valley to the west.

The structure of the Sangre de Cristo Range is extremely complex and inadequately known. The range is formed largely of rocks laid down in a trough in Pennsylvanian and Permian time during the tectonism that formed the Ancestral Rocky Mountains. Extensive faulting, both thrust and high-angle, formed mountains during the Laramide, and debris shed from these mountains formed major parts of the Eocene Huerfano Formation. Subdued topography is inferred by middle Tertiary time, when valleys bringing debris from the west crossed these mountains. Oligocene volcanic rocks and clastic beds containing volcanic debris of western provenance are found in remnants of these paleovalleys along the eastern foot of the range and the west side of the Wet Mountain Valley. Neogene uplift of the range is younger than the 29-m.y.-old Gribbles Park Tuff found in two of the paleovalleys, but neither the beginning nor ending of this tectonism can be dated.

Although the total Neogene uplift of the east side of the Sangre de Cristo Range on the Alvarado fault cannot be determined, a minimum figure of 3 km is required by the difference in elevation of the mountain tops and of the base of the Miocene fill in the Wet Mountain Valley graben. No remnants of Tertiary rocks are found on the Sangre de Cristo Range except for intrusive igneous rocks, including the Oligocene stock at Cottonwood Peak.

Upper Arkansas Valley

Van Alstine (1968) has demonstrated that the north end of the San Luis Valley and the south end of the upper Arkansas Valley were continuous in Miocene time when sediments of the Dry Union Formation (Santa Fe Group equivalent) filled a valley 4 to 5 km in width which was cut in Oligocene volcanic rocks

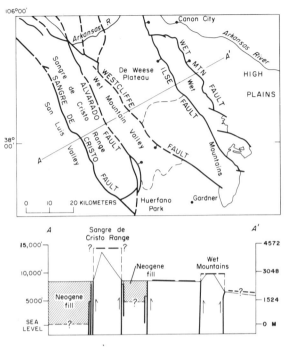

Figure 5. Neogene fault blocks south of the Arkansas River. Neogene faults outline modern basins and ranges from the San Luis Valley to Canon City. Inferred position of the prevolcanic surface is shown on cross section as a heavy dashed line; this late Eocene surface is same as modern surface on the De Weese Plateau and on a part of the Wet Mountains.

and Precambrian rocks west of Poncha Pass (Fig. 1). The Dry Union Formation is composed of poorly sorted and little consolidated light-brown, tan, or gray sandy mudstone, with interbedded clay, sand, and gravel, and thin layers of volcanic ash. These materials reflect several different source areas supplying volcanic, sedimentary, and metamorphic materials and indicate a distant source supplying fan gravels to a through going drainage system. The maximum thickness of the Dry Union Formation is not known; Tweto (1961, p. B133) suggested on the basis of geophysical data that it may be as much as 650 m locally.

The trough connecting the San Luis Valley and the upper Arkansas Valley was offset by a major west-northwest-trending Neogene fault that also cuts off the north end of the Sangre de Cristo Range. This fault shows the youngest major movement that can be identified on any basin-range type of break in central Colorado. South of Salida, south-dipping beds of the Dry Union Formation overlying the Precambrian crystalline basement contain a Valentine fauna, 12 m.y. old, based on fossils identified by Edward Lewis (1973, oral commun.). Higher beds in the Dry Union contain an Ash Hollow fauna indicating an age of about 7 m.y. Strata of both faunal zones are accompanied by well-rounded pebbles and cobbles of varied lithologies indicating distant sources, probably to the north and west. The Dry Union beds above the Ash Hollow fauna contain progressively more abundant coarse angular fragments; the proportion of coarse material derived from the north end of the Sangre de Cristo Range also increases upward in the section, and the far-traveled materials become sparse. The uppermost beds faulted against the Precambrian rocks of the mountains are almost entirely composed of distinctive Precambrian fragments derived from the area immediately south of the fault. The lithologic change in the south-dipping beds of the Dry Union Formation is interpreted to be evidence of the change in sedimentary basin geometry from a drainage system receiving sediment from the north and connecting with the San Luis Valley to the breakup of the trough and the rise of the Sangre de Cristo Range. The major movement on the west-northwest-trending cross fault is younger than the beds containing the 7-m.y.-old Ash Hollow fauna. The vertical component of movement cannot be accurately determined but is estimated at 1 to 2 km.

The cross fault south of Salida can be traced to the southeast where it splits; one branch forms the east flank of the Sangre de Cristo Range, and another forms the west side of a small Tertiary basin containing Miocene beds of Dry Union lithology. Traced southeastward, the latter fault joins the faults at the northern end of the Wet Mountain Valley, and there loses its identity.

On the east side of the upper Arkansas Valley north of Salida, step faults and tilted blocks characterize the valley margin. East-flowing paleovalley drainages were disrupted after deposition of the 29-m.y.-old Gribbles Park Tuff (Lowell, 1971), indicating that downfaulting of the valley graben began during early Miocene time. This faulting continued into Pliocene and Holocene time (Tweto, 1961, p. B135).

South Park

In southwestern South Park, a sequence of intertonguing calcareous, sandy, pinkish-brown to buff siltstone, mudstone, conglomeratic sandstone, and conglomerate of Miocene age was divided by Johnson (1937) into the coarse-grained Trump Formation and the fine-grained Wagontongue Formation. De Voto (1961, 1964) regarded these formations as facies of a single basin fill. About 200 m of fine-grained sediments in the center of the basin interfinger laterally with marginal coarser clasts representing alluvial fans. These beds rest with angular unconformity on the Oligocene Antero Formation of Johnson (1937) and locally are warped as much as 25° by Neogene folding. The northern outcrops (Fig. 1) outline a marked syncline and originally were regarded by Johnson (1937) as an upper member of the Antero Formation; De Voto assigned them to the Miocene on the basis of the angular unconformity. Miocene fossils found recently by Glenn R. Scott (1973, oral commun.) verify the latter conclusion. The folds represent continued deformation along earlier trends; the earlier folds in the Antero beds were accentuated in Neogene time.

Through going north-northwest-trending faults, including the Currant Creek fault (Fig. 1), transect the Thirtynine Mile and Guffey volcanic centers and have had major postvolcanic movements. As described by Epis and Chapin (1968), Precambrian shearing, late Paleozoic movement, and Laramide offset preceded the development of the late Eocene surface. Pennsylvanian, Permian, Jurassic, and Cretaceous rocks are preserved beneath the volcanic rocks. Epis and Chapin (1968) recognized more than 400 m of postvolcanic movement along these faults near the Thirtynine Mile center. The complex tilting of the structural block bearing Wall Mountain Tuff (ash flow I of Epis and Chapin, 1968) in this fault zone is due either to differential offset associated with volcanism or to Neogene movements on the regional fault system. This regional system can be traced to the south along Currant Creek, and it clearly joins the Ilse fault zone north of the Arkansas River. North of the Thirtynine Mile volcano the postvolcanic fault can be traced almost to Hartsel, where it appears to join the South Park fault as mapped by Stark and others (1949).

Fourmile Creek

The Fourmile Creek graben extends northward from the Canon City embayment, west of the Cripple Creek volcanic center. Paleozoic and Mesozoic sedimentary rocks are preserved within this graben, whereas Tertiary volcanic rocks rest upon Precambrian rocks in the uplifted blocks to the east and west, a situation indicative of Laramide faulting. The continuity of a prevolcanic surface of low relief across the Fourmile Creek blocks has been verified from remnants of a once more extensive volcanic cover. R. C. Epis (1973, oral commun.) has identified the 35-m.y.-old Oligocene Wall Mountain Tuff and andesitic conglomerates from western sources in the Thirtynine Mile volcanic field below locally derived phonolitic flows south of Cripple Creek (Fig. 6). Phonolite plugs cut the volcanic rocks from all sources. The Wall Mountain Tuff and andesitic conglomerate extend across the Fourmile Creek graben from the west, and phonolite flow remnants west of the graben are parts of more extensive outpourings that spread across the graben from the east. Thus the graben area appears to have been tectonically dormant during Oligocene time. Later reactivation of graben faults is indicated by an offset of at least 400 m along Fourmile Creek in the vicinity of Cripple Creek.

Uplift of the base of the volcanic rocks at Cripple Creek in relation to the Canon City level is suggested by the sharp faulted folds at the south end of the Front Range and by the more than 1 km difference in elevation between the base of the Oligocene volcanic deposits south of Cripple Creek and a comparable prevolcanic level just south of Canon City. Remnants of the exhumed late Eocene prevolcanic surface cut in Precambrian rocks at Cripple Creek indicate little local

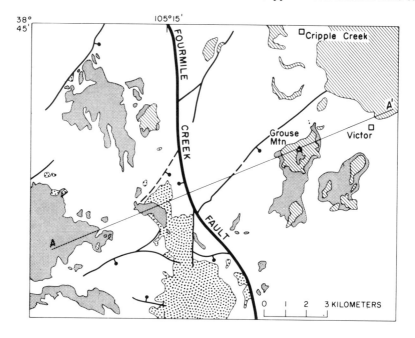

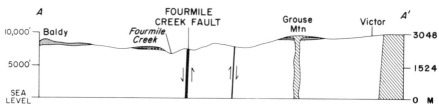

Figure 6. Geologic sketch map of the Fourmile Creek area. Volcanic rocks from the Cripple Creek center shown by diagonal line pattern, volcanic rocks of the Thirtynine Mile volcanic field shown by fine stipple, Cretaceous to Ordovician sedimentary rocks shown by coarse stipple. The late Eocene surface beneath the Oligocene volcanic rocks is displaced about 400 m across the Fourmile Creek fault. Geology from R. C. Epis (1970–72, unpub. data).

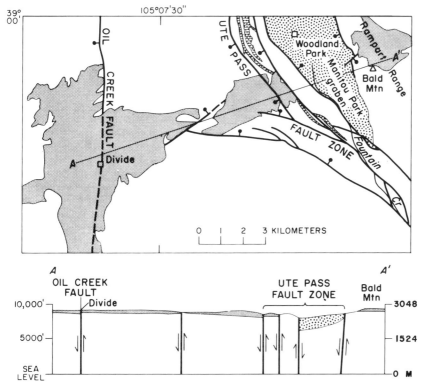

Figure 7. Geologic sketch map of the Divide area. Pliocene(?) and Miocene deposits shown by fine stipple pattern, Permian to Ordovician sedimentary rocks shown by coarse stipple, Cambrian sandstone dikes shown by coarse stipple with line pattern, and Precambrian rocks not patterned. Sedimentary rocks in graben were faulted down before cutting of the late Eocene surface; fault reactivation in the Neogene raised the Rampart Range nearly 400 m relative to the Divide area. Geology from G. R. Scott (1972, unpub. data).

relief, and comparably low relief probably existed on the softer sedimentary rocks to the south. Even allowing for considerable gradient on the surface, a postvolcanic offset between mountain block and plains of 350 to 1,000 m can be inferred.

Divide, Colorado, and the Rampart Range

Extensive deposits of probable Miocene-Pliocene(?) age cover broad flat areas in the vicinity of Divide, Colorado, 16 km north of Cripple Creek (Fig. 7). The deposits are composed of pinkish-tan and tan to gray, poorly consolidated siltstone, mudstone, and claystone with interlayered sandy and pebbly beds. The basal beds are coarse sandy gravels and conglomerates which lie in broad shallow channels cut into the Precambrian basement. The youngest identifiable volcanic clasts in the beds are of phonolite believed to have been derived from the Cripple Creek volcanic center. Fragments of phonolite from Cripple Creek and of volcanic materials from the Thirtynine Mile center establish northerly and easterly flow directions. Correlative gravels are found on the Rampart Range across the Manitou Park graben (Fig. 8). A Neogene uplift of more than 370 m for the Rampart Range relative to the upstream area around Divide is required to account for this distribution.

Faults associated with the Manitou Park graben are inferred to have Precambrian ancestry and to have been reactivated several times since then. The fault along the western side of the graben is followed by sandstone dikes of probable Cambrian age; the graben contains down dropped remnants of Paleozoic rocks, preserved below the level where the late Eocene (prevolcanic) surface was cut and which record post-Permian–pre-Eocene movements. Sandstone dike remnants in the graben now are higher than gravels in the western block, suggesting the Neogene elevation of the graben fill as well as of the Rampart Range. This Neogene faulting on the Ute Pass fault zone at the west side of the graben thus has a direction of movement opposite that of the older faulting that created the Manitou Park graben.

SUMMARY

South-central Colorado is riven by a series of north-northwest-trending to north-trending faults that cut the area into a series of structural blocks typical of many areas in the well-known Basin and Range province of the western United States. Geologic relations in many local areas demonstrate that these faults had major movement in Neogene time. High-angle faults separate blocks of greater or lesser uplift or subsidence, and sedimentation in local basins has been controlled by base levels established by tectonic movements.

Neogene tectonism began after an extended period of relative tectonic stability required for the cutting of a widespread late Eocene prevolcanic surface of low relief. The stability persisted through the time of Oligocene volcanic eruptions. Block faulting began early in Miocene time, and active tectonism may be inferred to have persisted into Holocene time in certain areas. Pulses of greatest activity are inferred for the early Miocene and for the latest Miocene or early Pliocene.

Figure 8. Manitou Park graben. View is northward along the Manitou Park graben. Boulder alluvium on grass-covered flats on the Rampart Range (on right) was deposited by a Miocene(?)-Pliocene(?) paleodrainage incised about 10 m into the late Eocene surface. Fault-line escarpment marks east boundary of Paleozoic rocks in the graben where they abut uplifted Precambrian rocks of the Rampart Range.

ACKNOWLEDGMENTS

For information reported here, I thank my co-workers Glenn R. Scott and Rudy C. Epis; the assistance of Carl Hedge, John Obradovich, and Charles Naeser with isotopic and fission-track dating, and of Edward Lewis for studies of Tertiary vertebrate fossils is gratefully acknowledge.

REFERENCES CITED

Boyer, R. E., 1962, Petrology and structure of the southern Wet Mountains, Colorado: Geol. Soc. America Bull., v. 73, no. 9, p. 1047-1069.

Burbank, W. S., and Goddard, E. N., 1937, Thrusting in Huerfano Park, Colorado, and related problems of orogeny in the Sangre de Cristo Mountains: Geol. Soc. America Bull., v. 48, no. 7, p. 931-976.

Christman, R. A., Brock, M. R., Pearson, R. C., and Singewald, Q. D., 1959, Geology and thorium deposits of the Wet Mountains, Colorado—A progress report: U.S. Geol. Survey Bull. 1072-H, p. 491-535.

De Voto, R. H., 1961, Geology of southwestern South Park, Park and Chaffee Counties, Colorado [D.Sc. thesis]: Golden, Colorado School Mines, 323 p.

——1964, Stratigraphy and structure of Tertiary rocks in southwestern South Park: Mtn. Geologist, v. 1, no. 3, p. 117-126.

Epis, R. C., and Chapin, C. E., 1968, Geologic history of the Thirtynine Mile volcanic field, central Colorado, in Cenozoic volcanism in the Southern Rocky Mountains: Colorado School Mines Quart., v. 63, no. 3, p. 51-85.

——1975, Geomorphic and tectonic implications of the post-Laramide late Eocene erosion surface in the Southern Rocky Mountains, in Cenozoic history of the Southern Rocky Mountains: Geol. Soc. America Mem. 144, p. 45-74.

Gaca, J. R., and Karig, D. E., 1966, Gravity survey in the San Luis Valley area, Colorado: U.S. Geol. Survey Open-File Rept.

Hills, R. C., 1888, The recently discovered Tertiary beds of the Huerfano River basin, Colorado: Colorado Sci. Soc. Proc., v. 3, p. 148-164.

——1900, Description of the Walsenburg quadrangle [Colorado]: U.S. Geol. Survey Geol. Atlas, folio no. 63.

Johnson, J. H., 1937, The Tertiary deposits of South Park, Colorado, with a description of the Oligocene algal limestone [abs. of thesis]: Colorado Univ. Studies Ser. Earth Sci., v. 25, no. 1, p. 77.

Lipman, P. W., and Mehnert, H., 1975, Late Cenozoic basaltic volcanism and development of the Rio Grande depression in the Southern Rocky Mountains, in Cenozoic history of the Southern Rocky Mountains: Geol. Soc. America Mem. 144, p. 119-154.

Lowell, G. R., 1971, Cenozoic geology of the Arkansas Hills region of the southern Mosquito Range, central Colorado, in James H. L., ed., Guidebook of the San Luis Basin, Colorado: New Mexico Geol. Soc. Guidebook, 22d Field Conf., 1971, p. 209-217.

Parker, R. L., and Hildebrand, F. A., 1963, Preliminary report on alkalic intrusive rocks in the northern Wet Mountains, Colorado, in Short papers in geology, hydrology, and topography: U.S. Geol. Survey Prof. Paper 450-E, p. E8-E10.

Scott, Glenn R., 1970, Quaternary faulting and potential earthquakes in east-central Colorado, in Geological Survey research 1970: U.S. Geol. Survey Prof. Paper 700-C, p. C11-C18.

——1975, Cenozoic surfaces and deposits in the Southern Rocky Mountains, in Cenozoic history of the Southern Rocky Mountains: Geol. Soc. America Mem. 144, p. 227-248.

Shawe, D. R., and Parker, R. L., 1967, Mafic-ultramafic layered intrusion at Iron Mountain, Fremont County, Colorado: U.S. Geol. Survey Bull. 1251-A, 28 p.

Singewald, Q. D., 1966, Description and relocation of part of the Ilse fault zone, Wet Mountains, Colorado, in Geological Survey research 1966: U.S. Geol. Survey Prof. Paper 550-C, p. C20-C24.

Stark, J. T., Johnson, J. H., Behre, C. H., Jr., Powers, W. E., Howland, A. L., Gould D. B., and others, 1949, Geology and origin of South Park, Colorado: Geol. Soc. America Mem. 33, 188 p.

Steven, T. A., 1975, Middle Tertiary volcanic field in the Southern Rocky Mountains, *in* Cenozoic history of the Southern Rocky Mountains: Geol. Soc. America Mem. 144, p. 75-94.

Tweto, Ogden, 1961, Late Cenozoic events of the Leadville district and upper Arkansas Valley, Colorado, *in* Short papers in the geologic and hydrologic sciences: U.S. Geol. Survey Prof. Paper 424-B, p. B133-B135.

Van Alstine, R. E., 1968, Tertiary trough between the Arkansas and San Luis Valleys, Colorado, *in* Geological Survey research 1968: U.S. Geol. Survey Prof. Paper 600-C, p. C158-C160.

MANUSCRIPT RECEIVED BY THE SOCIETY MAY 8, 1974

Geological Society of America
Memoir 144
© 1975

Cenozoic Surfaces and Deposits in the Southern Rocky Mountains

GLENN R. SCOTT

U.S. Geological Survey
Federal Center
Denver, Colorado 80225

ABSTRACT

The Cenozoic erosional history of the eastern part of the Southern Rocky Mountains is documented by ancient surfaces and deposits. During Laramide tectonism, streams eroded the sedimentary cover, exhuming pre-Mesozoic surfaces and beveling Precambrian rocks. By late Eocene time, erosion had produced the only widespread surface of low relief developed in the mountains. In early Oligocene time, broad shallow channels were cut into this surface, and these were partly filled with alluvial deposits and overlapped by Oligocene and Miocene volcanic rocks. Faulting in early Miocene through Pliocene time displaced mountain versus valley blocks as much as 12,000 m vertically. The late Eocene surface was fragmented, the mountains were deeply eroded, and adjacent grabens filled. Channels were again cut below the late Eocene surface and filled with equivalents of the Santa Fe Formation. Uplift accelerated in late Pliocene time, and concurrent erosion cut canyons 180 to 300 m deep at the mountain flanks.

Tertiary surfaces can be differentiated by their extent and degree of development, height above streams, and character of covering deposit. All widespread surfaces on Precambrian rocks are exhumed older surfaces or were cut during Tertiary time. Because the late Eocene surface is fragmented, reconstruction requires consideration of Neogene block faulting. Broad gentle channels cut into the late Eocene surface in early Oligocene and in Miocene or Pliocene time contain stream gravels that permit reconstruction of the geomorphic history. Some of these gravels have been called till, but they are not glacial and extensive icecap glaciation is unknown in the Southern Rockies. Quaternary surfaces are narrow and confined to valleys, and are not over 140 m above streams. Surfaces can be dated best where overlying deposits furnish direct evidence of age.

INTRODUCTION

This paper describes Tertiary and Quaternary erosion surfaces, explains how they formed, and gives criteria for recognizing and differentiating them and for making reasonable estimates of their ages. The geomorphic events described apparently took place throughout the Southern Rocky Mountains (Fig. 1), and the resulting landforms are broadly similar throughout the region.

Geologic literature contains many references to Tertiary and Quaternary surfaces in the Front Range, the chief articles being those by Lee (1917), Van Tuyl and Lovering (1935), Rich (1935), and Wahlstrom (1947). Most of these articles concentrate on the altitudes and ages of the surfaces, but give little attention to the covering deposits, which in general were erroneously called till. Very little consideration has been given to the processes forming the broad gentle surfaces or to associated tectonic and volcanic events, which were then less well known. In 1953 Knight described a geomorphic history of southern Wyoming almost identical to the sequence of events given here.

As used here, the Paleocene Epoch ended 53 to 54 m.y. ago, the Eocene, 37 to 38 m.y., the Oligocene, 28 m.y., the Miocene, 5 m.y., and the Pliocene, 1.8 m.y. ago.

STAGES IN GEOMORPHIC DEVELOPMENT OF THE SOUTHERN ROCKIES

The geomorphic development of the Southern Rockies is divided into six stages: (1) Laramide uplift, erosion, and deposition; (2) cutting of the Eocene surface; (3) Oligocene deposition; (4) early Miocene through Pliocene uplift, erosion, and deposition; (5) Pliocene canyon cutting; and (6) Quaternary glaciation and cutting of pediments and terraces.

Laramide Uplift, Erosion, and Deposition

Active geomorphic development of the Southern Rockies started in Late Cretaceous time and is still in progress. The Laramide orogeny began in Late Cretaceous time, about 67 m.y. ago, when the mountains began to rise and the sea began to withdraw. Laramide as used here is the orogenic time between Late Cretaceous and middle Eocene. The mountains were elevated about 3,600 m, enough to lift the Precambrian rocks above sea level. The sedimentary rock cover was eroded from the Precambrian rocks of the uplifts, and fine-grained coal-bearing beds equivalent to the Laramie Formation and overlying coarse-grained beds equivalent to the Dawson Formation were deposited in adjacent basins. In Paleocene time major stream systems, some of them 120 km long, developed in the rising mountains and carried debris from sources in the mountains to form sedimentary rocks and volcaniclastic deposits in bordering basin fills (Fig. 2). Deposits containing volcanic detritus at the distal ends of these stream channels are preserved on the plains near Golden (Denver Formation), near Colorado Springs (two zones in the Dawson Formation), and south of Canon City (in Poison Canyon Formation). Other basin fills lie in South Park (Denver? Formation), Middle Park (Middle Park Formation), and North Park (Coalmont Formation). Cretaceous and Paleocene volcanic flows and lahars also helped fill the Denver basin and South and Middle Parks. In addition, Laramide erosion exhumed widespread and well-formed pre-Paleozoic and pre-Mesozoic surfaces (Fig. 3) that had been cut on Precambrian rocks before deposition of the sedimentary rocks.

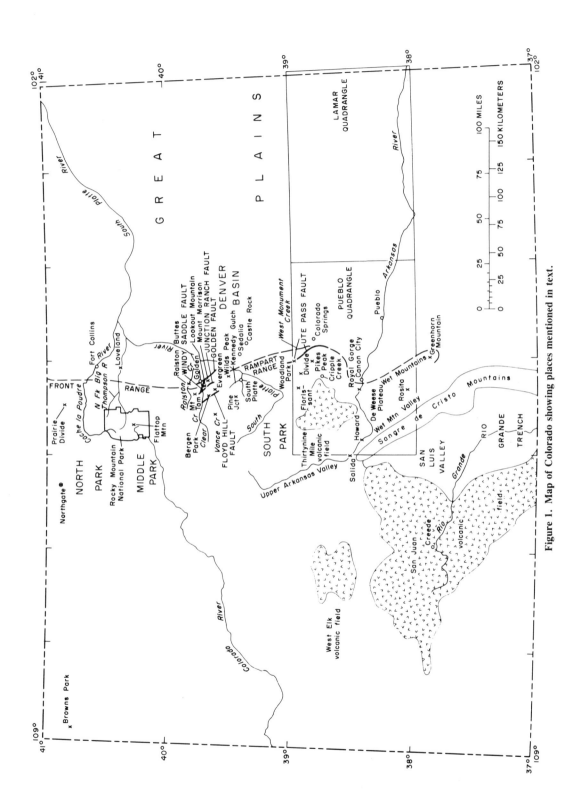

Figure 1. Map of Colorado showing places mentioned in text.

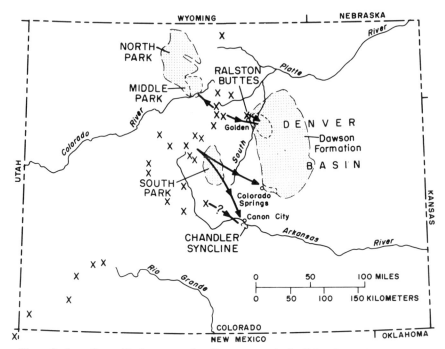

Figure 2. Some Laramide igneous and sedimentary rocks in Colorado. X, intrusive bodies of known or inferred Laramide age (from Tweto, 1975); short dashed lines within basins show rocks containing abundant volcanic detritus (not shown in South Park and Chandler syncline). Arrows show inferred channels of transport of volcanic flows and volcaniclastic rocks.

Cutting of the Eocene Surface

Erosion continued through both the Paleocene and Eocene Epochs, but apparently uplift nearly ceased after early Eocene time. In Eocene time, sediments that formed the Huerfano Formation, Echo Park Alluvium, and equivalent formations were carried in channels and across a developing pediment surface on the flanks of the uplifts, where local veneers still remain, and were deposited to thicknesses of hundreds or locally more than 1,000 m in the deepening grabens and basins. Deposition kept pace with basin development; therefore the deformation apparently did not appreciably disrupt throughgoing streams. The regional base level became stable in late Eocene time and remained nearly stable through Oligocene time. Prolonged stability led to the development of what has been called the late Eocene surface (Epis and Chapin, 1975). This late Eocene surface is the only widespread Cenozoic surface of low relief in the mountains (Fig. 4). It was cut in many areas along the eastern front of the Rocky Mountains from Greenhorn Mountain in the Wet Mountains northward along the Front Range to Wyoming where it was reported by Knight (1953). It possibly correlates with an Eocene surface described by Soister (1968), Harshman (1968), and Denson and Harshman (1969) in central Wyoming. The surface is overlain in many places by Eocene, Oligocene, or Miocene alluvial and volcanic deposits.

Locally this surface nearly coincides with parts of pre-Paleozoic or pre-Mesozoic surfaces cut on Precambrian rocks before deposition of the sedimentary rocks (Fig. 4). Thus part of it is an exhumed surface. In most places, however, it truncates all earlier surfaces and structures and was independently developed.

At the end of the Eocene Epoch, the landscape was probably similar to that of the western part of the Great Plains today. Pediments were the most widespread landforms. Owing to later structural deformation, the original inclination and direction of slope of these pediments can only locally be determined. Bajadas existed along the margins of basins bordering the larger ranges. In addition, mountain highlands having ridges and peaks as much as 100 m high (or possibly more) remained in the cores of earlier uplifts and probably resembled features that exist today. Widespread pediments were cut in areas of equigranular igneous rocks, and smaller pediments 100 or more m wide were formed between resistant ridges in some metamorphic terranes.

The landforms suggest that the early Tertiary streams were similar to Quaternary streams except for having broader flood plains, gentler gradients, and possibly larger amounts of water. The differences probably resulted from a warmer and wetter climate and a lower average altitude. The mid-Eocene (early Bridger time) of the Rocky Mountain region is known to be the hottest and wettest part of the entire Tertiary record. Following that, in the final phases of the Green River lakes, the climate first remained hot (average annual temperature 15° to 21°C and frost-free), but was drier than before (inferred rainfall 600 to 780 mm annually at lower altitudes). The late Eocene (Uinta) was considerably drier and cooler, resulting in a warm, temperate, mixed conifer-hardwood vegetation. The altitude in late Eocene time probably averaged less than 900 m (Estella Leopold, 1973, oral commun.; MacGinitie, 1969; Leopold and MacGinitie, 1972).

How and Where the Eocene Surface Is Preserved. The Eocene surface is well preserved only where it is still buried, as beneath the Oligocene deposits of the San Juan, Thirtynine Mile, and West Elk volcanic fields, all in southwestern Colorado

Figure 3. Royal Gorge west of Canon City, Colorado, showing excellent pre-Mesozoic surface largely stripped of former cover of Morrison Formation. Precipitous canyon was cut in Pliocene time.

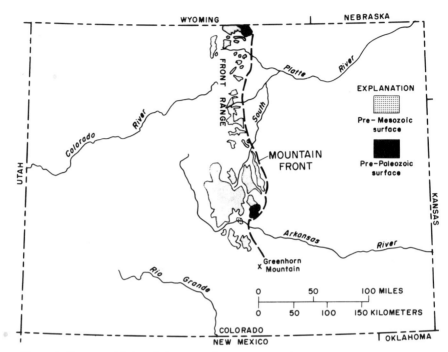

Figure 4. Known and inferred remnants of the Eocene surface in and near the Front Range. Two areas of known pre-Paleozoic surface (one pre-Manitou Limestone and one pre-Fountain Formation) and one area of pre-Mesozoic surface (pre-Morrison Formation) are shown.

(Epis and Chapin, 1975). Elsewhere, covering deposits remain only in channels cut below the general level of the surface. These deposits are best preserved in channels that parallel grabens, were dammed by volcanic rocks, or were blocked by faults. In the mountains, most of the deposits probably were removed from the Eocene surface during Miocene or more recent erosion, whereas on the plains of southeastern Colorado the erosion probably took place in both Chadron (early Oligocene) and part of Ogallala (late Miocene) time.

Other small unburied parts of the original surface were protected from erosion by their positions behind rising mountain blocks; streams behind these blocks generally managed to maintain their courses and drain the protected areas but were unable to cut deeply into them. One of the best examples of such a protected area is the De Weese Plateau, west of the Wet Mountains. Less than one-fifth of the streams that originally crossed the future site of the Wet Mountains were able to hold their courses as the mountains rose, and then only because they gained extra water from other streams that were blocked. Another example of a tectonically protected area can be identified at the east flank of South Park where a large area of the Eocene surface was well preserved by downfaulting and by being thinly buried.

In many places, the Eocene surface is clearly recognizable in the modern topography even though unprotected parts have been eroded to a greater or lesser degree. With experience, an observer can recognize remnants of the surface even where they have been severely modified by erosion or disrupted by faulting. Generally the surface has been dissected by a drainage network that cut sharp stream canyons but did not appreciably lower the intervening ridge crests. Over large areas, such as the Rampart Range (Fig. 5), the ridge crests are nearly accordant and represent an upland surface that is possibly less than a few tens of meters below its original level, as indicated by remnants of the early Oligocene Wall Mountain Tuff that filled shallow swales cut below the main flats of the surface. The higher parts of the Eocene surface were generally eroded the most, especially in areas that were glaciated during the Pleistocene. Even in these areas, however, flat areas are preserved that have long been inferred to be geomorphic surfaces, for example, the Flattop surface in Rocky Mountain National Park and the shoulders at 3,750 m on Pikes Peak. These surface remnants are seen only on the broader ranges; narrow ranges, such as the north half of the Sangre de Cristo, have lost all recognizable remnants of the surfaces that once existed there.

In the Front Range north of the South Platte River, most outflowing drainage was not blocked as it was behind the Wet Mountains and Rampart Range. Apparently, few of the late Cenozoic faults in that part of the range had sufficient movement down on the west as did faults in the Wet Mountains. The late Eocene surface to the north therefore generally has been more eroded than the same surface to the south. Only two places are known where drainage was blocked: one west of the Kennedy Gulch fault at Kennedy Gulch and another at Bergen Park. Tertiary gravel deposits show that the Kennedy Gulch fault moved up on the east nearly 300 m since Eocene time (Fig. 6).

The late Eocene surface at Bergen Park is now 600 m below the same surface on a block to the west across the major Floyd Hill fault (Sheridan and others, 1972), but is only about 100 m below the late Eocene surface on a block to the east across a small fault.

Oligocene Deposition

Oligocene rocks in the Southern Rocky Mountains consist largely of volcanic and related intrusive rocks and local alluvial and lake deposits. Most of these

Figure 5. Late Eocene surface preserved on Rampart Range north of Pikes Peak (highest peak on skyline). This surface is overlain by patches of Wall Mountain Tuff and post-Wall Mountain Tertiary alluvium.

Figure 6. View to west across north end of Rampart Range west of Sedalia. The late Eocene surface at or near skyline lies just east of the Kennedy Gulch fault. Alluvium lies on this surface west of the Kennedy Gulch fault but was stripped from the segment of surface shown here.

Figure 7. Tallahassee Creek Conglomerate of Oligocene age covers the foreground at High Park west of Cripple Creek. The Fourmile Creek graben lies between the conglomerate-covered grassy flat and the timbered flat-crested Front Range forming the skyline. The Front Range was elevated more than 400 m after deposition of this conglomerate. The conical peak on left is Mount Pisgah, a phonolite extrusive which poured out on the late Eocene surface on the Front Range.

rocks are in the San Juan, Thirtynine Mile, and West Elk volcanic fields (Fig. 1); other, small remnants exist elsewhere. Steven (1975) and Steven and Epis (1968) have shown that these once were part of a large composite volcanic field that covered most of the Southern Rocky Mountains in middle Tertiary time.

The Oligocene rocks covered the late Eocene surface (Fig. 7), filling and overlapping extensive shallow channels that had been cut below the general surface level. The floors of some of the channels were covered by meager deposits of Eocene prevolcanic alluvium probably equivalent to the Huerfano Formation or the Echo Park Alluvium. The channels were well developed when volcanism began in early Oligocene time and some remained open through Neogene time. In the southern Front Range area the channels were partly filled and locally overtopped by early Oligocene ash flows, gravel, and tuffaceous sedimentary beds that are in part equivalent to the White River Group. These were overlapped by volcanic rocks that erupted later in Oligocene and part of Miocene time. Channel cutting continued in Oligocene time, forming such features as the channel in the Wall Mountain Tuff at Castle Rock (subsequently filled with Castle Rock Conglomerate). Some Oligocene channels can still be traced or inferred for more than 160 km (Fig. 8).

Oligocene deposits changed the topography and modified the drainage systems. In the large volcanic fields, all features of the pre-existing landscape were buried. On the peripheries of the large fields, and in smaller fields, large parts of the late Eocene surface were buried, but peaks and ridges projected through the Oligocene deposits (Fig. 9). Many streams were dammed by volcanic flows or lahars and forced into new courses, in part cutting across the volcanic piles and in part following around the margins of the volcanic accumulations. Lakes locally

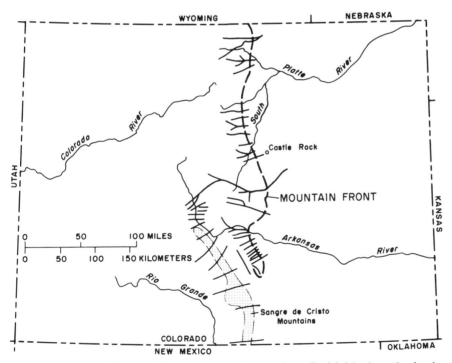

Figure 8. Inferred Oligocene-Miocene stream system based on alluvial, laharic, and volcanic flow deposits.

Figure 9. Oligocene(?) to Miocene(?) alluvium at Wixson Divide 13 km east of Rosita. View is to the west up the paleochannel that lies between valley walls composed of Precambrian rocks. The paleochannel headed near an andesitic vent, and the alluvium is rich in andesite clasts. A paleochannel just over the ridge to the north originated in a nonvolcanic area, and its alluvium is exclusively nonvolcanic.

formed behind the volcanic dams, and the lake beds deposited in them contain abundant fossils which tell much about the Oligocene climate. The most important lake deposits are at Creede (Miocene) in the San Juan Mountains, and at Florissant (Oligocene) in the southern Front Range. According to Leopold and MacGinitie (1972), the climate became warm, verging on a subtropical dry climate during Florissant time. The altitude probably averaged 900 m.

Early Miocene Through Pliocene Uplift, Erosion, and Deposition

A second episode of major Tertiary uplift, erosion, and deposition that started in early Miocene and continued through Pliocene time profoundly disrupted the Eocene surface and the overlying Oligocene deposits (Taylor, 1975). Vertically uplifted mountain blocks were deeply eroded, and the resulting debris filled adjacent grabens and basins. Offset may have exceeded 12,000 m along the west flank of the Sangre de Cristo Range (Gaca and Karig, 1965) and lesser amounts are suggested elsewhere. Eocene and Oligocene tectonic stability ended in early Miocene time when most ranges of the Southern Rocky Mountains began to rise. Basal sediments deposited in related grabens have been dated as early Miocene by vertebrate fossils (Edward Lewis, 1970, 1972, written commun.), indicating that the deformation began before Harrison or Marsland (early Miocene) time and continued through the Miocene.

Important evidence on the Neogene uplift comes from offset of Oligocene volcanic deposits and of drainage elements. Once-continuous deposits are now separated by as much as 1,500 m of vertical movement. The paleovalleys that formerly crossed the Sangre de Cristo Range and the upper Arkansas Valley (Fig. 8) were extensively disrupted during this period of deformation, and the deposits on the upthrown blocks were largely removed. For example, an excellent section of Oligocene rocks exists in the Arkansas Valley near Howard, but the channel extensions across the mountains on both sides have been removed by erosion.

An especially informative exposure of faulted Oligocene deposits lies on the plains south of Canon City where Gribbles Park Tuff (29 m.y.) and associated lahars lie at the same elevation (105 m) above the Arkansas River (30 m on projected profile) as the Slocum Alluvium of Illinoian or Sangamonian age. Equivalent rocks in the neighboring mountains are as much as 1,080 m higher as a result of Neogene tectonism. Evidently the Oligocene rocks here were displaced by this amount, buried by later Tertiary deposits, and then exhumed during the Quaternary.

Miocene deposits provide the strongest evidence for Miocene erosion because stream-cut surfaces are rare or difficult to date. Miocene alluvium was deposited in valleys formed by grabens, in basins or bajadas bordering the mountains, and in channels (Figs. 10, 17). Graben-fill deposits are largely confined to the San Luis Valley and upper Arkansas Valley segments of the Rio Grande trench and to the Wet Mountain Valley. The basin fills or bajadas are in the south end of South Park (Fig. 11), in Middle and North Parks, on the Great Plains, and at Browns Park. Channel-filling deposits of inferred Miocene age are known on the plateau west of the Wet Mountains, at Howard (which in part are also graben filling), at Divide (Fig. 12), west of South Platte (Peterson, 1964), near Wilds Peak (Bryant and others, 1973), at Pine Junction, west of Sedalia, at many places along Clear Creek west of Golden (J. C. Reed, Jr., R. B. Taylor, and D. M. Sheridan, 1972, oral commun.), north of the Big Thompson River (W. A. Braddock, 1972, written commun., Fig. 13), north of the Cache la Poudre River (W. A. Braddock, 1972, written commun.; Figs. 14 and 15), at Prairie Divide (M. E. McCallum, 1972, oral commun.), and in the Northgate district where the North Park Formation lies in channels cut into the White River Group (Steven, 1956). Other gravel deposits of possible Miocene age have been listed by Ives (1953) and by Wahlstrom (1947).

The Miocene deposits range widely in thickness and in height above nearby major modern streams. The graben fills are several hundred meters thick, but very little fill lies above modern stream level; the bajada deposits are as much as 120 m thick and commonly lie 150 m or more above major streams; most

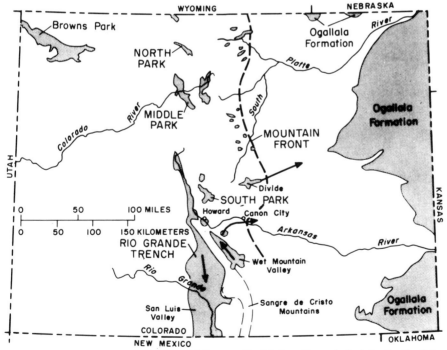

Figure 10. Some known and inferred Miocene alluvial deposits equivalent to the Ogallala Formation and inferred directions of flow in some channels (arrows).

Figure 11. Basin-fill deposits of the Wagontongue Formation of Johnson (1937) in the southern part of South Park. Fossils of late Miocene age were found at this outcrop.

Figure 12. Miocene(?) alluvium (grassy area) in paleochannel on upthrown east side of Manitou Park graben east of Woodland Park. Paleochannel was incised only several to a few tens of meters below the bordering tree-covered late Eocene surface to left and right. Woodland Park is in foreground.

of the channel deposits are 100 or 200 m thick or less and generally lie more than 150 m above adjacent major streams. The deposits range from boulder alluvium to silt and sand; the coarsest deposits are nearest the mountains and the finest deposits are near the centers of the graben fills. Volcanic ash layers have been found in the fine-grained deposits, but not in the coarse-grained deposits. Fossils generally are preserved only in the fine-grained deposits. No ash beds or fossils have been found in the channel deposits; their Miocene ages are therefore inferred.

I infer that, during Miocene time, the altitude of the lower mountains probably was similar to what it is now; the higher mountains probably were not so high as today. The Miocene pollen is similar to an impoverished Cordilleran flora of modern aspect (Leopold and MacGinitie, 1972, p. 163).

Preservation of the Miocene deposits depended largely on position relative to modern drainage. Deposits on elevated blocks were largely destroyed by erosion. The graben fills are almost completely preserved because of their protected positions relative to the bordering uplifts. In the San Luis Valley the Miocene-Pliocene deposits are almost completely buried by Quaternary alluvium and outwash and are hardly dissected. The basin-fill or bajada deposits are locally well preserved. Probably very few of the original channel deposits have been preserved; most occur now in channels abandoned because of stream captures, in channels along or across grabens, and in areas where channels were segmented and lowered by faults.

Erosion of the uplifted blocks and deposition in the basins were contemporaneous with Miocene deformation. Faults mark the margins of nearly all the graben fills and can be well documented along both flanks of the Sangre de Cristo Range and along the borders of the upper Arkansas Valley. Channels were disrupted by faults or fault systems. Many streams were tectonically dammed or were disrupted and rerouted; valleys were abandoned, and stream captures were commonplace. An example is a channel that was disrupted by a north-trending fault at Divide and to the east broken again by the Ute Pass fault at Woodland Park.

Figure 13. Oligocene(?) to Miocene(?) alluvium (grassy area) in paleovalley north of the Big Thompson River (foreground) and behind the sharp crags of Palisade Mountain. A high remnant of a geomorphic surface is preserved beyond an apparent fault scarp in the upper right part of the photograph.

Figure 14. Miocene(?) alluvium in paleochannel north of Cache la Poudre River. This channel can be traced for about 40 km. Axis of channel shown by arrow.

Pliocene Canyon Cutting

Uplift apparently was greatly accelerated in Pliocene (through Blanco Formation) time, and the resulting accelerated erosion cut the deep canyons that characterize the mountain flanks (Figs. 16, 17). These canyons were cut to 135 m above modern stream levels by the time the Nussbaum Alluvium was deposited in early Pleistocene time.

Most of the deeper parts of the modern canyons are cut near the mountain front close to the Great Plains, but major drainages extend as deep canyons many miles back into the mountains where they commonly grade to the base levels occupied by Pleistocene glaciers. These narrow, steep-walled canyons were incised either in the Eocene surface or in the floors of Miocene channels. Along Clear Creek canyon west of Golden the Pliocene canyon was cut 405 to 435 m below the Eocene surface and about 135 m below the base of Miocene(?) alluvium (Scott, 1972).

The initial positions of many major canyons seem to have been largely controlled by superposition as the streams cut through the sedimentary cover into the underlying Precambrian rocks. These positions were locally modified as erosion was later influenced by faults, rock structure, and rock hardness. Superposition of some canyons, such as Clear Creek canyon, West Monument Creek canyon, and the Royal Gorge, apparently started in Late Cretaceous time when streams cut through the sedimentary rock cover into Precambrian rocks, establishing courses across the rising mountains and carrying coarse volcanic and nonvolcanic detritus out onto the plains. These courses later became main drainage elements on the Eocene surface. The Eocene surface flanking Clear Creek is about 16 km wide at the mountain front between Mount Tom and Mount Morrison, but decreases in width to the west. In Miocene time ancestral Clear Creek cut a channel 1.5 km or more wide and about 240 m deep. In Pliocene time the canyon was deepened by 135 m and in Quaternary time by another 105 to 135 m to its present depth. Some stretches of the creek obviously were localized for short distances by shattered zones along faults, but in general, rock structure seems to have had little influence on the course of this drainage.

Although locally modified, the Pliocene canyons are still preserved throughout the mountains. Most of the detritus that resulted from canyon cutting apparently was transported across the plains and out of Colorado, but some was deposited as a thin widespread sheet of alluvium, the older part of which possibly now comprises the Kimball Member (of Lugn, 1938) of the Ogallala Formation on the eastern Great Plains, and the younger part of which possibly now comprises the lower beds of the Nussbaum Alluvium near the mountains (Scott, 1963a).

In the mountains, the upper part of the Santa Fe Formation in the San Luis Valley, the Santa Fe(?) Formation in the Wet Mountain Valley, and the upper part of the Dry Union Formation in the upper Arkansas Valley south of Salida probably include some of the Pliocene material resulting from canyon cutting. South of Salida the upper part of the Dry Union Formation of Miocene and Pliocene age consists of slightly rounded and poorly sorted Precambrian boulders, cobbles, and pebbles; it dips southward from the basin into a major fault that truncates the north end of the Sangre de Cristo Range. This upper part overlies beds, rich in volcanic detritus, which are of late Pliocene age and correlative with the Ash Hollow Member (of Moore and others, 1951) of the Ogallala Formation. These gravelly deposits apparently formed after great local Pliocene uplift of the north end of the Sangre de Cristo Range accelerated the erosion of the Precambrian rocks in the core of this range.

Pliocene uplift of the northern Sangre de Cristo Range appears to have been more than 1,200 m. This uplift, and concurrent downfaulting of comparable scale in the San Luis Valley, had a major influence in shaping the modern landscape, perhaps equaling the effects of Quaternary events.

Quaternary Pediments, Terraces, and Glacial Features

Canyon cutting and pedimentation continued in Quaternary time, but a major climatic cooling brought on glaciation. Episodic base level changes are recorded by pediments and by alluvial deposits. These changes are thought to be caused

Figure 15. Miocene(?) bouldery alluvium cemented by calcium carbonate and deeply weathered, in paleochannel north of Cache la Poudre River.

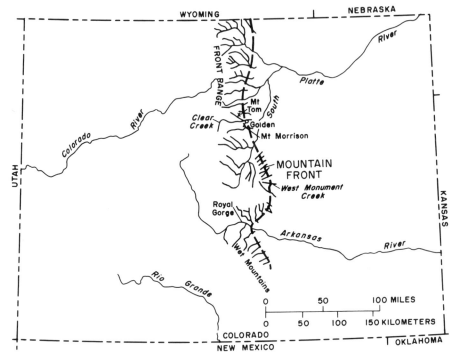

Figure 16. Major Pliocene canyons along the eastern flank of the Front Range and Wet Mountains. All canyons are the same as the modern stream valleys.

mainly by cyclic changes in the climate, but could partly be a result of uplift that continued into Holocene time. Local Quaternary uplift probably was nowhere more than 100 m. In describing the Quaternary surfaces and deposits, I present only the most important information here because these surfaces and deposits have been described several times before (Scott, 1960, 1963a, 1963b).

Nussbaum Alluvium. Pliocene canyon cutting apparently continued into Quaternary time until a stable base level was achieved and the Nussbaum Alluvium was deposited (Gilbert, 1897; Scott, 1963a; Soister, 1967). The Nussbaum Alluvium is an alluvial cover above a pediment surface and is the only deposit of this early canyon-cutting cycle that has been definitely identified. It caps the oldest of a set of pediments closely spaced in elevation above major streams, but differs in two ways from

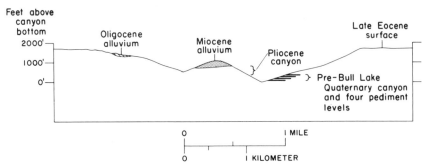

Figure 17. Typical Front Range canyon showing the form resulting from Cenozoic geomorphic processes. The four marks in the lower part of the canyon represent the levels of the four pre-Bull Lake pediments.

the younger pediment alluviums. First, the alluvium commonly is as much as 30 m thick, and thus is four to five times as thick as the alluvium on younger pediments; second, the alluvium is composed of a double fill separated by a buried soil, rather than the single fill characteristic of the younger pediments. The lower part of the Nussbaum is about 10 m thick; it probably is the cover genetically related to the pediment cutting, since bedrock lies beneath it within the possible depth of scour. The upper part is more than 10 m thick and is probably an alluvial fan deposit because it buried the underlying bedrock to depths beyond the possible range of scour. Similar changes from pedimentation to alluviation took place during each of the subsequent Quaternary pedimentation episodes, but during the younger alluviations the depth of scour was never exceeded. This suggests that the base level rose during deposition of the Nussbaum, promoting fan development.

The Nussbaum Alluvium in the type locality north of Pueblo is believed to be Quaternary in age, on the basis of a limb bone of *Camelops* sp. indet. (Edward Lewis, 1970, written commun.). The position of the Nussbaum within the Quaternary is unknown; the possibility remains that it might be equivalent to the Blanco Formation (3.5 to 1.4 m.y.—late Pliocene and early Pleistocene) in age.

The largest area of Nussbaum Alluvium still preserved is in the Lamar and Pueblo 1:250,000 quadrangles. On the plains the Nussbaum lies about 150 m below the base of the Ogallala Formation (Scott, 1963a) and about 135 m above modern major streams. In the mountains its base lies only about 105 m above modern stream level, owing to the greater resistance of Precambrian rocks to erosion. On the eastern plains the formation consists largely of fine-grained debris derived from the Ogallala Formation, the Dawson Formation, and the Pikes Peak Granite, and few pebbles larger than 2 cm in diameter are found. Most of the larger stones in the Nussbaum were derived from the Castle Rock Conglomerate and from the Wall Mountain Tuff.

Rocky Flats, Verdos, and Slocum Alluviums. The Rocky Flats, Verdos, and Slocum Alluviums are all alluvial covers on pediments. These units have been correlated from their type localities near Denver by use of geomorphic position, paleosols, and lithology with deposits to the south across the South Platte River basin and into the Arkansas River basin. Each major pediment has locally been found to contain more than one level. These probably result from a stream that created and did not destroy an intermediate level in the process of cutting down from one major geomorphic level to the next lower major level. These intermediate levels generally are found along only a few kilometers of stream length in the areas of greatest base-level and tectonic instability only a few miles from the mountain front. The intermediate-level pediment deposits are best mapped as genetic parts of the next younger formally named pediment alluvium.

Typically, in the plains areas, the Rocky Flats Alluvium (Nebraskan or Aftonian in age) is about 105 m above modern stream levels, the Verdos Alluvium (Kansan or Yarmouth) about 75 m, and the Slocum Alluvium (Illinoian or Sangamonian) 30 m above stream level. Along major drainages in the mountains the heights above streams are tens of meters less than on the plains because of the greater hardness of the Precambrian rocks. Local geomorphic and structural conditions in the mountains cause great variations in the positions of these alluviums above streams; for instance, the Verdos Alluvium at Westcliffe lies only 6 m above Grape Creek.

Louviers and Broadway Alluviums and Holocene Alluvium. These formations are all valley fills lying respectively about 21 m, 12 m, and <12 m above the major streams. Along some streams in the plains just east of the mountains, the episodic progression of downward cutting took place from the beginning of Quaternary time only through Louviers time; thereafter these streams merely reworked the upper part of the thick Louviers Alluvium. Along other streams, however, the downward cutting progressed through Holocene time to the present.

Glaciation. A general cooling of climate in Quaternary time caused cyclic glaciation. Most Quaternary glaciers in the Southern Rockies were confined to valleys and the pre–Bull Lake glaciers never extended far beyond the limits of Bull Lake Glaciation. The time of earliest glaciation in the Southern Rockies is unknown, for glacial deposits of early Pleistocene age are poorly known. The earliest recognized pre–Bull Lake till is possibly Nebraskan in age. The obvious Bull Lake and younger moraines are, from oldest to youngest: lower Bull Lake and upper Bull Lake (equivalent to Louviers Alluvium), lower Pinedale, middle Pinedale, and upper Pinedale (equivalent to Broadway Alluvium), and very minor younger moraines. Outwash terraces can be traced away from each of the two Bull Lake moraines, but they merge into one terrace within a few kilometers. Similarly, Pinedale deposits merge into one terrace, and the Holocene outwash equivalent to neoglaciation apparently either underlies the flood plains of modern streams or forms very low flanking terraces.

Glacial erosion was the chief cause of the destruction of the Eocene surface in the higher parts of the mountains. No remnant of the surface remains in narrow ranges such as the northern half of the Sangre de Cristo. In the glaciated parts of the broader ranges, such as the Front Range, small remnants of a postulated old surface are found (Fig. 18), as at Flattop Mountain in Rocky Mountain National Park (Lee, 1917, p. 28). No glaciation took place in lower mountain ranges such as the Wet Mountains, and consequently the Tertiary surfaces and deposits were more widely preserved.

Throughout the glaciated mountains in addition to destruction of the surfaces, erosion related to glaciation also removed most of the Tertiary alluvial deposits (even in the lower parts of middle Tertiary channels) that had not already been removed in Pliocene time. In addition, along the Arkansas River, melt water removed most of the older Quaternary alluvial deposits; they remain only where the canyon parallels a major fault far enough to have a wide valley where terraces could be preserved. Many of the preserved older Quaternary alluvial deposits were cemented by travertine and so they were not easily eroded. Most of the alluvium remaining in the mountain valleys is Bull Lake and Pinedale outwash. These remnants help to extend and correlate the outwash sequence with the nonglacial terrace sequence on the Great Plains.

RECOGNITION OF TERTIARY SURFACES AND DEPOSITS

Some surfaces in the mountains are very difficult to recognize and differentiate. Proving that a local flat or semi-flat area is a geomorphic surface is difficult; but determining whether it is an exhumed pre-Paleozoic or pre-Mesozoic surface or a primary Eocene, Oligocene, Miocene, or Quaternary erosion surface is virtually impossible in many places. The only certain way of recognizing and identifying a surface is by means of overlying datable deposits, and these are all too scarce. Many factors combine to make the surfaces difficult to recognize and differentiate:

1. Stream or glacial erosion has removed most or all of the overlying deposits, severely modified the surfaces, and reduced the extent of both surfaces and deposits.

2. Faulting has segmented the surfaces and changed the vertical positions of both the surfaces and deposits.

3. Deposits do not contain datable materials.

4. The sources of the rocks in the deposits cannot be identified.

Figure 18. Surfaces at crest of Front Range in Rocky Mountain National Park. View from Specimen Mountain eastward across Cache la Poudre River, toward Forest Canyon Pass, Forest Canyon (forested), and high peaks along continental divide. Bare talus slope at Forest Canyon Pass is formed by Miocene ash-flow tuff; high mountains by Precambrian rocks. Faults have influenced the modern topography. The high flat areas should not be considered remnants of a Tertiary geomorphic surface without definitive local evidence.

5. Surfaces and deposits are buried under basin fills and cannot be seen.

Generally it is easier to tell Quaternary from Tertiary surfaces and deposits than it is to differentiate the Tertiary surfaces and deposits. The following criteria are helpful in differentiating the Tertiary and Quaternary:

1. Tertiary surfaces are commonly extensively developed on Precambrian crystalline rocks, whereas only extremely small areas of Quaternary surfaces were cut on crystalline rocks. The only widespread surface of low relief was cut in Eocene time. No widespread montane Quaternary stream-cut surface exists at any level.

2. Normally, no montane Quaternary alluvium lies higher than 108 m above major streams. However, Tertiary alluvium can be 108 m or less above stream levels in downfaulted areas or where later entrenchment has been minor.

3. No high-level gravel deposit far from areas of Bull Lake Till is glacial in origin. All known pre–Bull Lake till is adjacent to Bull Lake Till and was deposited by valley glaciers. Furthermore, no icecap glaciers that could have deposited the high-level gravels are known in Colorado or New Mexico.

4. Precambrian rock beneath the Eocene surface is weathered to depths of about 10 to more than 40 m, whereas under Quaternary surfaces in the lower parts of the canyons weathering is less than 3 m deep.

Although it generally is possible to tell whether a surface or a deposit is Tertiary or Quaternary in age, it is more difficult to distinguish which Tertiary surface is present unless a datable overlying deposit is found. Following are some suggestions for differentiating the Eocene, Oligocene, and Miocene surfaces.

EOCENE SURFACE

1. The Eocene surface was very widespread, and much of it was nearly flat, although many high ridges and peaks persisted locally.
2. Underlying rocks were weathered to depths of about 10 to more than 40 m.
3. The Eocene surface was widely covered by alluvial or volcanic deposits.
4. The Eocene surface may lie at any height in relation to modern streams, depending on local structure.
5. Amount of offset by faults is greater than it is for younger surfaces.

OLIGOCENE CHANNELS

1. Minor Oligocene channels have been recognized in central Colorado, but major channels existed in Wyoming.
2. Underlying rocks are weathered like those under the Eocene surface.
3. The Oligocene-channeled surface was widely covered by volcanic deposits.
4. Oligocene channels lie a little lower than the Eocene surface, where both can be recognized.

MIOCENE CHANNELS AND GRABEN FILLS

1. Minor Miocene channels are cut into the Eocene surface; they tend to be narrower than Oligocene channels.
2. The bedrock is weathered about 10 m deep beneath the channels.
3. The associated alluvial gravels contain mainly fragments of Precambrian rocks, but locally contain some volcanic rocks. The gravels are generally not covered by volcanic deposits.
4. Local structure determines height above modern major streams.
5. There is a characteristic reddish-brown (Santa Fe Formation) color for deposits in graben fills.
6. Pliocene canyons are entrenched into these channels and deposits.

As a result of the recent work done on Cenozoic surfaces and deposits, I believe that the following previously stated or implied ideas in the geologic literature should be discarded.

1. That there are many levels of high surfaces, each level representing a separate major episode of cutting.
2. That remnants of surfaces can be correlated on the basis of their altitudes. The old surfaces were offset in many places by Cenozoic faulting.
3. That fault movement since Laramide time is minor. Local offset of as much as 12,000 m is probable since Laramide time.
4. That early Pleistocene glaciation was icecap glaciation. Icecap glaciation did not take place in Colorado or New Mexico. Pre–Bull Lake glaciers occupied valleys.

The modern landscape thus appears to have its origins in a widespread surface of low relief that was completed by late Eocene time; it has since been modified by minor broad alluvium-filled valleys cut in Oligocene and Miocene time, and further modified by deep canyons cut in Pliocene time and by glaciation and pedimentation in Quaternary time. These surfaces and valleys were segmented by block faulting mainly from early Miocene through Pliocene time, with the greatest local displacement during the Pliocene; some faults moved as late as Holocene time.

ACKNOWLEDGMENTS

I am particularly indebted to my co-workers in the Pueblo, Colorado, 1:250,000 quadrangle: R. C. Epis and R. B. Taylor; together we distinguished surfaces, paleovalleys, and associated deposits, and mapped faults that displace these features. Fossil identifications by Edward Lewis have permitted us to assign ages to several Tertiary basin-fill deposits. Glen A. Izett, Ogden Tweto, Thomas A. Steven, and Peter W. Lipman of the U.S. Geological Survey have discussed Tertiary geomorphology with me and have supplied unpublished information. W. A. Braddock and his students at the University of Colorado have mapped deposits associated with surfaces and paleovalleys in many 7½′ quadrangles northwest of Loveland, and kindly shared the data with me. Information about some paleovalleys and deposits west of Fort Collins has been provided by M. E. McCallum of Colorado State University. Photographs presented herein were taken by R. B. Taylor and K. L. Pierce.

REFERENCES CITED

Bryant, Bruce, Miller, R. D., and Scott, G. R., 1973, Geologic map of the Indian Hills quadrangle, Jefferson County, Colorado: U.S. Geol. Survey Geol. Quad. Map GQ-1073.

Denson, N. M., and Harshman, E. N., 1969, Map showing areal distribution of Tertiary rocks, Bates Hole-Shirley Basin area, south-central Wyoming: U.S. Geol. Survey Misc. Geol. Inv. Map I-570.

Epis, R. C., and Chapin, C. E., 1975, Geomorphic and tectonic implications of the post-Laramide, late Eocene erosion surface in the Southern Rocky Mountains, in Curtis, Bruce, ed., Cenozoic history of the Southern Rocky Mountains: Geol. Soc. America Mem. 144, p. 45-74.

Gaca, J. R., and Karig, D. E., 1965, Gravity survey in the San Luis Valley area, Colorado: U.S. Geol. Survey open-file report.

Gilbert, G. K., 1897, Description of the Pueblo quadrangle, Colorado: U.S. Geol. Survey Geol. Atlas, Folio 36, 7 p.

Harshman, E. N., 1968, Geologic map of the Shirley Basin area, Albany, Carbon, Converse, and Natrona Counties, Wyoming: U.S. Geol. Survey Misc. Geol. Inv. Map I-539.

Ives, R. L., 1953, Anomalous glacial deposits in the Colorado Front Range area, Colorado: Am. Geophys. Union Trans., v. 34, no. 2, p. 220-226.

Johnson, J. H., 1937, The Tertiary deposits of South Park, Colorado, with a description of the Oligocene algal limestones [abs.]: Colorado Univ. Bull., v. 25, no. 1, Gen. Ser. 403, p. 77.

Knight, S. H., 1953, Summary of the Cenozoic history of the Medicine Bow Mountains, Wyoming, in Wyoming Geol. Assoc. Guidebook, 8th Ann. Field Conf., Laramie Basin, Wyoming, and North Park, Colorado, 1953: p. 65-76.

Lee, W. T., 1917, The geologic story of the Rocky Mountain National Park, Colorado: U.S. Dept. Interior, National Park Service, 89 p.

Leopold, E. B., and MacGinitie, H. D., 1972, Development and affinities of Tertiary floras in the Rocky Mountains, in Aham, A. G., ed., Floristics and paleofloristics of Asia and Eastern North America: Amsterdam, Elsevier Pub. Co., chap. 12, p. 147-200.

Lugn, A. L., 1938, The Nebraska State Geological Survey and the "Valentine problem": Am Jour. Sci., 5th Ser., v. 36, no. 213, p. 220-227.

MacGinitie, H. D., 1969, The Eocene Green River flora of northwestern Colorado and northeastern Utah: California Univ. Pubs. Geol. Sci., v. 83, 203 p.

Moore, R. C., and others, 1951, The Kansas rock column: Kansas Geol. Survey Bull. 89, 132 p.

248   G. R. Scott

Peterson, W. L., 1964, Geology of the Platte Canyon quadrangle, Colorado: U.S. Geol. Survey Bull. 1181-C, p. C1-C23.

Rich, J. L., 1935, Physiographic development of the Front Range [disc.]: Geol. Soc. America Bull., v. 46, p. 2046-2051.

Scott, G. R., 1960, Subdivision of the Quaternary alluvium east of the Front Range near Denver, Colorado: Geol. Soc. America Bull., v. 71, no. 10, p. 1541-1543.

——1963a, Nussbaum Alluvium of Pleistocene(?) age at Pueblo, Colorado: U.S. Geol. Survey Prof. Paper 475-C, p. C49-C52.

——1963b, Quaternary geology and geomorphic history of the Kassler quadrangle, Colorado: U.S. Geol. Survey Prof. Paper 421-A, p. 1-70.

——1972, Geologic map of the Morrison quadrangle, Jefferson County, Colorado: U.S. Geol. Survey Misc. Geol. Inv. Map I-790 A.

Sheridan, D. M., Reed, J. C., Jr., and Bryant, B. H., 1972, Geologic map of the Evergreen quadrangle, Jefferson County, Colorado: U.S. Geol. Survey Misc. Geol. Inv. Map I-786 A.

Soister, P. E., 1967, Relation of Nussbaum Alluvium (Pleistocene) to the Ogallala Formation (Pliocene) and to the Platte-Arkansas divide, southern Denver basin, Colorado, in Geological Survey research 1967, Chap. D: U.S. Geol. Survey Prof. Paper 575-D, p. D39-D46.

——1968, Stratigraphy of the Wind River Formation in south-central Wind River basin, Wyoming: U.S. Geol. Survey Prof. Paper 594-A, p. A1-A50.

Steven, T. A., 1956, Cenozoic geomorphic history of the Medicine Bow Mountains near the Northgate fluorspar district, Colorado: Colo. Sci. Soc. Proc., v. 17, no. 2, p. 35-55.

——1975, Middle Tertiary volcanic field in the Southern Rocky Mountains, in Curtis, Bruce, ed., Cenozoic history of the Southern Rocky Mountains: Geol. Soc. America Mem. 144, p. 75-94.

Steven, T. A., and Epis, R. C., 1968, Oligocene volcanism in south-central Colorado, in Epis, R. C., ed., Cenozoic volcanism in the Southern Rocky Mountains: Colorado School Mines Quart., v. 63, no. 3, p. 241-258.

Taylor, R. B., 1975, Neogene tectonism in south-central Colorado, in Curtis, Bruce, ed., Cenozoic history of the Southern Rocky Mountains: Geol. Soc. America Mem. 144, p. 211-226.

Tweto, Ogden, 1975, Laramide (Late Cretaceous-early Tertiary) orogeny in the Southern Rocky Mountains, in Curtis, Bruce, ed., Cenozoic history of the Southern Rocky Mountains: Geol. Soc. America Mem. 144, p. 1-44.

Van Tuyl, F. M., and Lovering, T. S., 1935, Physiographic development of the Front Range: Geol. Soc. America Bull., v. 46, p. 1291-1350.

Wahlstrom, E. E., 1947, Cenozoic physiographic history of the Front Range, Colorado: Geol. Soc. America Bull., v. 58, no. 7, p. 551-572.

MANUSCRIPT RECEIVED BY THE SOCIETY MAY 8, 1974

Geological Society of America
Memoir 144
© 1975

Late Cretaceous and Cenozoic History
of Laramie Basin Region,
Southeast Wyoming

D. L. Blackstone, Jr.

University of Wyoming
Laramie, Wyoming 82071

ABSTRACT

Earliest deformation is recorded by conglomerates of late Campanian-earliest Maestrichtian age derived from distant sources. Early Paleocene deformation defined the northern Front Range, Medicine Bow Mountains, and the Sierra Madre-Park Range, which attained limited height as erosion matched uplift. Late Paleocene gravels unconformably overlapped the margins of the uplifts.

By early Wasatch time major northwest-trending structural features were outlined, the Paleocene conglomerates had been folded, and locally derived sediments again overlapped uplift margins. During middle Eocene time, sediments were deposited in the Shirley basin area. A mature landscape was produced in late Eocene time, and northeast-trending structural features developed.

During Oligocene time, volcanic debris intertongued with local conglomerates and was unconformably deposited on older rocks. Aggradation continued until deposition extended across the northern Laramie Mountains and central Medicine Bow Mountains.

Miocene rocks in the Saratoga valley, Hanna basin, and near Cheyenne suggest areas of nondeposition lying south of these localities. Present high-level erosion surfaces probably coincide with the level of basin fill in early Pliocene time.

Volcanic activity near Specimen Mountain, Colorado, in late Oligocene time (28 m.y.) may have continued into Pliocene(?) time.

Post-Miocene normal faulting and folding were active in the Laramie and Medicine Bow Mountains, Saratoga valley, North Park basin, Rawlins uplift, and along the Wheatland fault zone. Pliocene(?) and Pleistocene boulder deposits reflect late structural and climatic changes.

INTRODUCTION

The Laramie basin is the structural depression in southeastern Wyoming that lies between the Laramie Mountains on the east and the Medicine Bow Mountains on the west (Beckwith, 1938). In map view the depression is V-shaped with the apex to the south and the open end of the V roughly limited by the northeast-trending Como Bluff anticline which passes near Medicine Bow, Wyoming. The depression is rimmed on two flanks by rocks of Precambrian age and on the north by rocks of Paleozoic and Mesozoic age and is a compound syncline containing numerous well-defined folds of both large and small magnitude. The position of the basin within the regional framework is shown in Figure 1, which also shows some details of geologic structure within the basin. The area considered in detail is shown in Figure 2.

The events reviewed here fall within the time interval beginning with the last widespread incursion of marine waters associated with the Western Interior seaway during late Campanian–early Maestrichtian time and ending with the Holocene. This history includes marine regression as represented in part by the Lewis Shale–Fox Hills Sandstone and Medicine Bow Formation, the local documentation of the classic Laramide orogeny, the pre-Oligocene unconformity, the middle and upper Tertiary depositional and structural history, and finally Pleistocene events which are given limited but necessary attention.

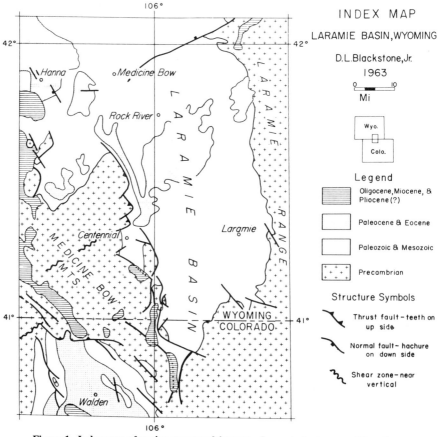

Figure 1. Index map of major structural features, Laramie basin area, Wyoming.

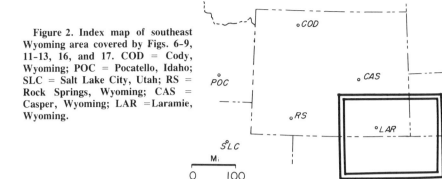

Figure 2. Index map of southeast Wyoming area covered by Figs. 6-9, 11-13, 16, and 17. COD = Cody, Wyoming; POC = Pocatello, Idaho; SLC = Salt Lake City, Utah; RS = Rock Springs, Wyoming; CAS = Casper, Wyoming; LAR = Laramie, Wyoming.

CRETACEOUS HISTORY

Campanian-Maestrichtian

Lewis Sea. The area under discussion was inundated by marine waters in Late Cretaceous time (Fig. 3) and remained under marine waters until late Maestrichtian time. In this sea, rocks were laid down that are now referred to as the Niobrara Formation, the Frontier Formation, and the Steele Shale. Regression occurred during deposition of the beds in the Mesaverde Formation followed by marine incursion through the time of deposition of the Lewis Shale and sandstone beds and intermittent marine episodes during deposition of the Medicine Bow Formation. It has become ever more apparent that a very large amount of the interior Cretaceous sediment was deposited in the relatively short time span outlined above. The Late Cretaceous and early Tertiary rocks deposited in the Laramie basin area have been fully documented by Gill and others (1970).

The level of the Lewis sea provides the last positive widespread structural datum for the Laramie basin area by which relative movement can be judged. By the close of Cretaceous time (Blackstone, 1963), the contact between the subsedimentary rocks and the Precambrian basement had been depressed to at least 6,100 m below sea level in the Hanna basin area, but to only about 150 m below sea level in the deeper part of the Laramie Basin.

Medicine Bow Formation. The regression of marine waters to the east and southeast following deposition of the Lewis Shale and sandstone beds and the Fox Hills Sandstone equivalents was followed by deposition of sediments in a generally paludal continental environment (Fig. 4). These sediments are designated the

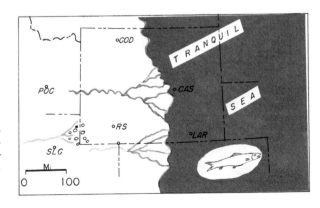

Figure 3. Extent of Cretaceous sea in late Campanian–early Maestrichtian time in Wyoming area.

Medicine Bow Formation and are considered to be late Maestrichtian in age. Fox (1970) reported foraminifera from parts of the Medicine Bow Formation, indicating that marine conditions persisted intermittently into late Maestrichtian time.

The Medicine Bow Formation along the western edge of the Hanna basin at the type locality (Bowen, 1918) is approximately 1,900 m thick, contains sandstone characterized by angular to subangular grains, and includes numerous coal beds in the lower part but contains no conglomerate.

The southernmost outcrops of the Medicine Bow Formation are located on the south flank of the Mill Creek syncline, a subsidiary fold on the west side of the Laramie basin. The outcrops are in sec. 29, T. 16 N., R. 77 W., near an old coal prospect pit known as Citizen's Coal Mine. Here about 120 m of coals and carbonaceous shales, claystones, and shales containing a brackish-water oyster fauna conformably overlie the marine Lewis sandy shales. Where exposed, the sequence is nearly vertical due to drag associated with a reverse fault and is overlain by the basal conglomerate beds of the Hanna Formation with a 90° unconformity.

Within this 120-m interval of basal Medicine Bow Formation there is a lens of conglomerate about 6 m thick that crops out along the strike for about 150 m. The conglomerate was originally reported by Knight (1953), later described in detail by Houston and others (1968), and subsequently mapped in detail by Blackstone (1970). The conglomerate contains recognizable clasts of Precambrian, Paleozoic, and Mesozoic age rocks. Knight (1953) interpreted the conglomerate as having been derived from an immediate local source and reflecting the earliest elevation of the Medicine Bow Mountains above sea level.

In my estimation the conglomerate was derived from a more distant source to the southwest, and the Medicine Bow Formation was continuous across the

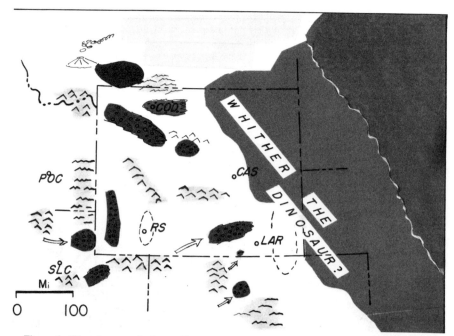

Figure 4. Wyoming area in latest Cretaceous (Lance) time. Dark shading = marine water; light-gray shading = marine and paludal areas. Deposits of conglomerate shown by open circles; volcanic debris shown by triangles.

present site of the mountains during Maestrichtian time. The Medicine Bow Formation, which crops out in discontinuous fashion from Citizen's Coal Mine to the type locality, has been drastically truncated by Paleocene erosion, so that only the basal 120 to 150 m of the formation remain in the southern part of the Laramie basin.

Ferris Deposition. In the Hanna basin, which is contiguous to the Laramie basin to the northwest, about 1,200 m of conglomerate, sandstone, siltstone, and coal are assigned to the Ferris Formation of latest Cretaceous (Lance) and Paleocene age. The unit crops out only in the present Hanna and Carbon basins. In terms of tectonic history the conglomerate is the most interesting part of the succession. The conglomerate is described by Bowen (1918):

The pebbles consist mainly of chert, and white quartz or quartzite: then follow in about the order named red and gray quartzite, porphyries, and conglomerate, with subordinate amounts of other constituents. Three facts impress the student of this conglomerate: (1) It is made up only of the most resistant kinds of rocks; (2) it is lacking in several kinds of rock now exposed in the surrounding mountains—for example, granite, limestone, sandstone, Cloverly conglomerate, and Mowry shale; (3) it contains materials that seem foreign to this region, namely, quartzite, porphyry (quartz latite), and rhyolite. These features lead to the inference that the conglomerate and also the associate sandstone were derived from some distant source rather than from the nearby mountain.

The conglomerate occurs in the lower 240 m of the Ferris Formation and was deposited in a west-trending trough essentially parallel to the present axis of the Hanna basin (Love and others, 1963; Blackstone, 1963). No detailed studies concerning the clast-size distribution and paleocurrent directions have been published for the conglomerate of the Ferris Formation. The lithology and gross pattern of distribution suggest a remote source to the southwest. Reynolds (1971) and Love (1960) pointed out that the Sweetwater arch or uplift in central Wyoming was initially elevated in part in Lance time. The lack of locally derived clasts in the conglomerate of the Ferris Formation suggests that the source will be found in the general area of the Uinta Mountains, approximately 250 km to the southwest.

Considerable discussion of the age of the Ferris Formation as defined by Bowen (1918) and as mapped by Dobbin and others (1929a) appears in the geologic literature. Knight (1951) presented an interpretation in which the Ferris and Hanna Formations are not separated and are considered as a single unit that spans the Cretaceous-Tertiary time boundary. Recent palynological investigations reported by Gill and others (1970) and by petroleum organizations indicate that the lower and conglomeratic part of the Ferris Formation is Late Cretaceous in age and that the upper coal-bearing nonconglomeratic part of the formation is early Paleocene in age, as Blackstone (1973) pointed out. The stratigraphic level of the time boundary is approximately at the Dana coal series as mapped near the Dana siding on the Union Pacific Railroad by Dobbin and others (1929a).

Houston and others (1968) used the term "Ferris-Hanna undivided" to describe rocks lying above the Medicine Bow Formation and below the Wind River Formation on the north, northeast, and east flanks of the Medicine Bow Mountains. I consider this to have been a poor choice and believe that the Ferris Formation is present only in the adjoining Hanna and Carbon basin synclines and not in the mountain-flank facies (see later discussion).

TECTONIC SIGNIFICANCE

The evidence derived from the conglomerate beds of the Ferris Formation relative to tectonic activity is that little local uplift had occurred in the Laramie basin area at this time, that uplift to the west was well advanced, and that in the depositional trough on the site of the present Hanna basin the Late Cretaceous–early Paleocene boundary lies within a conformable sequence of strata.

CENOZOIC HISTORY: TERTIARY

Paleocene

Hanna Formation. Within the Laramie basin there are rocks of late Paleocene age (Figs. 5 and 6) that correlate with the type Hanna Formation within the Hanna basin. The Hanna Formation near the type locality and within the central part of the depositional basin is approximately 2,100 m thick and so far as can be determined is conformable upon the Ferris Formation along the axis of the basin. Around the margins of the Hanna, Carbon, and Laramie basins, the Hanna Formation lies with marked unconformity upon older rocks. Knight (1951) mapped a small area on the north flank of the Hanna basin showing this relation in some detail. Here large granitic boulders derived from the Precambrian age core of the Shirley Mountains, which border the basin on the north, are common in the Hanna Formation.

In the Laramie basin area the formation contains much poorly exposed sandstone that weathers to a deep-greenish limonitic brown. The base of the formation is marked by conglomerate beds whose boulders increase in size toward the mountain flanks. The clasts in the conglomerate beds are roundstones of boulder to pebble size and are largely of local derivation. Near the Medicine Bow Mountains the roundstones are predominantly light-colored quartzite derived from the Medicine Peak Quartzite of the northern axial remnant of the mountains. The roundstones tend to be percussion marked and to have a deep-brown iron stain. The conglomerate crops out near and around Kennaday Peak on the north flank of the Medicine Bow Mountains and along the east flank of the mountains in a series of well-defined

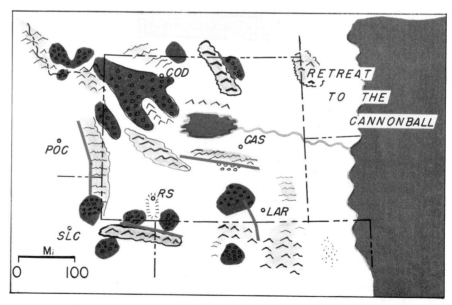

Figure 5. **Wyoming area in Paleocene time. Cannonball sea in dark shading. Deposits of coarse clastic rocks shown by open circles. Major faults as solid lines.**

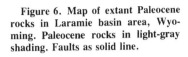

Figure 6. Map of extant Paleocene rocks in Laramie basin area, Wyoming. Paleocene rocks in light-gray shading. Faults as solid line.

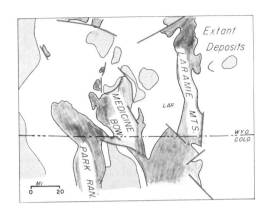

hogbacks. The more easterly outcrops of the conglomerate contain only pebble-size material.

Rocks of equivalent age that occur in the North Park basin of north-central Colorado have been called the Coalmont Formation (Beekly, 1915); they attain a thickness of 2,100 m according to Hail (1965). These rocks were apparently once continuous between the Hanna and North Park basins along the axis of the present Saratoga valley syncline, which lies on the southwest side of the Medicine Bow Mountains. The evidence reported by Montagne (1957) consists of limited exposures of yellow-green sandstones and waxy clays in the Saratoga valley near the Big Creek Ranch in T. 13 N., Rs. 81 and 82 W. No rocks of Paleocene age are known on the east flank of the Laramie Mountains, but they are present east of the Front Range in the Denver basin.

During Late Cretaceous and early Paleocene time a depositional trough existed across Wyoming at about the latitude of the Hanna basin (Love and others, 1963; Blackstone, 1963; Reynolds, 1971) (Fig. 5). The existence of Waltman Lake in central Wyoming with a probable connection with the Cannonball sea to the east during this time interval suggests that the topographic relief in southeastern Wyoming was quite low and that the Laramie Mountains may have not been extant.

Very shortly thereafter tectonic activity increased markedly, and the Sierra Madre, Medicine Bow Mountains, and the Sweetwater arch to the north were strongly elevated (Fig. 7). In the Laramie basin secondary folds along the mountain flanks were defined and in the early stages of the uplift marginal thrusting was very important in achieving elevation of the Precambrian cores of the mountain masses. Erosion was at a maximum resulting in the stripping of all Phanerozoic rocks from the crests of the ranges and also in breaching the cores of many auxiliary folds along the mountain flanks. As uplift and erosion went on, the debris fans grew headward toward their source areas and overlapped the folded and faulted margins of the uplift; this is particularly well shown in the Medicine Bow Mountains.

Examples of the relation of the Paleocene clastic rocks to the older rocks are discussed in the following sections.

Kennaday Peak Area. Kennaday Peak, 3,294 m summit elevation, located in sec. 8, T. 17 N., R. 81 W. in the upper reaches of Pass Creek drainage basin, is an isolated mountain carved in part from a quartzite-rich conglomeratic sequence of strata that rest unconformably upon rocks of all ages down to and including the Precambrian basement (Fig. 7). In this mountain, about 600 m of sandy conglomeratic beds (originally mapped by Neely, 1934) are exposed. Gries (1964) and Houston and others (1968) further contributed to the mapping of the extent of the conglomerate. The conglomerate is composed dominantly of quartzite in well-rounded clasts and is interbedded with coarse-grained sandstones.

Southeast of Kennaday Peak a long ridge of the same conglomeratic type of material extends for about 15 km up the north flank of the Medicine Bow Mountains to an elevation of approximately 3,350 m near Phantom Lake. The same material also caps an adjoining ridge between the North and South Forks of Brush Creek about 3 km northwest of Phantom Lake. The beds in these areas are identified as the Hanna Formation which overlaps southward onto the Precambrian core of the range. A second ridge, lying about 11 km east of Kennaday Peak, extends from Turpin Park (sec. 10, T. 17 N., R. 80 W.) to Pine Butte about 5.5 km to the southeast and is underlain by conglomeratic sediment of the Hanna(?) Formation.

These two areas of outcrop represent progressive overlap of the late Paleocene debris southward across the folded and faulted rocks in the Pass Creek syncline (which extends beneath Kennaday Peak) and finally onto the Precambrian core of the range.

Medicine Bow River Drainage and Adjoining Area. At lower elevations: (1) southeast of the townsite (sec. 20, T. 20 N., R. 80 W.) of Elk Mountain, Wyoming, in the Medicine Bow River valley; (2) just west of the southern Medicine Bow valley in the headwaters of Pass Creek; and (3) along Interstate 80 near Bear and Wagonhound Creeks (near the northwest corner of T. 19 N., R. 79 W.), the basal conglomerate of the Hanna Formation is well exposed and lies with strong angular unconformity (up to 90°) on rocks as young as Lewis Shale (or possibly Medicine Bow Formation) and as old as the Precambrian. The basal conglomerate along the Medicine Bow River contains a large percentage of gneissic and amphibolitic

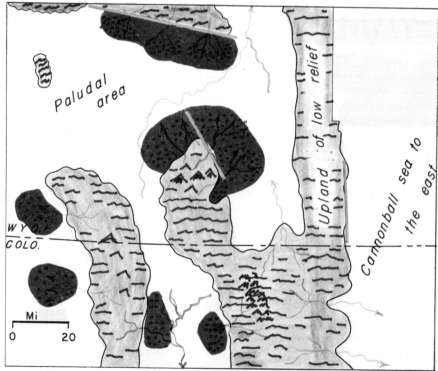

Figure 7. Paleogeography of Laramie basin area, Wyoming, near the close of Paleocene time. Coarse clastic debris shown by open circles.

cobbles and boulders reflecting the lithology of the Precambrian basement closest to the site of deposition.

Both east and west of the East Fork of the Medicine Bow River in Tps. 18 and 19 N., Rs. 79 and 80 W., there are extensive exposures of a nearly flat-lying sequence of conglomeratic sediments. The basal conglomerate is particularly well developed and contains abundant well-rounded cobbles of amphibolite. Good exposures occur at the head of Sedrun Slide (secs. 9, 10, and 14, T. 18 N., R. 80 W.). The conglomeratic sequence was mapped by Houston and others (1968) as Ferris-Hanna undivided and is shown as extending up the north flank of the Medicine Bow Mountains to about the south line of T. 18 N.

Palynological material collected from these rocks in sec. 10, T. 18 N., R. 80 W. (Hyden and others, 1968; USGS Paleobotany Locality D3608) and identified by R. H. Tschudy, was considered to be Eocene(?). However, on the basis of the stratigraphic relations and the lithology of the sediments, I consider these rocks to be the basal Hanna Formation.

Wick Ranch. East of Bear Creek and along Wagonhound Creek near the Old Wick Ranch (sec. 8, T. 19 N., R. 79 W.), there are excellent exposures of the basal conglomerate of the Hanna Formation. The cobbles are predominantly quartzite, probably derived from the Medicine Peak Quartzite. The Hanna Formation overlies steeply dipping to overturned Steele Shale, Mesaverde Formation, and Lewis Shale on the northeast flank of the Wagonhound anticline at about 90° unconformity. Near the Wagonhound exit of U.S. Interstate 80 reverse faulting places Frontier Sandstone upon overturned Steele Shale, and in turn the basal conglomerate of the Hanna Formation overlaps the fault trace and thus dates the major faulting and folding as Paleocene.

Arlington Fault. In the eastern parts of Tps. 16 and 17 N., R. 78 W., east of the Arlington thrust fault (Darton and Siebenthal, 1909; Blackstone, 1970), the Hanna Formation crops out in well-defined hogbacks, and it is present in a narrow syncline associated with the south-trending Rock Creek–Seven Mile line of folding, which lies a few kilometers farther east. In these areas the Hanna rests with angular unconformity upon rocks at least as old as the Niobrara Formation and as young as basal Medicine Bow Formation. In addition the basal conglomerate of the Hanna Formation can be seen to overlap the trace of the Arlington fault near Corner Mountain (Blackstone, 1970).

Sheep Mountain. Reverse faults that bound the east flank of Sheep Mountain (T. 15 N., R. 77 W.) extend northward to the Citizen's Coal Mine area (sec. 29, T. 15 N., R. 77 W.). There vertical to overturned beds in the footwall east of these thrust faults are overlain at about 90° unconformity by the basal conglomerate of the Hanna Formation (Blackstone, 1970), which indicates that the subsidiary structural features of the Medicine Bow Mountains are also Paleocene in age.

VARIATION IN THICKNESS OF THE HANNA FORMATION

The maximum thickness of the Hanna Formation in the Laramie basin area is about 370 m, based on exploratory drilling for oil and gas and for uranium ores; and in many places the formation is less than 150 m thick. This is much less than the 1,200 m predicted by Knight (1953), and very much less than the 2,100 m reported in the nearby Hanna basin (Dobbin and others, 1929a).

The great variation in thickness has puzzled workers in the area and led to misinterpretations of both stratigraphy and structure. The answer to the puzzle lies in the relative ages of the Paleocene rocks exposed in the two areas. The central part of the Hanna basin reflects continuous deposition from Late Cretaceous through Paleocene time. On the other hand, the Paleocene rocks exposed in the

Laramie basin in the Mill Creek syncline (northwestern T. 16 N., R. 77 W.) and at Cooper Cove anticline (western T. 18 N., R. 77 W.) have been dated by Tschudy (Houston and others, 1968; Blackstone, 1973) as very late Paleocene. The age determinations are based on palynological material collected from carbonaceous shales lying less than 15 m above the basal conglomerate of the Hanna Formation which in turn rests with angular unconformity upon lower Medicine Bow Formation in one place and upon sandy upper Lewis Shale in another. Thus, the variation in thickness is the result of overlapping relations around the basin margins as mountains were elevated. The erosional unconformity varies in degree from place to place and in many places is not well dated.

TECTONIC SIGNIFICANCE OF PALEOCENE RELATIONS

It appears that the major northwest-trending faulting and folding in the Laramie basin occurred in early to middle Paleocene time concomitant with deposition in the Hanna, Carbon, Laramie, and North Park basins (Fig. 6). Evidence is less clear as to the behavior of the site of the present Laramie Mountains. The absence of Paleocene sediments along the east flank of the Laramie Mountains and 50 to 100 km farther east in Goshen Hole (though rocks of Lance age exist in both places) allows for two hypotheses.

1. The Laramie Mountains or at least their central part was not elevated in Paleocene time, and an irregular stratigraphic section was deposited eastward to the Cannonball sea.

2. The Laramie Mountains were barely elevated and provided a barrier to deposition to the east.

Hail (1965) stated that the northern Front Range was elevated and stripped of cover by the time the Coalmont Formation was deposited in North Park:

The oldest outcropping beds of the Coalmont Formation contain detritus from Precambrian rocks. The evidence indicates that the Mesozoic rocks had been eroded and the Precambrian core of the Park Range was exposed when the basal gravels of the Coalmont were deposited.

Eocene

Wind River Formation. Nace (1936) reported that Knight originally recognized that the variegated beds in the Cooper Creek drainage of the central Laramie basin are of Eocene age and equivalent to the Wind River Formation of central Wyoming. The variegated claystones crop out over several townships near Cooper Lake on the floor of the basin and in isolated areas along the mountain front (Fig. 8). In the Shirley basin about 65 km to the north there are extensive exposures of lower and middle Eocene rocks (Harshman, 1968).

Prichinello (1971) described all the material in the University of Wyoming collections that had come from exposures of the sequence along Cooper Creek in T. 18 N., R. 77 W. The study of this fauna indicated that the rocks are of Greybullian (lowest Wasatchian) provincial age.

The Wind River Formation consists of arkosic sandstone that is largely whitish, brown cross-bedded sandstone rich in dark minerals, and much red, purple, and green variegated mudstone. The vertebrate fauna occurs in the latter. The sandstone units that contain "cannonball" concretions about 10 to 15 cm in diameter might prove to be useful marker beds. The conglomerate on both flanks of the basin contains clasts of pebble size with occasional cobble-sized fragments.

Considerable confusion has existed concerning the distribution of Eocene rocks in the Laramie basin and distinguishing between Wind River and Hanna Formations in some localities. The most characteristic aspect of the Wind River Formation is a purple to maroon color; no other rocks of Tertiary age in the basin exhibit such hues. The basal conglomerate of the Wind River Formation is less distinctive and because of poor exposure is difficult to differentiate from the conglomerate of the Hanna Formation.

Several areas are discussed here in order to point out the physical relation between the Eocene and Paleocene rock units.

Hanna Basin. In the central part of the Hanna syncline, T. 22 N., R. 81 W., where there is active strip mining of coal, the strata lying above the Hanna No. 1 coal seam are considered to be Eocene on the basis of palynology. The relations indicate continuous deposition and a conformable sequence (Blackstone, 1973). Knight (1951) mapped in detail an area in T. 24 N., Rs. 80 and 81 W. along the Medicine Bow River, where the contact relation of Eocene(?) rocks is shown. The Eocene strata are conformable with the Hanna Formation near the basin axis, but overlap unconformably northward toward the sediment source.

Wagonhound Creek to Arlington. Hyden and McAndrews (1967) on the T L Ranch 7 1/2' quadrangle (Tps. 19 and 20 N., Rs. 79 and 80 W.) mapped extensive areas of Quaternary colluvium (Qc) which are actually bedrock exposures of the Hanna and other formations. The mapping was further complicated by the introduction of two unnecessary new formation names—Dutton Creek and Foote Creek (Hyden and others, 1965). Both names have subsequently been abandoned (Gill and others, 1970; Blackstone, 1973). Revised mapping indicates that in the southern half of the quadrangle, the dip of the conglomerate beds at the base of the Hanna Formation increases to about 40° NE. As the dip increases the conglomerate beds crop out along the crests of prominent hogbacks, paralleled on the south by a strike valley in which U.S. Interstate 80 is routed. Within this valley are found several areas of variegated claystone of the Wind River Formation which lie with angular unconformity upon the Hanna Formation and older rocks. These areas of the Wind River Formation were included in the mapped Quaternary colluvium by Hyden and McAndrews (1967).

Arlington Area. Rather extensive excavations have been made in the vicinity of the Arlington exit from U.S. Interstate 80, which is at the northeast edge of the Medicine Bow Mountains near Rock Creek. In the large cut extending northwestward for about 1.5 km from the exit, gently dipping variegated claystone beds of the Wind River Formation are exposed on both sides of the highway. They rest with angular unconformity on rocks as old as Precambrian and also overlap the trace of the Arlington fault. East of the highway the Wind River Formation overlaps the steeply dipping conglomerate of the Hanna Formation. Mapping by

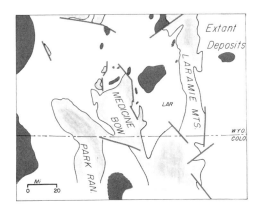

Figure 8. Map of extant Eocene rocks in Laramie basin area, Wyoming. Eocene rocks in dark gray. Faults as solid lines.

Hyden and others (1967) preceded the construction of U.S. Interstate 80 and does not show the critical details.

Along the highway east and southeast of Arlington interchange, essentially flat-lying variegated beds of the Wind River Formation lie on beds of brown sandstone and carbonaceous shale of the Hanna Formation which dip gently eastward. Hyden (1966a) indicated the existence of Wind River Formation beneath the Arlington pediment to a point about 6.5 km northeast of the interchange. Palynological material collected from these rocks in sec. 5, T. 19 N., R. 78 W. (USGS Paleobotany Locality D3020-B) and identified by Tschudy is considered to be Eocene. Unfortunately Hyden did not recognize the extent of the Wind River Formation in this area or its overlapping unconformable relation to the older folded and faulted rocks.

Bengough Hill. The crest of the Cooper Cove anticline in western T. 18 N., R. 77 W. is breached by erosion to create a topographic basin bounded on both east and west flanks by well-defined hogbacks reflecting the basal conglomerate of the Hanna Formation. Maps prepared by Hyden (1966b) show these rocks as Quaternary colluvium. On the east flank of the anticline, along the west face of Bengough Hill, east of U.S. Interstate 80 in sec. 28, T. 18 N., R. 77 W., the basal conglomerate of the Hanna Formation rests unconformably upon the upper Lewis Shale. I collected palynological material from a carbonaceous shale unit about 12 m above the basal conglomerate. R. Tschudy (1969, written commun.) identified the material (Table 1) and assessed the results.

The presence of *Carya, Pistillipollenites* (P3-pl), *Momipites tenuipolis* (P3-smlB), and BCP3-r11B indicates that these samples are from the upper part of the Hanna Formation of late Paleocene age. I could find no clear evidence of an Eocene age, although I examined the samples carefully with this possibility in mind. I am convinced that these samples are from the uppermost part of the Paleocene. The assemblages resemble most closely assemblages from the Tongue River Member of the Fort Union Formation.

TABLE 1. POLLEN CONTENT OF TWO SAMPLES FROM FRESH ROAD CUTS ON I-80 NORTHWEST OF LARAMIE

Code species	Lower sample*	Upper sample*
Carva 2 species	x	x
C3-p7	x	x
BCP4-r11B	x	x
C3-rt6	x	x
P4-sm4	x	x
P3-sm1B	x	x
Tax-sm1	x	x
P3-sm16D	x	x
P3-sm 111	x	x
P3-p1	x	
C3-rt1B	x	x
Tax-r7 var	x	x
P4-foss 1		x
Sequoia-1		x
P3-sm21		x
P3-foss 1		x
Pa4-sm1		x

*The samples were productive and were given USGS paleobotany numbers. Lower sample: DB-A D4286-A; upper sample: DB-B D4286-B.

A second and parallel hogback about 0.4 km to the east is supported by the basal conglomerate of the Wind River Formation. The conglomerate is similar to those of the Hanna Formation but contains less well-rounded cobbles and pebbles, and the clasts are much smaller. Traced northward from Bengough Hill, the basal conglomerate of the Wind River Formation transgresses westward across the Hanna Formation with a low angularity and rests upon the Lewis Shale (Blackstone, 1973). A similar overlap of the Hanna Formation by the Wind River Formation lies near James Lake, which is in the northeast part of T. 17 N., R. 76 W. in the central Laramie basin.

Mill Creek Syncline. The most southerly known exposures of the Wind River Formation in the Laramie basin are in sec. 20, T. 16 N., R. 77 W. At this locality variegated claystone is preserved in the downdropped side of a normal fault that extends northwestward into the northeast-trending Mill Creek syncline (Blackstone, 1970).

Sybille Springs Area. On the west flank of the Laramie Mountains in sec. 1, T. 20 N., R. 73 W., about 1.5 km north of the Sybille Springs anticline, a small area of variegated claystone and yellow sandstone beds was found in the summer of 1972. The claystone lies in a shallow depression and rests upon Casper Formation that dips westward at a low angle (Tudor, 1953). The claystone and yellow sandstone beds contain boulders and cobbles of granite and occasional clasts of sandstone. The granitic boulders are so deeply weathered that they may be broken with one's fingers. I assigned these rocks to the Wind River Formation on the basis of color and stratigraphic position.

Lookout Siding Area. East of Lookout siding (sec. 34, T. 20 N., R. 75 W.) on the Union Pacific Railroad, the basal sandstone of the Wind River Formation is conglomeratic and is overlain by variegated claystone that crops out to the west of the railroad around the rim of the Cooper Lake depression. The U.S. Geological Survey drilled a core hole in sec. 26, T. 20 N., R. 75 W. in the course of coal evaluation studies (McAndrews, 1965). An assemblage of palynological material (USGS Paleobotany Locality D3199), recovered from the lower part of the hole, was dated by Robert Tschudy as Paleocene in age. On the basis of this identification, the entire drilled section was assigned to the Hanna Formation, whereas in fact the upper part of the hole penetrated the Wind River Formation. The contact between the Wind River Formation and the underlying Hanna Formation is unconformable, and the Wind River overlaps the Hanna Formation to the southeast.

About 5 km southeast of Lookout siding, the conglomeratic basal beds of the Wind River Formation are exposed in low bluffs on the west bank of the Laramie River. The conglomerate contains rather abundant clasts of pebble to cobble size of anorthosite whose source must be the large mass of anorthosite exposed in the Laramie Mountains south of Sybille Canyon (Newhouse and Hagner, 1957).

TECTONIC SIGNIFICANCE

The conclusions to be drawn from the areal distribution and contact relations of the Eocene Wind River Formation are listed below (Figs. 9, 10).

1. Major folding and faulting in the Laramie basin area is of pre-Greybull (pre-Eocene) age.

2. Sediment was derived from the Medicine Bow, Shirley, and Laramie Mountains, and the cores of the ranges had been stripped of sedimentary cover by this time.

3. The large volume of variegated claystone indicates a deeply weathered and mature landscape of low relief in the uplands.

4. The Hanna basin continued to subside, giving a continuous record of deposition from Late Cretaceous to at least middle Eocene time.

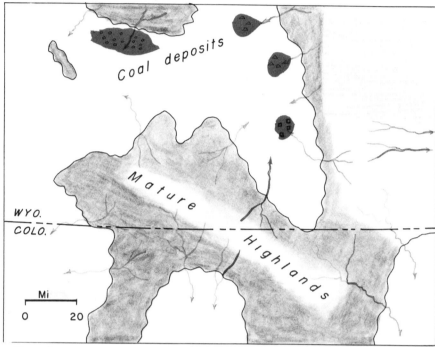

Figure 9. Paleogeography of Laramie basin area in early Eocene (Wasatch) time. Coarse granitic debris shown by circles; arkose by triangles; anorthosite pebbles by squares.

Figure 10. Diagrammatic cross section across the east flank of Medicine Bow Mountains, Wyoming, showing unconformable relations of the Paleocene Hanna and Eocene Wind River Formations.

5. Uplift of the mountains continued in early Eocene with minor folding on the basin margins.

6. The Cooper Lake basin area subsided, and the Eocene rocks were warped into a broad shallow syncline.

7. Northeast-trending features appear to have become structurally active in Eocene time.

Oligocene

The Oligocene rocks of the Laramie basin (Fig. 11), which have been assigned to the White River Formation undivided (McGrew, 1953), consist of fine-grained siltstone and bentonitic claystone with limited amounts of conglomerate and sandstone. The rocks have been dated as Oligocene on the basis of collections of vertebrate fossils made near Albany, Wyoming, in sec. 12, T. 14 N., R. 77 W.; near Kings Canyon in the northern part of North Park basin (Steven, 1956); and near Wheatland Reservoir in the northern part of the Laramie basin (Brooks, 1957). Extensive outcrops of Oligocene rocks occur in Bates Hole to the north (Harshman, 1968); along the east flank of the Laramie Mountains (Denson and Bergendahl, 1961; Toots, 1965); along the northeast-trending Wheatland fault zone

(McGrew, 1963); and in scattered areas across the northern Laramie Mountains along the old Fort Fetterman road which approximately follows the center line of R. 74 W.

Undated rocks of similar lithology and stratigraphic position occur widely throughout the upper Big Laramie River valley as far south as Glendevey, Colorado (sec. 29, T. 10 N., R. 76 W.) (Beckwith, 1942). Similar rocks have been observed about 24 km south of there at Chambers Lake near the junction of the Medicine Bow Mountains and the Front Range. Corbett (1966) reported volcanic activity about 16 km farther south in the Mount Richthofen-Iron Mountain area of the Never Summer Range, which is dated on radiometric evidence as late Oligocene.

The age of the Oligocene White River rocks in the Laramie basin area, on the basis of their vertebrate fauna, is early Oligocene (Chadronian) according to McGrew (1953), though middle Oligocene (Orellan) age rocks are reported by Harshman (1972) in the Shirley basin area, and by Denson and Bergendahl (1961) in the area east of the Laramie Mountains.

No sedimentary rocks of Oligocene age have been reported in the Saratoga valley west of the Medicine Bow Mountains by Ashley (1948), Ebens (1966), Montagne (1957), or Vine and Prichard (1959). In view of the rather extensive mapping that has been done there, it is doubtful if any are present, though they may have been deposited and removed.

Within the Laramie basin there is at present no known instance of White River Formation resting upon strata of Eocene age. It is possible, since mapping is incomplete, that such a situation may be found in the vicinity of Wheatland Reservoir at the north end of the basin, but localities at which Eocene and Oligocene vertebrate fossils have been found are in close proximity geographically.

The White River Formation lies with angular unconformity upon rocks as young as Cretaceous and as old as Precambrian in the vicinity of Albany, Wyoming, at the south end of the Centennial valley, the structural indentation on the east flank of the Medicine Bow Mountains. The White River is essentially flat lying; and the uppermost strata reach an elevation of about 2,800 m in the railroad cuts above Albany. Similar situations exist along the valley of the Big Laramie River upstream to Glendevey, Colorado. Likewise in the northern part of the Laramie basin the White River Formation overlaps onto the Precambrian rocks of the Laramie Mountains (Tudor, 1953; Brooks, 1957).

An extensive zone of normal faults named the Wheatland and Whelan fault systems trends about N. 45° E. across southeastern Wyoming and passes through the towns of the same names. The Wheatland zone has been mapped as far to the southwest as the east flank of the Laramie Mountains where the fault zone

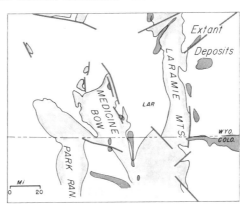

Figure 11. Map of extant Oligocene rocks in Laramie basin area, Wyoming. Oligocene rocks in dark gray. Faults as solid lines.

roughly parallels Wyoming Highway 34, which crosses the mountains along Sybille Creek (McGrew, 1963). The fault zone has been observed and mapped farther to the southwest in the James Lake quadrangle in the Laramie basin (Blackstone, 1973) and it appears to continue southwestward to the southern end of the Quealy anticline (eastern part of T. 17 N., R. 77 W.). At Quealy the fault is normal in character, down to the south, and has about 430 m displacement. The entire fault zone aligns with the Nash Fork–Mullen Creek shear zone described by Houston and others (1968).

The fault movement involves the Eocene Wind River Formation in the James Lake area and both Oligocene and Miocene age rocks in the Wheatland area, indicating a post-middle Miocene age for the faulting (McGrew, 1963). The movement probably reflects reactivation of northeast-trending fracture zones in the Precambrian basement.

It is striking that the Oligocene sedimentary rocks in southeastern Wyoming are essentially flat lying, composed of bentonitic siltstone, and rest with marked angular unconformity upon a surface of strong topographic relief. The source of the fine-grained White River Formation was volcanic (Denson and Chisholm, 1971), but the precise location of the vents has not been established. Along the east and northeast flanks of the Laramie Mountains large amounts of arkose and conglomerate of local derivation are found in the White River Formation (Moore, 1959; McGrew, 1963; Stanley, 1971).

By the beginning of Oligocene time the Laramie basin had been elevated and eroded into a mature landscape which was at least reminiscent of the present topography. During Oligocene time the earlier landscape (Fig. 12) was literally drowned in fine-grained volcanically derived sediments which probably originated outside of Wyoming. The deposition may have been in part accomplished by free fall of volcanic debris from the atmosphere, but in part must have been debris reworked by aggrading streams. The lower areas of the topography were filled first and as deposition continued only the highlands stood above the plain of aggradation. The deposition of locally derived arkoses and conglomerates containing clasts up to cobble size along the east flank of the Laramie Mountains may reflect only climatic variation with little or no tectonic activity. (The problem of climatic versus tectonic influence is discussed further under the Pleistocene events.)

At present, rocks of Oligocene age rest upon rocks of Precambrian age within the mountain uplifts at relatively high elevations. In the long north-trending open valley across the northern end of the Laramie Mountains, approximately along the center line of R. 74 W. (known locally as the old Fort Fetterman road), the

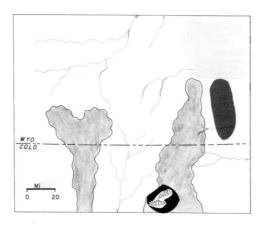

Figure 12. Paleogeography of Laramie basin area, Wyoming, in late Oligocene time. Deposits of coarse clastic debris east of the Laramie Mountains shown in dark gray.

present drainage divide rests upon Oligocene rocks at an elevation of approximately 2,400 m, giving a minimum value for the height of fill during Oligocene time. Steven (1956) reported that Oligocene rocks near the Kings Canyon fluorite deposits in North Park lie at an elevation as high as 2,840 m and reflect a southward drainage pattern in Oligocene time.

The possibility of general epeirogenic uplift in later Tertiary time cannot be ignored; thus, the present elevations do not reflect the sea-level datum elevation of filling during Oligocene time, but actually the degree of filling relative to the exposed cores of the mountains at that time, in the absence of later differential movement.

The unconformity between middle and late Eocene rocks and the early Oligocene rocks as reported by Harshman (1972) in the Shirley basin–Bates Hole area represents about 6 to 8 m.y., rather a long time span within the Cenozoic. Many investigators consider this time interval and the accompanying unconformity as the logical termination of the Laramide orogeny (Love, 1960; Tweto and Sims, 1963). A. J. Eardley informally proposed that the subsequent geological history be described as the Absarokan disturbance.

TECTONIC SIGNIFICANCE

The relation of the Oligocene White River Formation to the older rocks, and the nature of the sediments, lead to the following conclusions.

1. No major orogenic folding or faulting occurred during Oligocene time.

2. Aggradation of the area by material from a remote source commenced and continued into late Cenozoic time.

3. Some further reduction of the relief occurred, resulting in locally derived sediments in favorable areas.

4. A certain degree of crustal stability was attained; this allowed sedimentation to obscure the earlier landscape.

5. Differential elevations of the base of the Oligocene rocks suggest a gradient of deposition to the east but may also reflect post-Oligocene tilting.

Miocene

The existence of Miocene rocks in the Laramie basin (Fig. 13) has not been documented on the basis of faunal evidence; deposits that have been assigned to the Miocene North Park Formation by McCallum (1968) must be interpreted with this fact in mind.

Strata definitely of Miocene age are known in the Miller Hill area north of the Park-Sierra Madre Range (Vine and Prichard, 1959), in the Saratoga valley (Montagne, 1957), in North Park (Hail, 1965), in the Shirley basin (Harshman, 1972), and on the High Plains east of the Laramie Mountains (Sato and Denson, 1967; Moore, 1959; Minick, 1951).

In the vicinity of Walcott Junction, Wyoming, along the south flank of the Hanna basin, there are extensive outcrops of light-colored tuffaceous sedimentary rocks that have been correlated with the North Park Formation of the Saratoga area 50 km farther south (Dobbin and others, 1929a; Chadeayne, 1966; McGrew, 1953). No vertebrate fauna has been found in these strata, but on the basis of lithology and stratigraphic position the correlation appears reasonable.

McCallum (1968) summarized data for the east-central Medicine Bow Mountains and presented a map showing the distribution of rocks he considered to be Miocene North Park(?) equivalents. Unfortunately he did not include in his study the exposures in Dry Park at the head of the Albany slide area in the eastern Medicine Bow Mountains. In the headwall scar of this large area of mass gravity movement

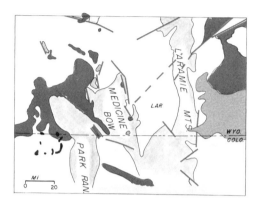

Figure 13. Map of extant Miocene and Pliocene rocks in Laramie basin area, Wyoming. Miocene rocks in light gray; Pliocene rocks in dark gray. Faults as solid lines.

in secs. 21 and 22, T. 14 N., R. 78 W., a thick conglomerate sequence containing abundant cobbles of mafic Precambrian rocks is well exposed. The conglomerate overlies about 240 m of fine-grained, pale salmon-pink siltstones near the base of which vertebrate fossils of Chadron (lower White River) age have been collected. The conclusion that this conglomerate probably represents the base of the Miocene succession in this area is based on the stratigraphic position and on similar successions to the west of the Medicine Bow Mountains. In several areas around the Sierra Madre and the western Medicine Bow Mountains, a persistent conglomerate occurs (at the base of the sequence mapped as North Park Formation) that marks change in the tectonic activity during Miocene time. In these areas the basal conglomerate marks an unconformity representing extensive erosion in the highlands and the possibility that Oligocene rocks were removed from some areas. Denson and Bergendahl (1961) provided an excellent diagram of the relation found east of the Laramie Range.

The Miocene strata assigned to the Arikaree Formation by Sato and Denson (1967) and Denson and Bergendahl (1961) contain large amounts of fine-grained gray sandstone and an almost universally present accessory of minute grains of bluish-gray magnetite. The source of the widespread sandstone of the Arikaree Formation has not been clearly established, but Chisholm (1963) indicated that the sands on the west side of the Sierra Madre were derived from a southern source.

The limited outcrops in the central-eastern Medicine Bow Mountains, assigned by McCallum (1968) to the Miocene North Park Formation(?), are a dozen meters wide and consist of very white, fine-grained, calcareous sediment containing angular fragments of the underlying or immediately adjacent rocks. The matrix fills fractures in the underlying material to a depth of about one meter. There is no basal conglomeratic layer, probably because the material lies near the ancient drainage divide of the mountains.

Rock Creek Valley. The upper valley of Rock Creek in the northern Medicine Bow Mountains is very straight and appears to be controlled by a late- or post-Tertiary normal fault, down to the west. In this valley there is light-colored sediment assigned to the Tertiary by Knight (1953), who stated:

A conspicuous remnant of ash clay, some 300 feet thick, partially fills the valley of Rock Creek at elevations from 9300 to 9600 feet. Some coarse clastic material from the valley sides is interbedded with the clay at this locality. It is not likely that the volcanic ashes were extensively transported by wind and water during the time of deposition . . . .

Houston and others (1968) considered these strata to be infaulted Miocene North

Park(?) Formation, but no fossils have been found to positively date the material. From Knight's description it is apparent that he considered the strata to be Oligocene in age and similar to White River at the southern end of Centennial valley, about 30 km to the south.

Mount Richtofen–Iron Mountain Area. Corbett (1966) reported Tertiary volcanic activity in the Mount Richtofen–Iron Mountain area on the northwest boundary of Rocky Mountain National Park. The area lies near the heads of the Cache la Poudre, Big Laramie, Michigan, and Colorado Rivers on the north flank of the Never Summer Range (Fig. 14). The volcanic activity here has an age of 27 to 28 m.y., which is about the Oligocene-Miocene boundary as defined by Evernden and others (1964) and roughly synchronous with volcanic eruptions in the Rabbit Ears Range and the Hot Sulphur Springs quadrangle in north-central Colorado as reported by Izett (1968).

The extrusions of andesite, trachyte, rhyolite, and latite in the Mount Richtofen–Iron Mountain area provided a source for volcanic rock clasts that are found in the Platte River drainage on both east and west sides of the Medicine Bow, Front Range, and Laramie Mountains. Montagne (1957) reported clasts of volcanic rock types in the North Park Formation of the Saratoga valley, at Pick Ranch north of Saratoga, Wyoming.

TECTONIC SIGNIFICANCE

The widespread occurrence of coarse conglomerate beds at the base of the Arikaree-North Park Formations, which lie in part on rocks of Oligocene age and then overlap onto older rocks down to and including the Precambrian core of the ranges, indicates that only the higher parts of the uplifts were exposed

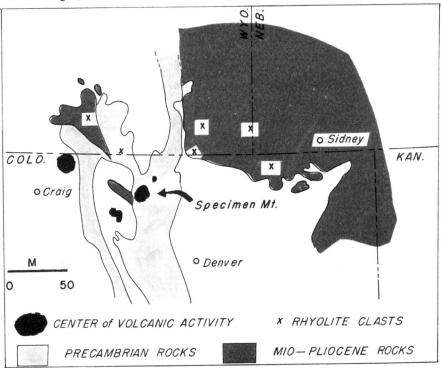

Figure 14. Specimen Mountain, Colorado, is the source for clasts of rhyolitic rock types found in the Ogallala Formation at points marked by X.

above the plain of aggradation that had been developing throughout Oligocene and early Miocene time. During the Miocene time interval erosion continued to reduce the elevation of the highlands along the crests of the mountain ranges, and in some areas it may have effected removal of a considerable volume of Oligocene age rocks.

The surface of aggradation sloped gently eastward and probably crossed the Laramie Mountains in the sag between the Sherman Mountain remanent mass and the Laramie Peak highland to the north, approximately in the vicinity of the present Sybille Canyon drainage. At the west side of the Laramie basin the streams were cutting headward and beginning the development of the high-level erosion surface on the Medicine Bow Mountains.

Pliocene

Any discussion of Pliocene tectonic history must take note of the fact that the absolute age to be assigned to the Miocene-Pliocene boundary is in a state of flux. The relation between the North American and European provincial time scales is being re-evaluated and the trend seems to be toward shortening the time assigned to the Pliocene in North America. If the European viewpoint is accepted, several rock units in Wyoming that are now considered Pliocene in age will be reassigned to the Miocene.

No rocks of Pliocene age are extant in the Laramie basin area, which is to be expected since they would have been deposited last and eroded first during the late Pliocene-early Pleistocene period of denudation of the entire region. However, east of the Laramie Mountains near Cheyenne, the Ogallala Formation of late Miocene (Barstovian) and early Pliocene (Clarendonian) age is widespread on the High Plains as shown by Denson and Chisholm (1971), Lowry and Crist (1967), Moore (1959), and Minick (1951). Fossil assemblages of Pliocene age have been collected at Trail Creek, north of Cheyenne in sec. 29, T. 17 N., R. 66 W. (P. O. McGrew, 1974, oral commun.), and Minick (1951) reported fossil seeds, establishing the Pliocene age of strata at Chalk Buttes in T. 11 N., R. 65 W. in Weld County, Colorado.

McCallum (1968) assigned certain scattered deposits on the Medicine Bow upland surface at about 2,900 m elevation to early Pliocene(?), middle to late Pliocene(?), and late Pliocene(?) on the basis of local lithologic characteristics and relative chronology, but without faunal evidence for dating the rocks. Steven (1956) assigned the rocks in the Northgate fluorspar district of North Park basin to the North Park Formation and considered the age to be late Miocene or early Pliocene(?).

In the Saratoga valley southwest of the Medicine Bow Range, the youngest faunally dated rocks are of Barstovian (late Miocene) age (Montagne, 1957; P. O. McGrew, 1974, oral commun.). The Saratoga valley sequence can be divided into two units: a lower one that has been correlated with the Browns Park Formation and an upper one that has been correlated with the North Park Formation. The faunal evidence indicates a Barstovian age for the Browns Park Formation. The upper unit in this sequence dips eastward on the eastern side of Saratoga valley, and these higher beds may well be Pliocene.

The Moonstone Formation, preserved in the shallow central graben of the Granite Mountains area in central Wyoming, is considered on the basis of rather limited faunal evidence to be Pliocene in age (Love, 1961).

On the basis of the known localities in which deposition appears to have been continuous from late Miocene (Barstovian) to early Pliocene (Clarendonian), as exemplified by the Ogallala Formation and other scattered occurrences, certain reconstructions of paleogeography have been made. Love and others (1963) indicated that Wyoming was essentially covered by Pliocene deposits by the close of that epoch. The reconstruction is based on a minimal amount of outcrop data, but it is plausible in view of the widespread aggradation that had occurred earlier in the Cenozoic and the extensive distribution on the High Plains of Pliocene rocks, which must have been derived from the west (Fig. 16).

One source of data for the reconstruction of the Pliocene landscape is available in southeastern Wyoming and north-central Colorado. As noted in the discussion of Miocene rocks, Wahlstrom (1944, 1947), Gorton (1953), Ward (1957), and Corbett (1966) reported on the intrusive and extrusive igneous activity centering around Specimen Mountain, Mount Richthofen, Nokku Craigs, and Iron Mountain in the northwest area of Rocky Mountain National Park. The 27- to 28-m.y.-old Specimen Mountain volcanic rocks are a probable source for the volcanic clasts in the late Tertiary sediments. A reasonable interpretation of source and distribution of the sediments may be based on this relation.

The Ogallala Formation of Miocene-Pliocene age crops out over a wide area in Laramie County, Wyoming, and to the south in Larimer and Weld Counties, Colorado, as well as extending eastward into Kimball, Banner, Scotts Bluff, and Sioux Counties, Nebraska. The formation was characterized by Denson and Chisholm (1971).

Notable lithologic dissimilarity from area to area. Extreme heterogeneity of composition and sorting both laterally and vertically. Unit composed of poorly cemented calcareous claystone, siltstone, sandstone, and conglomerate of fluviatile origin. Thin lenticular beds of algallike limestone suggest local areas of lacustrine deposition. Blue-gray rhyolitic ash beds as much as 30 feet thick in intermontane basin areas. Stringers, nodules, and concretionary masses of black to milk-gray chert near base. Lenticular hard dense lithographic limestone as much as 50 feet thick locally at contact with underlying lower Miocene. Rocks are porous and permeable. Thicknesses range from 0 to 2,500 feet.

One distinctive aspect of the Ogallala Formation in southeastern Wyoming is the presence of extrusive felsic volcanic rock fragments best described as rhyolite. Minick (1951) reported the existence of cobbles of rhyolite at Chalk Buttes in T. 11 N., R. 65 W., Weld County, Colorado; at Pawnee Buttes in T. 14 N., R. 60 W., Weld County, Colorado; and as far east as Pine Bluffs on the Wyoming-Nebraska border. Moore (1959) reported cobbles and boulders of rhyolitic extrusive rocks in the Ogallala Formation at the Gangplank west of Cheyenne, Wyoming, and P. O. McGrew (1974, oral commun.) noted the existence of clasts of volcanic rock types at Trail Creek Quarry in T. 17 N., R. 16 W. about 32 km north of Cheyenne, Wyoming. I originally noted the presence of rhyolite cobbles near Mountain Home, Wyoming, T. 12 N., R. 78 W., at an elevation of 2,600 m on the Medicine Bow upland surface.

Figure 14 shows the Specimen Mountain source area and the localities noted above. The rhyolitic clasts are now found at least 200 km from the potential source area and the distribution pattern is less than simple. It is possible in view of the pattern of distribution of ash flows and (or) ignimbrites in other parts of the west that the original area covered by the ejecta from the Specimen Mountain vents was much larger than at present. In this case the boulders found in the Ogallala Formation might have been transported shorter distances than the map suggests. The reported age of the volcanic activity in the Specimen Mountain area is late Oligocene–early Miocene, allowing for transport of clasts from this area in all subsequent time up to and including Holocene. The cobbles in the High Plains area occur within what is currently considered to be the Pliocene part of the Ogallala Formation. The net result, however, suggests that the transport of the cobbles was across a surface of aggradation, probably of low relief and moderate slope, which extended from the higher parts of the Front Range and the Medicine Bow Mountains eastward to the High Plains and which crossed the Laramie Mountains near the Gangplank (Fig. 15).

This relation of high-level erosion surfaces in the central Rocky Mountain region has been discussed by many geologists including Blackwelder (1909), Lee (1923), Atwood and Atwood (1938), Knight (1953), Van Tuyl and Lovering (1934), Wahlstrom (1947), Moore (1959), and Eggler and others (1969). Some of the well-known surfaces located within this area are (1) the Sherman surface, (2) Green Mountain surface, and (3) the Medicine Bow upland. The development of the surfaces may have been almost synchronous and appears to best fit events in late Cenozoic time.

The Sherman surface was named by Blackwelder (1909), but the extent of the surface was poorly defined. Eggler and others (1969) discussed the origin of the surface and in effect limited it to the area of exposure of the Precambrian Sherman granite. I consider the surface to be much more extensive and roughly coincident with the outcrop area of both the Sherman granite and the Laramie Mountains anorthosite. The low-relief area along the Laramie Mountains, lying generally between the 7,000- and 8,000-foot contours as shown on the Army Map Service Cheyenne Code No. Nk 13-8, 1:250,000 sheet, is not a smooth surface, but is coextensive with the surface typically developed in the area about the old Sherman station on the Union Pacific Railroad.

Moore (1959) clearly documented the development of the Sherman surface (restricted sense) relative to the deposition of the Ogallala Formation. If the Ogallala is at least partly of Pliocene age, then the Sherman surface was developed during that time interval. I believe that the term Sherman surface should be extended to cover the area of low relief extending as far north as the canyon of Sybille Creek, and perhaps across that canyon to the area near the headwaters of Blue Grass Creek in Tps. 21 and 22 N., R. 72 W.

The presence of a conglomerate containing rhyolite clasts at an elevation of 2,650 m near Mountain Home Post Office (T. 12 N., R. 78 W.) near the Colorado-Wyoming border in the Medicine Bow Mountains indicates that the Medicine Bow surface may be contemporaneous with the beds containing such cobbles in the High Plains area.

TECTONIC SIGNIFICANCE

The existence of rocks of Pliocene age east of the Laramie Mountains strongly suggests that such deposits existed at one time on the west side of the mountains as well (Fig. 15). In order for the rhyolitic cobbles in the Ogallala Formation derived from the Specimen Mountain area to have reached the Plains region, they must have been transported across the Front Range–Laramie Mountain area. During this time the region was relatively stable tectonically.

In the process of transporting the cobbles eastward the streams reduced parts of the ranges to low relief and formed the present high-level erosion surfaces such as the Sherman. Subsequently these same streams were in part superimposed upon the older rocks in the core of the ranges (Fig. 16).

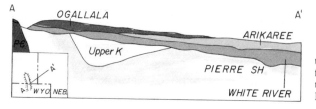

Figure 15. Digrammatic cross section on east flank of the Laramie Mountains (from Denson and Bergendahl, 1961).

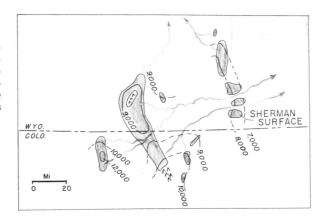

Figure 16. Map of paleography in Miocene-Pliocene time. Contours on present high-level erosion surfaces with possible drainage patterns. Erosion surfaces shaded light gray.

The profound denudation of the region that has taken place since the deposition of the Pliocene rocks would seem to require regional elevation to provide in part the erosive power of the streams that denuded the region.

QUATERNARY

Pleistocene

Active glaciers occupied the axial remnant areas of the northern and southern Medicine Bow Mountains, the Sierra Madre-Park Range, and the northern Front Range during late Pleistocene time (Fig. 17).

The extent and character of the late Pleistocene glaciation has been reported by Atwood (1937), Eschman (1955), Hail (1965), Jones and Quam (1944), Kiver (1968), McCallum (1962), Mears (1953), Price (1973), Ray (1940), Saulnier (1968), and Van Tuyl and Lovering (1934).

The investigators have concerned themselves in general with late Wisconsin history and problems of regional correlation of glacial stages. However, geologic mapping on the flanks of the Medicine Bow Mountains requires discrimination between varying types of Quaternary deposits. As a result of this sort of mapping, in which I have participated, it has been established that late Wisconsin ice on the east flank of the Medicine Bow Mountains was somewhat more extensive than previously reported. Ice overtopped Rock Creek Ridge (in T. 17 N., R. 78 W.) rather than moving entirely northward down the Rock Creek Valley as indicated by Atwood (1937). The front of the ice extended eastward over the ridge and down to an elevation of about 2,700 m along a trend about 16 km north to south between the headwaters of Cooper Creek and the south fork of Mill Creek. An ill-defined moraine accumulated at the ice front. Extensive areas of solifluction later developed behind the moraine. The drainages that led eastward from the ice front are characterized by boulder trains of large quartzite clasts deposited by the melt waters. The relations are shown in Blackstone (1973, Plate 1).

Prior to the Wisconsin period of glaciation, here considered to be equivalent to the Pinedale Stage (Mears, 1953; McCallum, 1962), there had been a period of extensive pedimentation along the east and west flanks of the Medicine Bow Mountains. The most prominent of these surfaces is the Arlington terrace (Dobbin and others, 1929b), which is actually a pediment under present terminology. The surface extends northeastward from Arlington (in sec. 30, T. 19 N., R. 78 W.) for more than 32 km toward the town of Rock River. Other extensive surfaces of similar character are found both northeast and northwest from Lookout siding

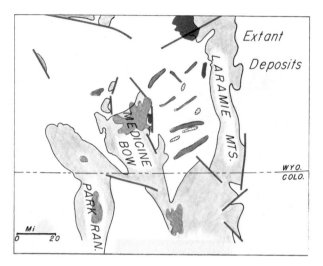

Figure 17. Map of Pleistocene features. Areas occupied by Wisconsin ice in dark gray; pre-Wisconsin pediments in light gray; Wisconsin deflation hollows hachured. Two areas of bouldery debris in northeast of area also shown in dark gray.

(T. 20 N., Rs. 74 and 75 W.); south of Cooper Lake (T. 19 N., R. 75 W.); along the north rim of Big Hollow where it is followed by Wyoming Highway 130, 8 to 21 km west of Laramie; and along Wyoming Highway 230 southwest of Laramie toward Harmony (sec. 19, T. 14 N., R. 75 W.) (Cleven, 1956; Montagne, 1953). These surfaces slope eastward at gradients ranging from 4 to 22 m per km and were controlled by the Laramie River drainage near its present position along the eastern side of the basin.

The interpretations of the structural evolution in the Laramie basin area (Knight, 1953; Houston and others, 1968) have been based in part on the distribution of conglomerates that have been assigned varying ages. The ages of some of the conglomerate sequences in the Medicine Bow, Hanna, and Wind River Formations can be demonstrated clearly in those areas where outcrops are reasonably continuous as previously described. Other deposits of bouldery material, most of which are not well indurated or not indurated at all, occur at high elevations in the Medicine Bow Mountains and in certain localities on the flanks of the Laramie Mountains. Some of these deposits have been assigned a Cenozoic age when a Pleistocene age assignment would be at least as logical if not more so.

Several examples will be discussed.

Rock Creek Ridge. The crest of Rock Creek Ridge, a north-trending topographic feature underlain by Precambrian age rocks, reaches an elevation of 3,240 m at Woolf triangulation station in sec. 16, T. 17 N., R. 78 W. The ridge parallels the valley of Rock Creek for about 24 km and forms the eastern topographic front of the Medicine Bow Mountains as viewed from the east. As shown on the Morgan quadrangle (Blackstone, 1973), the higher parts of the ridge near Woolf and Rock Creek Point (sec. 28, T. 17 N., R. 78 W.), and other unnamed high points over 3,050 m in elevation farther south are capped by accumulations of large quartzite boulders. A similar deposit caps a rounded summit (elevation 3,100 m) in sec. 15, T. 16 N., R. 78 W., a few kilometers to the southeast.

The large quartzite boulders are generally fragmented by frost action and their surfaces have been extensively wind-blasted. The boulders range up to 3 m or more in diameter but generally are about 1 m in diameter. No clean exposures have been seen, and the matrix if any is unknown. Knight (1953) considered the boulders to be a mountainward flank phase of the Hanna Formation and included them in an interpretation of the local tectonic history. McCallum (1968) assigned a small area of bouldery material at an elevation of 2,900 m near Centennial, Wyoming, to the Hanna Formation.

I believe that these high-level boulder deposits are pre-Wisconsin Pleistocene in age and that they exist only in scattered isolated patches along the mountain front. All conglomerate beds of definite Paleocene age in the Laramie basin area contain roundstones, the largest of which may reach 2 m in diameter but .5 m is the usual maximum and most of the material is cobble size or smaller. The bouldery debris on Rock Creek Ridge does not contain roundstones, nor are the boulders deeply iron stained as are many of the Hanna boulders. If my interpretation is correct, some modification of earlier interpretations of Paleocene history are necessary.

Kennaday Peak Area. As discussed under Paleocene strata, the upper 600 m of the isolated Kennaday Peak summit on the northwest flank of the Medicine Bow Mountains is composed of conglomeratic Hanna Formation. The summit of the mountain and the spurs radiating from the peak are strewn with rather angular boulders up to 4 m in diameter. Similar large boulders are found scattered through the timber at lower elevations on the north flank of the mountain and concentrated on a northwest-trending ridge northwest of Cedar Pass, which is 4.5 km northwest from Kennaday Peak. The largest boulders are granitic gneiss; amphibolite is common (more than 20 percent), but quartzite is essentially lacking and occurs only as small pebbles that appear to be 5 percent or less of the deposit. Neely (1934) and later Gries (1964) assigned these boulders to the Hanna Formation. An excellent picture of them appears in Neely's report.

I consider this boulder deposit to be Pleistocene(?) in age and probably equivalent to the Rock Creek Ridge material. There are no roundstones on the summit surface such as are found throughout the underlying 460 m of Hanna Formation which makes up the mass of the mountain; and quartzite is essentially lacking. Since the 6,100-m-thick Medicine Peak Quartzite that underlies the crest of the northern Medicine Bow Mountains is the obvious source for most of the Hanna roundstones, the absence of quartzite in the Kennaday Peak boulder deposit is puzzling. The nearest possible source of the gneiss and amphibolite boulders is about 10 km to the northwest on Pennock Mountain whose present summit elevation is only 3,062 m. If this was the source of these Pleistocene(?) boulders, an implication of some Pleistocene or later tectonic adjustment cannot be avoided.

Garrett–North Laramie Area. In the northern Laramie Range centered around Laramie Peak (roughly within a circle of 32 km radius), there are deposits of "king"-sized boulders. Individual well-rounded to subrounded boulders of granite up to 6 m in diameter are common. The most northern deposit is in the drainage of Horseshoe Creek in sec. 18, T. 28 N., Rs. 70 and 71 W., as reported by Pennington (1947). A second sequence extends for about 24 km in a west-to-east alignment capping the divide between Horseshoe and Cottonwood Creeks about 1.5 km north of the line between T. 27 and 28 N. Bretz and Horberg (1952) assigned this deposit a late Tertiary (probably Pliocene) age and stated that it is "certainly much older than early Pleistocene" because of the induration of the boulder beds.

On the west side of the Laramie Mountains, boulders occur at several localities from Marshall Post Office (sec. 17, T. 27 N., R. 75 W.) south and southeast to the canyon of the North Laramie River near the old Garrett Post Office (sec. 29, T. 25 N., R. 73 W.). These boulders occur in a fanlike arrangement opposite the mouths of southwestward flowing drainages or in elongate patches extending westward some kilometers out onto the basin floor. The size of the boulders decreases rather regularly away from the mountain front, but no well-exposed bedded deposits have been observed.

The age of these boulder deposits has been variously interpreted as Paleocene, Eocene, Oligocene, and Pleistocene. The deposits described by Bretz and Horberg (1952) overlie rocks as young as the Miocene Arikaree Formation, and those near Marshall Post Office overlie tuffaceous sediments that can be traced to dated Oligocene rocks nearby. Those near Garrett overlie strata that appear to be Eocene on the basis of color and stratigraphic position. If all the boulders represent one episode of deposition they are younger than Miocene. The distribution of the boulder beds around the Laramie Peak topographic high, and the concentration of the boulders opposite the mouths of present valleys, suggests that they are related to the existing stream pattern. It is my conclusion that these boulders were deposited during a period of high precipitation (in Pleistocene time), which has not yet been correlated with the standard Wisconsin glacial chronology.

Long Lake–Wheatland Reservoir No. 2 Area. A discontinuous but extensive boulder-capped surface extends through Tps. 19–23 N., and Rs. 73 and 74 W., Albany County, Wyoming, roughly occupying the area from Long Lake to the north end of Wheatland Reservoir No. 2 in the northeastern Laramie basin. The bouldery debris lies on a gentle west-sloping surface of very low relief which bevels west-dipping strata ranging in age from Triassic to Late Cretaceous. The deposit cannot be more than a few tens of meters in thickness, probably 6 to 12 m.

The boulders on this surface range up to 9 m in maximum dimension, though most of them are cobble size. The lithologies present include norite and anorthosite, white milky quartz, granitic gneiss, amphibolite, sandstone of the Casper Formation, and fossiliferous gray Mississippian(?) chert. There is no doubt that the source was from the east; this conclusion, based on the presence of anorthosite, is reinforced by the lack of quartzites from the Medicine Bow Mountains to the west.

The geologic age of the deposit is in doubt, but I consider it to be Pleistocene(?). The evidence is tenuous and is based on the character of the boulders, relation to source, elevation, and the local geomorphic history. Further investigations are in progress.

TECTONIC SIGNIFICANCE: CLIMATE VERSUS TECTONISM

The presence of deposits of bouldery debris on the flanks of mountainous areas always gives rise to questions concerning their origin and tectonic significance. Some of the deposits can be uniquely related to sharp pulses of tectonic uplift in the immediate area; others are less satisfactorily explained. Given an area of moderately strong topographic relief and no *new* active tectonic event, it is still possible to generate large quantities of very coarse clastic debris.

There is no evidence, in the Laramie basin region, of tectonic activity during the Quaternary; yet there are boulder deposits that can be logically assigned to this age. I conclude that the particular boulder deposits described in this section of the paper reflect sharp climatic variation during Pleistocene time, which provided sufficient volumes of water to enhance the transporting power of the streams and to generate fanglomerate type deposits.

Removing these various bouldery deposits from the Tertiary and placing them in the Quaternary simplifies the structural interpretation necessary to account for their distribution. The concept that all deposits containing "king"-sized boulders that are found on the flanks of mountain uplifts are evidence of Eocene uplift no longer seems tenable.

Drainage adjustment by stream capture and by deflation during late Pleistocene time has been complex and widespread. In general the headward erosion of the tributaries of the Medicine Bow River has diverted waters northwestward from the Laramie basin into the North Platte River south of the Ferris–Seminoe Mountains.

These drainages formerly flowed eastward across the Laramie Mountains and joined the North Platte River farther to the east. Tectonic activity does not seem to have influenced the drainage changes.

Holocene

Seismic activity has occurred periodically during historic time in the Medicine Bow Mountains, centering southwest of Laramie near the Fox Park and Kings Canyon communities. Several local events occurred during the middle 1950s. None of these events could be directly related to known faults, and all were of minor magnitude.

Denudation continues to be the order of the day.

SUMMARY

The Laramide orogenic event in the Laramie basin area began after the deposition of the Maestrichtian Medicine Bow Formation (Lancian provincial age) and reached a climax before the deposition of the youngest part of the Paleocene Hanna Formation.

Deformation continued in early Eocene time, resulting in the sharp folding of the conglomerate of the Hanna Formation along the Rock River–Seven Mile–Quealy dome line of folds (from T. 20 N., R. 78 W. to T. 17 N., R. 77 W.). The early Eocene Wind River Formation (Greybull provincial age) lies with angular unconformity across major folds and faults and is in turn folded, indicating mild deformation of probable late Eocene age. The primary elevation of the Medicine Bow Mountains began during Paleocene time and continued after the deposition of the youngest part of the Paleocene Hanna Formation, since the Hanna is folded to dips as high as 30° along the Mill Creek syncline (in the central western part of T. 16 N., R. 77 W.) and along the line of folding mentioned above.

Following the deposition of the Wind River Formation a period of erosion ensued, probably during late Eocene time, which etched out the topography so that the cores of the mountain uplifts stood above the basin areas. Fine-grained sediment, in large part of volcanic derivation, began to accumulate during Oligocene time and to cover the topography at lower elevations. The source of the fine-grained Oligocene sediment has not been fully established, but it must have been in part to the south in Colorado and in part farther west in the Great Basin. Local conglomeratic debris derived from the exposed Precambrian rocks in the core of the mountains is interbedded with the material from the remote sources. The fine-grained material continued to accumulate throughout Oligocene time and filled the areas of lower relief.

Little change occurred in Miocene time, except for the nature of the sediment. The usual lithology of the Miocene is one of fine-grained gray sandstone, which is commonly strongly cross-bedded. The base of the Miocene section is marked in many places by coarse conglomeratic debris derived from the Precambrian core of the uplifts. The conglomerate indicates a bevelling across the uplifts and reflects that fact that the lowland areas must have been filled with fine-grained material. The aggradation continued into Pliocene time until only the higher parts of the mountain masses stood above the plain of aggradation, and these were being bevelled by stream action to form what are now the high-level erosion surfaces of the present mountains. Clasts of extrusive felsic rock types derived from the Specimen Mountain–Iron Mountain area in Colorado are found incorporated in the Ogallala Formation at least 120 km east of the present mountain front. During the transport of this material across the Laramie Mountains the Sherman erosion surface must have been developed.

Normal faulting apparently occurred in post-middle Miocene time, but is not well documented. Regional denudation began in late Pliocene and has continued to the present.

During Pleistocene time, mountain glaciers were active and somewhat more widespread than previously reported. Bouldery debris was deposited during at least two stages of glacial history. Pedimentation of pre-Wisconsin age is widespread in the Laramie basin.

Minor seismic activity in the Medicine Bow Mountains during the past 20 years documents continued tectonic activity.

REFERENCES CITED

Ashley, W. H., 1948, Geology of the Kennaday Peak-Pennock Mt. area, Carbon County, Wyoming [M.S. thesis]: Laramie, Wyoming Univ., 71 p.

Atwood, Wallace W., Jr., 1937, Records of Pleistocene glaciers in the Medicine Bow and Park Ranges: Jour. Geology, v. 45, no. 2, p. 113-140.

Atwood, W. W., and Atwood W. W., Jr., 1938, Working hypothesis for the physiographic history of the Rocky Mountain region: Geol. Soc. America Bull., v. 49, p. 957-980.

Beckwith, R. H., 1938, Structure of the southwest margin of the Laramie basin, Wyoming: Geol. Soc. America Bull., v. 49, p. 1515-1544.

——1942, Structure of the upper Laramie River valley, Colorado-Wyoming: Geol. Soc. America Bull., v. 53, p. 1491-1532.

Beekly, A. K., 1915, Geology and coal resources of North Park, Colorado: U.S. Geol. Survey Bull. 596, p. 1-121.

Blackstone, D. L., Jr., 1963, Development of geologic structure in central Rocky Mountains, in The backbone of the Americas-A symposium: Am. Assoc. Petroleum Geologists Mem. 2, p. 160-179.

——1970, Structural geology of the Rex Lake quadrangle, Laramie basin, Wyoming: Wyoming Geol. Survey Prelim. Rept. 11, p. 1-17.

——1973, Structural geology of the Strous Hill quadrangle, James Lake quadrangle and part of the Morgan quadrangle, Albany and Carbon Counties, Wyoming: Wyoming Geol. Survey Prelim. Rept. 13, p. 1-45.

Blackwelder, E., 1909, Cenozoic history of the Laramie region, Wyoming: Jour. Geology, v. XVII, p. 429-445.

Bowen, C. F., 1918, Stratigraphy of the Hanna basin, Wyoming: U.S. Geol. Survey Prof. Paper 108-L, p. 227-241.

Bretz, J. Harlen, and Horberg, Leland, 1952, A high-level boulder deposit east of the Laramie range, Wyoming: Jour. Geology, v. 60, p. 480-488.

Brooks, Billy G., 1957, The geology of the Wheatland Reservoir area, Albany County, Wyoming [M.S. thesis]: Laramie, Wyoming Univ., 42 p.

Chadeayne, Dennis K., 1966, Geology of Pass Creek Ridge, St. Marys, and Cedar Ridge anticlines, Carbon County, Wyoming [M.S. thesis]: Laramie, Wyoming Univ., 87 p.

Chisholm, Wayne A., 1963, Effect of climate and source area location on Browns Park petrology [abs.]: Am. Assoc. Petroleum Geologists Bull., v. 47, p. 353.

Cleven, Gale W., 1956, A statistical analysis of erosion surfaces in the vicinity of Laramie, Wyoming [M.S. thesis]: Laramie, Wyoming Univ., 64 p.

Corbett, Marshall K., 1966, The geology and structure of the Mt. Richtofen-Iron Mt. region, north-central Colorado: Mountain Geologist, v. 3, p. 3-21.

Darton, N. H., and Siebenthal, C. E., 1909, Geology and mineral resources of the Laramie basin, Wyoming: U.S. Geol. Survey Bull. 364, p. 1-81.

Denson, N. M., and Bergendahl, M. H., 1961, Middle and upper Tertiary rocks of southeastern Wyoming and adjoining areas: U.S. Geol. Survey Prof. Paper 424-C, p. C168-C172.

Denson, N. M., and Chisholm, W. A., 1971, Summary of mineralogical and lithologic characteristics of Tertiary sedimentary rocks in the middle Rocky Mountains and the northern Great Plains: U.S. Geol. Survey Prof. Paper 750-C, p. 117-126.

Dobbin, C. E., Bowen, C. F., and Hoots, H. W., 1929a, Geology and coal and oil resources of the Hanna and Carbon basins, Carbon County, Wyoming: U.S. Geol. Survey Bull. 804, p. 1-88.

Dobbin, C. E., Hoots, H. W., and Dane, C. H., 1929b, Geology of the Rock Creek oil field and adjacent areas, Carbon and Albany Counties, Wyoming: U.S. Geol. Survey Bull. 806-D, p. 131-153.

Ebens, Richard J., 1966, Stratigraphy and petrography of Miocene volcanic sedimentary rocks in southeastern Wyoming and north-central Colorado [Ph.D. thesis]: Laramie, Wyoming Univ., 129 p.

Eggler, D. H., Larson, E. E., and Bradley, W. C., 1969, Granites, grusses, and the Sherman erosion surface, southern Laramie range, Colorado-Wyoming: Am. Jour. Sci., v. 267, p. 510-522.

Eschman, Donald F., 1955, Glaciation of the Michigan River basin, North Park, Colorado: Jour. Geology, v. 63, p. 197-213.

Evernden, J. F., Savage, D. E., Curtis, G. H., and James, G. T., 1964, Potassium-argon dates and the Cenozoic mammalian chronology of North America: Am. Jour. Sci., v. 262, p. 145-198.

Fox, James E., 1970, Foraminifera in the Medicine Bow Formation, south-central Wyoming: Contr. Geology, v. 9, p. 98-101.

Gill, J. R., Merewether, E. A., and Cobban, W. A., 1970, Stratigraphy and nomenclature of some Upper Cretaceous and lower Tertiary rocks in south-central Wyoming: U.S. Geol. Survey Prof. Paper 667, p. 1-53.

Gorton, Kenneth A., 1953, Geology of the Cameron Pass area, Grand, Jackson, and Larimer Counties, Colorado, in Wyoming Geol. Assoc. Guidebook 8th Ann. Field Conf., Laramie Basin, Wyoming, and North Park, Colorado, 1953, p. 87-98.

Gries, John C., 1964, Structure and Cenozoic stratigraphy of the Pass Creek basin area, Carbon County, Wyoming [M.S. thesis]: Laramie, Wyoming Univ., 69 p.

Hail, William J., Jr., 1965, Geology of northwestern North Park, Colorado: U.S. Geol. Survey Bull. 1188, p. 1-133.

Harshman, E. N., 1968, Geologic map of the Shirley basin area, Albany, Carbon, Converse, and Natrona Counties, Wyoming: U.S. Geol. Survey Misc. Geol. Inv. Map I-539.

——1972, Geology and uranium deposits, Shirley basin area, Wyoming: U.S. Geol. Survey Prof. Paper 745, p. 1-82.

Houston, R. S., and others, 1968, A regional study of rocks of Precambrian age in that part of the Medicine Bow Mountains lying in southeastern Wyoming—With a chapter on the relationship between Precambrian and Laramide structure: Wyoming Geol. Survey Mem. 1, 167 p.

Hyden, Harold J., 1966a, Geologic map of the McFadden quadrangle, Carbon County, Wyoming: U.S. Geol. Survey Geol. Quad. Map GQ-533.

——1966b, Geologic map of the Bengough Hill quadrangle, Albany and Carbon Counties, Wyoming: U.S. Geol. Survey Geol. Quad. Map GQ-579.

Hyden, Harold J., and McAndrews, Harry, 1967, Geologic map of the T L Ranch quadrangle, Carbon County, Wyoming: U.S. Geol. Survey Geol. Quad. Map GQ-637.

Hyden, Harold J., McAndrews, Harry, and Tschudy, Robert H., 1965, The Foote Creek and Dutton Creek Formations, two new formations in the north part of the Laramie basin, Wyoming: U.S. Geol. Survey Bull. 1194-K, p. K1-K12.

Hyden, Harold J., King, John S., and Houston, Robert S., 1967, Geologic map of the Arlington quadrangle, Carbon County, Wyoming: U.S. Geol. Survey Geol. Quad. Map GQ-643.

Hyden, Harold J., Houston, Robert S., and King, John S., 1968, Geologic map of the White Rock Canyon quadrangle, Carbon County, Wyoming: U.S. Geol. Survey Geol. Quad. Map GQ-789.

Izett, Glen A., 1968, Geology of the Hot Sulphur Springs quadrangle, Grand County, Colorado: U.S. Geol. Survey Prof. Paper 586, p. 1-79.

Jones, W. D., and Quam, Louise O., 1944, Glacial land forms in Rocky Mountain National Park, Colorado: Jour. Geology, v. 52, p. 217-234.

Kiver, Eugene P., 1968, Geomorphology and glacial geology of southern Medicine Bow Mountains [Ph.D. thesis]: Laramie, Wyoming Univ., 129 p.

Knight, S. H., 1951, The Late Cretaceous-Tertiary history of the northern portion of the Hanna basin, Carbon County, Wyoming, in Wyoming Geol. Assoc. Guidebook 6th Ann. Field Conf., p. 45–53.

——1953, Summary of the Cenozoic history of the Medicine Bow Mountains, Wyoming, in Wyoming Geol. Assoc. Guidebook 8th Ann. Field Conf., Laramie Basin, Wyoming, and North Park, Colorado, 1953, p. 65–76.

Lee, W. T., 1923, Peneplains of the Front Range and Rocky Mountain National Park, Colorado: U.S. Geol. Survey Bull. 730–A, p. 1–17.

Love, J. D., 1960, Cenozoic sedimentation and crustal movement in Wyoming: Am. Jour. Sci., v. 258A, p. 204–214.

——1961, Split Rock Formation (Miocene) and Moonstone Formation (Pliocene) in central Wyoming: U.S. Geol. Survey Bull. 1121–I, p. I1–I37.

Love, J. D., McGrew, Paul O., and Thomas, H. D., 1963, Relationship of latest Cretaceous and Tertiary deposition and deformation to oil and gas in Wyoming, in The backbone of the Americas—A symposium: Am. Assoc. Petroleum Geologists Mem. 2, p. 196–208.

Lowry, Marlin E., and Crist, Marvin A., 1967, Geology and ground-water resources of Laramie County, Wyoming: U.S. Geol. Survey Water-Supply Paper 1834, p. 1–71.

McAndrews, Harry, 1965, Geologic map of the Cooper Lake quadrangle, Albany County, Wyoming: U.S. Geol. Survey Geol. Quad. Map GQ-430.

McCallum, M. E., 1962, Glaciation of Libby Creek canyon, east flank of Medicine Bow Mountains, southeastern Wyoming: Contr. Geology, v. 1, p. 78–88.

——1968, Cenozoic history of the east-central Medicine Bow Mountains, Wyoming: Mountain Geologist, v. 5, p. 69–81.

McGrew, Laura W., 1963, Geology of the Fort Laramie area, Platte and Goshen Counties, Wyoming: U.S. Geol. Survey Bull. 1141–F, p. F1–F39.

McGrew, Paul O., 1953, Tertiary deposits of southeastern Wyoming, in Wyoming Geol. Assoc. Guidebook 8th Ann. Field Conf., Laramie Basin, Wyoming, and North Park, Colorado, 1953, p. 61–64.

Mears, Brainerd, Jr., 1953, Quaternary features of the Medicine Bow Mountains, Wyoming, in Wyoming Geol. Assoc. Guidebook 8th Ann. Field Conf., Laramie Basin, Wyoming, and North Park, Colorado, 1953, p. 81–84.

Minick, J. N., 1951, Tertiary stratigraphy of southeastern Wyoming and northeastern Colorado [M.S. thesis]: Laramie, Wyoming Univ., 53 p.

Montagne, John, 1953, Geomorphology of the Centennial–Big Hollow area, southeastern Wyoming, in Wyoming Geol. Assoc. Guidebook 8th Ann. Field Conf., Laramie Basin, Wyoming, and North Park, Colorado, 1953, p. 77–80.

——1957, Cenozoic structural and geomorphic history of northern North Park and Saratoga valley, Colorado and Wyoming, in Rocky Mtn. Assoc. Geologists Guidebook to the Geology of North and Middle Park Basins, Colorado, 1957, p. 36–41.

Moore, Fred., 1959, The geomorphic evolution of the east flank of the Laramie Range, Colorado and Wyoming [Ph.D. thesis]: Laramie, Wyoming Univ., 123 p.

Nace, Raymond L., 1936, Summary of the Late Cretaceous and early Tertiary stratigraphy of Wyoming: Wyoming Geol. Survey Bull. 26, p. 175.

Neely, Joseph, 1934, The geology of the north end of the Medicine Bow Mountains, Carbon County, Wyoming: Wyoming Geol. Survey Bull. 25, p. 1–15.

Newhouse, W. H., and Hagner, A. F., 1957, Geologic map of anorthosite areas, southern part of Laramie Range, Wyoming: U.S. Geol. Survey Minerals Inv. Map MF-119.

Pennington, Jack J., 1947, Stratigraphy and structure of the Horse Draw area, Albany and Platte Counties, Wyoming [M.S. thesis]: Columbia, Missouri Univ., 85 p.

Price, Chadderdon, 1973, Glacial and drainage history of the Upper Cow Creek drainage, Sierra Madre Range, Wyoming [M.S. thesis]: Laramie, Wyoming Univ., 82 p.

Prichinello, Kathryn A., 1971, Earliest Eocene mammalian fauna from the Laramie basin of southeast Wyoming: Contr. Geology, v. 10, p. 73–87.

Ray, Louis, 1940, Glacial chronology of the Southern Rocky Mountains: Geol. Soc. America Bull., v. 51, p. 1851–1918.

Reynolds, Mitchell W., 1971, Geologic map of the Bairoil quadrangle, Sweetwater and Carbon Counties, Wyoming: U.S. Geol. Survey Geol. Quad. Map GQ-913.

Sato, Yoshiaki, and Denson, N. M., 1967, Volcanism and tectonism as reflected by distribution of nonopaque heavy minerals in some Tertiary rocks of Wyoming and adjacent states: U.S. Geol. Survey Prof. Paper 575-C, p. C42-C54.

Saulnier, George J., 1968, Ground water resources and geomorphology of the Pass Creek basin area, Albany and Carbon Counties, Wyoming [M.S. thesis]: Laramie, Wyoming Univ., 91 p.

Stanley, K. O., 1971, Tectonic implications of Tertiary sediment despersal on the Great Plains east of the Laramie Range, in Symposium on Wyoming Tectonics and Their Economic Significance, Wyoming Geol. Soc. Guidebook 23, p. 65-70.

Steven, T. A., 1956, Cenozoic geomorphic history of the Medicine Bow Mountains near the Northgate fluorspar district, Colorado: Colo. Sci. Soc. Proc., v. 17, p. 35-55.

Toots, Heinrich, 1965, Reconstruction of continental environments: The Oligocene of Wyoming [Ph.D. thesis]: Laramie, Wyoming Univ., 176 p.

Tudor, Mathew S., 1953, Structural geology of the west-central flank of the Laramie Range, in Wyoming Geol. Assoc. Guidebook 8th Ann. Field Conf., Laramie Basin, Wyoming, and North Park, Colorado, 1953, p. 101-102.

Tweto, Odgen, and Sims, P. K., 1963, Precambrian ancestry of the Colorado mineral belt: Geol. Soc. America Bull., v. 74, p. 991-1014.

Van Tuyl, F. M., and Lovering, T. S., 1934, Physiographic development of the Front Range: Geol. Soc. America Bull., v. 46, p. 1291-1350.

Vine, James D., and Prichard, George E., 1959, Geology and uranium occurrences in the Miller Hill area, Carbon County, Wyoming: U.S. Geol. Survey Bull. 1074-F, p. 201-239.

Wahlstrom, Ernest E., 1944, Structure and petrology of Specimen Mountain, Colorado: Geol. Soc. America Bull., v. 46, p. 77-90.

——1947, Cenozoic physiographic history of the Front Range, Colorado: Geol. Soc. America Bull., v. 58, p. 551-572.

Ward, Dwight E., 1957, Geology of the Middle Fork of the Michigan River, Jackson County, Colorado: Rocky Mtn. Assoc. Geologists Guidebook to the Geology of North and Middle Park Basins, 1957, p. 70-73.

MANUSCRIPT RECEIVED BY THE SOCIETY MAY 8, 1974

Printed in U.S.A.